Soil Science: A Comprehensive Approach

Soil Science: A Comprehensive Approach

Editor: Michael Higgins

Callisto Reference

www.callistoreference.com

Callisto Reference,
118-35 Queens Blvd., Suite 400,
Forest Hills, NY 11375, USA

Visit us on the World Wide Web at:
www.callistoreference.com

ISBN: 978-1-64116-272-2 (Hardback)

Printed in China.

Cataloging-in-Publication Data

Soil science : a comprehensive approach / edited by Michael Higgins.
p. cm.
Includes bibliographical references and index.
ISBN 978-1-64116-272-2
1. Soil science. 2. Soil management. 3. Agriculture. 4. Earth sciences. I. Higgins, Michael.
S591 .S65 2020
631.4--dc23

Table of Contents

Preface

This book aims to highlight the current researches and provides a platform to further the scope of innovations in this area. This book is a product of the combined efforts of many researchers and scientists, after going through thorough studies and analysis from different parts of the world. The objective of this book is to provide the readers with the latest information of the field.

Soil science is a field concerned with the study of the soil, its classification, formation and mapping. The analysis of the fertility of the soil as well as its biological, physical and chemical properties is also encompassed in this field. Pedology and edaphology are the two branches of soil science. Pedology studies soil in its natural setting while edaphology studies the interactions of soil. Empirical soil data is obtained from soil surveys that are acquired through field sampling and remote sensing. In the current scenario, due to the prospects of a crucial water crisis, escalating population growth, increased land degradation and high per capita food consumption, soil science is an important field of study. Current direction of soil research is therefore aimed at the study of soil in relation to carbon sequestration, climate change and greenhouse gases. This book is a compilation of chapters that discuss the most vital concepts and emerging trends in the field of soil science. Such selected concepts that redefine this field have been presented in this book. The readers would gain knowledge that would broaden their perspective about soil science.

I would like to express my sincere thanks to the authors for their dedicated efforts in the completion of this book. I acknowledge the efforts of the publisher for providing constant support. Lastly, I would like to thank my family for their support in all academic endeavors.

Editor

EDTA-induced phytoextraction of lead and barium by brachiaria (*B. decumbens* cv. Basilisk) in soil contaminated by oil exploration drilling waste

André Fernão Martins de Andrade[1], Nelson Moura Brasil do Amaral Sobrinho[2], Fabiana Soares dos Santos[3], Marcio Osvaldo Lima Magalhães[4*], Alfredo Tolón-Becerra[5] and Leilane da Silva Lima[6]

*[1]Instituto Federal de Educação, Ciência e Tecnologia do Rio de Janeiro, Pinheiral, Rio de Janeiro, Brazil. [2]Departamento de Solos, Instituto de Agronomia, Universidade Federal Rural do Rio de Janeiro, Seropédica, Rio de Janeiro, Brazil. [3]Departamento de Engenharia de Agronegócios, Universidade Federal Fluminense, Volta Redonda, Rio de Janeiro, Brazil. [4]Departamento de Agronomia, Universidade do Estado de Mato Grosso, Rodovia MT-358, km 7, 78300-000, Tangará da Serra, Mato Grosso, Brazil. [5]Departamento de Engenharia Rural, Universidade de Almería, Almería, Andaluzia, Spain. [6]Programa de Pós-graduação em Agronomia-Ciência do Solo, Universidade Federal Rural do Rio de Janeiro, Rio de Janeiro, Rio de Janeiro, Brazil. *Author for correspondence. E-mail: marciomagalhaes@gmail.com*

ABSTRACT. The phytoextraction of heavy metals using chelating agents has been widely studied for the remediation of contaminated soils. To evaluate the efficiency of EDTA-induced phytoextraction of Ba and Pb using *Brachiaria decumbens* for the remediation of soil contaminated by oil well drilling and exploration waste, an experiment was conducted by applying a single dose (6 mmol EDTA kg^{-1} soil) and split doses of EDTA (three applications of 2 mmol EDTA kg^{-1} soil). The samples were subjected to sequential extractions using the method proposed by Ure et al. (1993) as modified by Rauret et al. (1999).The application of EDTA did not influence the distribution of Ba in various chemical fractions of the soil. The dry matter production did not differ significantly between the treatments and the control, thereby demonstrating the tolerance of plants to the experimental conditions. The absorption of Pb by plants was influenced by the application of EDTA. The application of a single dose of EDTA influenced the absorption of Pb and its translocation to the aerial plant parts. The application of split doses favoured higher accumulation of Pb in roots. Because of its tolerance to heavy metals and EDTA, *B. decumbens* has the potential to be used in phytostabilisation.

Keywords: heavy metals, phytoremediation, chelating compounds.

Introduction

Exploration for oil generates waste that can potentially cause pollution. This waste arises from the drilling of prospecting wells and primarily consists of fragmented rock mixed with drilling fluid. This waste also contains hydrocarbons and heavy metals, such as Ba and Pb (DEUEL JR.; HOLLIDAY, 1997).

Contamination by heavy metals and other elements considered toxic to humans and the environment has occurred since the late eighteenth century, thereby raising concerns and warranting studies to search for effective low-cost technologies to reduce the toxic effects of such contamination (GRATÃO et al., 2005).

Phytoremediation has been considered a promising and relatively cost-effective method compared to other techniques for the remediation of areas contaminated with heavy metals (SAIFULLAH et al., 2009; SALT et al., 1998). Phytoextraction can be classified as natural or induced. Natural phytoextraction involves the use of plants known as hyperaccumulators of heavy metals. Induced phytoextraction involves the use of plants with a high biomass production that can accumulate elevated heavy metal concentrations in their tissues when grown in soils treated chemically with chelating agents that increase the bioavailability of metals and absorption by the plant (MEERS et al., 2005).

Among the many chelating agents used to enhance phytoextraction, ethylene-diamine-tetraacetic acid (EDTA) has been shown to be very efficient in forming soluble complexes with metals, especially for the removal of Pb (DOUMET et al., 2008; KIM et al., 2003; LESTAN et al., 2007; XIA et al., 2009).

The objective of the present study was to evaluate the use of EDTA in the induced phytoextraction of Ba and Pb using *Brachiaria decumbens* for the remediation of soil contaminated by waste from oil exploration drilling.

Material and methods

The contaminated soil used in the experiment consisted of a mixture of waste from oil prospecting wells; this waste was deposited more than 20 years ago in an area containing Latossolo Vermelho Distrófico típico.

The chemical (Table 1) and physical characteristics of soil samples were determined according to Embrapa (1997). A soil particle size analysis showed a silt content of 62%, a sand content of 38% and a negligible percentage of clay.

Contaminated soils were used to assess the ability of *Brachiaria decumbens* cv. Basilisk to phytoextract Ba and Pb in the presence of EDTA. The experiment was assembled and conducted in a greenhouse at the Federal Rural University of Rio de Janeiro ('Universidade Federal Rural do Rio de Janeiro'), and the contaminated soil samples were collected in Santa Maria do Oeste in the Brazilian State of Paraná.

Pots were filled with 4 kg of contaminated soil, and 24 seeds were placed in each pot. After sprouting, the number of plants was thinned to four per pot. The pots were maintained at a moisture content equivalent to 80% the field capacity by the addition of deionised water. After the first month, approximately 100 mL of a one-quarter-strength Hoagland solution (HOAGLAND; ARNON, 1950) were added every three days. Two treatments plus the control were used; one treatment received a split application of EDTA, and the other received a single dose of EDTA. Each condition had four replicates for a total of 12 experimental units.

Table 1. Chemical characterisation of the soil used in the experiment.

	Soil
pH H_2O	8.3
Ca^{2+} ($cmol_c$ dm^{-3})	9.5
Mg^{2+} ($cmol_c$ dm^{-3})	2.0
H + Al ($cmol_c$ dm^{-3})	1.3
Al^{3+} ($cmol_c$ dm^{-3})	0.0
S (%)	11.53
T (%)	12.83
V (%)	90
P (mg kg^{-1})	3
K (mg kg^{-1})	5
C (g kg^{-1})	1.4
Ba (mg kg^{-1})	6.700
Pb (mg kg^{-1})	350.4

pH in water (1:2.5); Ca and Mg, H+Al – extracted in 1 mol L^{-1} KCl; Al - extracted in 1 mol.L^{-1} calcium acetate at pH 7.0; P - extracted with North Carolina solution; K - extracted with 0.5 mol.L^{-1} potassium dichromate; C – Walkley – Black; Ba and Pb - aqua regia digestion (pseudototal) (ISO 11466, 1995).

The treatment with a split application of EDTA, applied via the surface pot, consisted of three doses of 2 mmol EDTA kg^{-1} soil (0.6 mg EDTA g^{-1} soil). The first dose was administered at 120 days after sowing the plants followed by two additional applications at intervals of seven days. The single EDTA dose was administered by the addition of 6 mmol EDTA kg^{-1} (1.8 mg EDTA g^{-1} soil) on the day of the last dose in the split application treatment.

Soil samples were collected at five days after the first application of 2 mmol EDTA kg^{-1} soil in the split application treatment and after the application of 6 mmol EDTA kg^{-1} soil in the single-dose treatment. The samples were subjected to sequential extractions using the method proposed by Ure et al. (1993) as modified by Rauret et al. (1999).

The plants were collected one month after the final application of EDTA, which completed a cycle of 164 days. After collection, the roots and leaves were separated, weighed, rinsed in deionised water and placed in an oven at 70 °C until a constant weight was obtained. The samples were ground and then digested using a 6:1 nitryl perchlorate solution according to the method of Tedesco et al. (1995).

The concentrations of metals in soil extracts were determined by inductively coupled plasma optical emission spectrometry (ICP-OES; Perkin Elmer model OPTIMA 3000). The limit of detection (LD) and limit of quantitation (LQ) for Ba were 0.036 mg kg^{-1} and 0.36 mg kg^{-1}, respectively, and the LD and LQ for Pb were 0.020 mg kg^{-1} and 0.067 mg kg^{-1}, respectively, in the soil digestion extracts. The LD was calculated as the mean of the blanks plus three standard deviations of the blanks from all analyses (10 replicates). The National Institute of Standards and Technology (NIST) certified standard reference material (SRM) 22709a (San Joaquin soil; Ba and Pb concentrations of 979 ± 28 and 17.3 ± 0.1 mg kg^{-1}, respectively) was used to validate the determination of the pseudototal contents of Ba and Pb in the soil. NIST certified SRM 1573a (tomato leaves; Ba concentration of 63 ± 0.7 mg kg^{-1}) was used to validate the Ba contents of the plant material, and NIST certified SRM 1547 (peach leaves; Pb concentration of 0.87 ± 0.03 mg kg^{-1}) was used to validate the Pb contents of the plant materials. All analyses of the certified samples showed a range of recovery of 93-95%, which was within the ranges accepted as normal by NIST for soil and plant samples.

To evaluate the potential of Brachiaria to extract Pb and Ba, we determined the bioaccumulation factor (AF) and the index of translocation (IT) of each element in the plant according to Araújo et al. (2011); The phytoextraction efficiency (PE) was determined according to Vysloužilová et al. (2003), as described below:

AF = M (aerial part) / M (soil)

IT = (accumulation of the element in the aerial part / accumulation of the element in the plant) * 100.

PE = [M(plant) x Plant weight / M(soil) x pot weight] x 100, where M is the concentration of the metal contaminant.

The data were evaluated by analysis of variance using the F test ($p < 0.05$), and the means were compared using Tukey's test ($p < 0.05$). The analyses were performed using the statistical software SAEG statistical software package, version 9.0 (Arthur Bernardes Foundation at the Federal University of Viçosa, Viçosa State, Minas Gerais).

Results and discussion

The geochemical fractionation of contaminated soil samples (Table 2) showed that the Ba was almost entirely in the most recalcitrant fraction (F4 - residual), thus resulting in low concentrations in the other soil fractions. The presence of EDTA, both in the split application and single-dose treatments, did not produce significant differences between the different factions. Studies conducted by Smeda and Zyrnicki (2002) also found higher levels of Ba in the residual fraction. The occurrence of Ba in the soil, especially in the residual fraction, and the Ba^{2+} ion in the soil tends to be precipitated and adsorbed over time in the form of oxides and hydroxides (PICHTEL et al., 2000).

Table 2. Geochemical fractionation of Ba and Pb in contaminated soils and soils with different EDTA concentrations.

Metal	EDTA concentration	Geochemical fraction			
		F1	F2	F3	F4
		mg kg^{-1}			
Ba	0	118 (± 1.5)	252(±1.7)	358 (± 10.9)	6082 (±6.2)
	3 x 2[1]	119(±6.2)	243(±4.0)	377(±12.4)	6091(±22.4)
	1 x 6[2]	115(±2.5)	257(±1.4)	374(±16.1)	6136(±130.8)
Pb	0	53 (± 4.0)	156 (± 2.2)	26 (± 1.5)	335 (± 3.3)
	3 x 2	95(±7.97)	151(±17.4)	27(±2.0)	303(±26.9)
	1 x 6	92(±5.9)	151(±7.4)	25(±2.2)	312(±15.0)

F1 - acid-soluble fraction; F2 - fraction bound to iron oxides and manganese oxides; F3 - fraction bound to organic matter and sulphides; F4 - Residual. The values correspond to an average of 4 treatments ± standard deviation. There was no significant difference (Tukey's test at 5% significance) between treatments. 1: Treatment with three applications of EDTA at 2 mmol kg^{-1} soil. 2: Treatment with the application of a single dose of EDTA at 6 mmol kg^{-1} soil.

In the control, approximately 10% of the total Pb content was detected in the exchangeable phase (F1), where electrostatic and carbonate bonds occur, and approximately 30% occurred in the fraction bound to oxides (F2). This binding of Pb to carbonates and oxides was also mentioned by Pichtel et al. (2000). The Pb has a high affinity for iron oxides and, in particular, manganese oxides (XIA et al., 2009).

The addition of EDTA caused a significant change in the distribution of Pb in the geochemical fractions of the soil. Significant increases in Pb concentrations were observed in the soluble acid fractions (F1). Studies by Grčman et al. (2001) and Saifullah et al. (2009) also showed a marked increase in Pb levels in the soil solution with the addition of 3 mmol L^{-1} EDTA.

The plants developed without any symptoms of stress until the addition of EDTA. The treatment that received the first of three doses of 2 mmol EDTA kg^{-1} soil began to show leaves with a more intense green colour at 125 days after sowing.

In the treatment that received a single dose of EDTA (6 mmol EDTA kg^{-1} soil), the plants showed a yellowing of the leaves at 140 days after sowing; this yellowing continued until the time of collection. These symptoms could be attributed to the solubilisation of metals in the soil at levels greater than the phytotoxicity limit in the region of the rhizosphere as well as the phytotoxicity of the EDTA itself (EVANGELOU et al., 2007).

Despite these symptoms, the dry matter production of *Brachiaria* was statistically similar between treatments (Table 3), thereby indicating that the plants were tolerant of the metals and

EDTA under the experimental conditions. In an evaluation of the effect of EDTA on the induced phytoextraction of Zn and Cd by *Brachiaria decumbens*, Santos et al. (2006) also found that this plant had a tolerance to high concentrations of Zn and Cd in the soil.

Table 3. Dry matter production in the different vegetative parts of *Brachiaria decumbens* grown in contaminated soil with varied concentrations of EDTA.

Treatments	Dry matter (g per pot)		
EDTA kg^{-1} soil	Aerial parts	Roots	Total
0	16.04 ^{NS} (±1.21)	6.43 ^{NS} (±0.34)	22.47 ^{NS} (±1.25)
3 x 2^1	19.70 ^{NS} (±4.02)	4.78 ^{NS} (±1.28)	24.48 ^{NS} (±4.95)
1 x 6^2	15.90 ^{NS} (±2.47)	4.19 ^{NS} (±0.18)	20.09 ^{NS} (±2.63)

Values correspond to the means of four treatments ± standard deviation. NS - no significant difference (Tukey's test at 5% significance) in column. 1: Treatment with 3 applications of EDTA at 2 mmol kg^{-1} soil. 2: Treatment with the application of a single dose of EDTA at 6 mmol kg^{-1} soil.

Neugschwandtner et al. (2008) studied the application of split and single doses of EDTA and showed that signs of phytotoxicity in corn plants were more visible after the application of a single dose; the single dose also led to the death of the plants before harvest in some cases. According to Barocsi et al. (2003), the addition of the same amount of EDTA in split doses can increase the plants' resistance to damage caused by heavy metals and EDTA compared to a single dose.

There was no significant difference in the concentration of Ba among the treatments (Table 4), which may have been caused by the low bioavailability of this element in the soil (Table 1).

Table 4. Concentrations of Ba and Pb in the leaves and roots of *Brachiaria decumbens* (mg kg^{-1}) grown in contaminated soil and under different concentrations of EDTA.

	Leaf			Root		
Metal	0	3 x 2^1	1 x 6^2	0	3 x 2	1 x 6
Ba	26.9 ^{NS} (±5.8)	27.3 ^{NS} (±2.8)	24.3 ^{NS} (±1.28)	17.6 ^{NS} (±1.03)	18.5 ^{NS} (±2.56)	20.9 ^{NS} (±4.7)
Pb	0.5 C (±0.0)	1.41 B (±0.07)	9.02 A (±3.18)	1.41 B (±0.18)	5.3 A (±1.11)	4.2 A (±0.90)

The values correspond to the means of four treatments ± standard deviation. Different letters along the rows indicate statistically significant differences (Tukey's test at 5% significance or less) between the plant parts. NS = Not significant. 1: Treatment with three applications of EDTA at 2 mmol kg^{-1} soil. 2: Treatment with the application of a single dose of EDTA at 6 mmol kg^{-1} soil.

However, there was a marked increase in the Pb concentrations in the leaves relative to the control plants, especially in the treatment with the application of a single dose of EDTA at 6 mmol kg^{-1} soil. The addition of EDTA also led to an increase in the concentration of Pb in the roots. The induction of Pb uptake by plant roots has also been observed by other authors (DOUMETT et al., 2008; KIM et al., 2003; LESTAN et al., 2007) and is in agreement with the geochemical fractionation results (Table 2). Although no significant difference was found in the geochemical fractionation of Pb among the different treatments (Table 2), increased Pb absorption may have occurred due to the acidification of the medium in the region of the rhizosphere, which would increase the formation of soluble complexes around the roots.

The treatment with the split addition of EDTA retained the greatest amount of Pb in the roots, thereby reducing the extent of Pb translocation to the aerial parts. This greater ability to retain Pb in the roots may be due to the limited mobility of Pb within the plant (PICHTEL et al., 2000) as well as the capacity of the plant to acquire mechanisms to adapt and overcome injury caused by EDTA at lower concentrations (VASSIL et al., 1998).

The accumulation of Ba in the roots and aerial plant parts (Table 5) was not significant. However, the accumulation of Pb in the leaves was significantly higher in the treatment with a single addition of EDTA. In the roots, the application of EDTA also significantly increased the accumulation of Pb relative to the control plants.

Table 5. Accumulation of metals in the leaves and roots of *Brachiaria decumbens* (mg per pot^{-1}) grown in contaminated soil and under different treatments with EDTA.

	Leaf			Root		
Metal	0	3 x 2^1	1 x 6^2	0	3 x 2	1 x 6
Ba	0.42 ^{NS} (±0.08)	0.56 ^{NS} (±0.14)	0.38 ^{NS} (±0.06)	0.11 ^{NS} (±0.01)	0.09 ^{NS} (±0.02)	0.09 ^{NS} (±0.02)
Pb	0.01 C (±0.00)	0.03 B (±0.01)	0.13 A (±0.05)	0.01 B (±0.00)	0.02 A (±0.00)	0.02 A (±0.00)

The values correspond to the mean of four replicates ± standard deviation. Different letters along the rows indicate significant differences (Tukey's test of 5% significance or less) in each plant part. 1: Treatment with three applications of EDTA at 2 mmol kg^{-1} soil. 2: Treatment with the application of a single dose of EDTA at 6 mmol kg^{-1} soil.

The IT and PE of Ba and Pb in *Brachiaria decumbens* in response to the application of EDTA are shown in Table 6.

Table 6. Index of translocation (IT) and phytoextraction efficiency (PE) in *Brachiaria decumbens* plants grown in contaminated soil and under different treatments with EDTA.

	IT		PE	
EDTA	Ba	Pb	Ba	Pb
(mmol kg^{-1}soil)	%			
0	77.0 (±3.69)	47.6 (±5.26)	0.0018	0.0009
3 x 2^1	84.7 (±4.17)	53.4 (±8.40)	0.0021	0.003
1 x 6^2	80.6 (±4.42)	86.9 (±1.77)	0.0016	0.005

The values correspond to the mean of four replicates ± standard deviation. 1: Treatment with three applications of EDTA at 2 mmol kg^{-1} soil. 2: Treatment with the application of a single dose of EDTA at 6 mmol kg^{-1} soil.

Pb demonstrated an IT greater than 85% in the treatment with the application of a single dose of EDTA. This finding contradicts that of Pulford and Watson (2003) who claim that there are two major limitations for the phytoextraction of Pb: 1) low bioavailability and 2) a low extent of translocation from roots to the leaves. Pichtel et al. (2000), working with various plant species, found limited mobility of Pb in the plants. However, Santos et al. (2006) report that the presence of EDTA favours an increase in the Pb content in the leaves.

Despite the increased efficiency of Pb phytoextraction with the addition of EDTA when compared to controls, especially in treatments with a single-dose application, the values obtained in the present study are much lower than those reported by Huang et al. (1997) who state that greater than 1% of the total metal in the soil should be absorbed by the aerial parts for phytoextraction to be economically feasible. Similar results were observed by Neugschwandtner et al. (2008) who studied EDTA-induced the phytoextraction of Pb in corn.

The results of the present study suggest that the application of a single dose of EDTA enhanced the efficiency of Pb accumulation and translocation to the aerial plant parts. Similar results were reported by Neugschwandtner et al. (2008).

Conclusion

The application of EDTA did not influence the absorption of Ba in plants.

The application of EDTA caused an increase in the concentration of Pb in plants.

The application of EDTA increased the rate of Pb translocation to the aerial parts.

The efficiency of Pb phytoextraction was low, and under the experimental conditions, *Brachiaria decumbens* does not have the potential to be used in phytoextraction programs.

References

ARAÚJO, S. A. A.; GUILHERME, L. R. G.; LOPES, G.; CAMPOS, M. L. Phytoremediation of arsenic-contaminated soils using Brachiaria grass. **Ciência Agrotécnica**, v. 35, n. 1, p. 84-91, 2011.

BAROCSI, A.; CSINTALAN, Z.; KACSANYI, L.; DUSHENKOV, S.; KUPERBERG, J. M.; KUCHARSKI, R.; RICHTER, P. I. Optimizing phytoremediation of heavy metal-contaminated soil by exploring plants' stress adaptation. **International Journal of Phytoremediation**, v. 5, n. 1, p. 13-23, 2003.

DEUEL JR., L. E.; HOLLIDAY, G. H. **Soil remediation for the petroleum extraction industry**. 2nd ed. Oklahoma: PennWell, 1997.

DOUMETT, S.; LAMPERI, L.; CHECCHINI, L.; AZZARELLO, E.; MUGNAI, S.; MANCUSO, S.; PETRUZZELLI, G.; DEL BUBBA, M. Heavy metal distribution between contaminated soil and *Paulownia tomentosa*, in a pilot-scale assisted phytoremediation study: Influence of different complexing agents. **Chemosphere**, v. 72, n. 10, p. 1481-1490, 2008.

EMBRAPA-Empresa Brasileira de Pesquisa Agropecuária. **Manual de métodos de análises de solos**. Rio de Janeiro: CNPS, 1997.

EVANGELOU, W. H. M.; EBEL, M.; SCHAEFFER, A. Chelate assisted phytoextraction of heavy metals from soil. Effect, mechanism toxicity and fate of chelating agents. **Chemosphere**, v. 68, n. 6, p. 989-1003, 2007.

GRATÃO, L. P.; PRASAD, V. V. M.; CARDOSO, F. P.; LEA, J. P.; AZEVEDO, A. R. Phytoremediation: green technology for the clean up of toxic metals in the environment. **Brazilian Journal of Plant Physiology**, v. 17, n. 1, p. 53-64, 2005.

GRČMAN, H.; VELINKONJA-BOLTA, S.; VODNIK, D.; LESTAN, D. EDTA enhanced heavy metal phytoextraction: metal accumulation, leaching and toxicity. **Plant and Soil**, v. 235, n. 1 , p. 105-114, 2001.

HOAGLAND, D. R.; ARNON, D. L. The water-culture method for growing plants without soil. **California Agricultural Experimental Station Circular**, v. 347, n. 2, p. 347-352, 1950.

HUANG, J. W.; CHEN, J.; BERTI, W. R. Phytoremediation of Pb-contaminated soils: role of synthetic chelates in lead phytoextraction. **Environmental Science and Technology**, v. 31, n. 3, p. 800-805, 1997.

ISO 11466. **Soil quality** – Extraction of trace elements soluble in aqua regia. Geneva: International Organisation for Standardisation, 1995.

KIM, C.; LEE, Y.; ONG, S. K. Factors affecting EDTA extraction of lead from lead-contaminated soils. **Chemosphere**, v. 51, n. 9, p. 845-853, 2003.

LESTAN, D.; LUO, C.-L.; LI, X.-D. The use chelating agents in the remediation of metal-contaminated soils: A review. **Environmental Pollution**, v. 153, n. 1, p. 3-13, 2007.

MEERS, E.; RUTTENS, A.; HOPGOOD, M. J.; SAMSON, D.; TACK, F. M. G. Comparison of EDTA and EDDS as potential soil amendments for enhanced phytoextraction of heavy metals. **Chemosphere**, v. 58, n. 8, p. 1011-1022, 2005.

NEUGSCHWANDTNER, W. R.; TLUSTOS, P.; KOMÁREK, M.; SZÁKOVÁ, J. Phytoextraction of Pb and Cd from a contaminated agricultural soil using different EDTA application regimes: Laboratory versus field scale measures of efficiency. **Geoderma**, v. 144, n. 3, p. 446-454, 2008.

PICHTEL, J.; KUROIWA, K.; SAWYERR, H. T. Distribution of Pb, Cd and Ba in soils and plants of two contaminated sites. **Environmental Pollution**, v. 110, n. 1, p. 171-178, 2000.

PULFORD, I. D.; WATSON, C. Phytoremediation of heavy metal-contaminated land by trees – a review. **Environmental International**, v. 29, n. 4, p. 529-540, 2003.

RAURET, G.; LOPEZ-SAANCHEZ, J. F.; SAHUQUILLO, A.; RUBIO, R.; DAVIDSON, C.; URE, A.; QUEVAUVILLER, P. H. Improvement of the BCR three step sequential extraction procedure prior to the certification of new sediment and soil reference materials. **Journal of Environmental Monitoring**, v. 1, n. 1, p. 57-61, 1999.

SAIFULLAH, M. E.; QADIR, M.; CARITAT, P.; TACK, F. M. G.; DU LAING, G.; ZIA, M. H. EDTA-assisted Pb phytoextraction. **Chemosphere**, v. 74, n. 10, p. 1279-1291, 2009.

SALT, D. E.; SMITH, R. D.; RASKIN, I. Phytoremediation. **Annual Review of Plant Physiology and Plant Molecular Biology**, v. 49, n. 1, p. 643-668, 1998.

SANTOS, F. S.; HERNÁNDEZ-ALLICA J.; BECERRIL, M. J.; AMARAL SOBRINHO, N.; MAZUR, N.; GARBISU, C. Chelate-induced phytoextraction of metal polluted soils with Brachiaria decumbens. **Chemosphere**, v. 65, n. 1, p. 43-50, 2006.

SMEDA, A.; ZYRNICKI, W. Application of sequential extraction and the ICP-AES method for study of the partitioning of metals in fly ashes. **Microchemical Journal**, v. 72, n. 1, p. 9-16, 2002.

TEDESCO, M. J.; GIANELLO, C.; BISSANI, C. A.; BOHNEN, H.; VOLKWEISS, S. J. **Análise de solo, plantas e outros materiais**. Porto Alegre: Brasil, 1995.

URE, A. M.; QUEVAUVILLER, P.; MUNTAU, H.; GRIEPINK, B. Speciation of heavy metals in soils and sediments. An account of the improvement and harmonization of extraction techniques undertaken under the auspices of the BCR of the Commission of the European Communities. **International Journal of Environmental Analytical Chemistry**, v. 51, n. 1, p. 135-151, 1993.

VASSIL, A. D.; KAPULNIK, Y.; RASKIN, I.; SALT, D. E. The role of EDTA in lead transport and accumulation by Indian mustard. **Plant Physiology**, v. 117, n. 2, p. 447-453, 1998.

VYSLOUŽILOVÁ, M.; TLUSTOŠ, P.; SZÁKOVÁ, J. Cadmium and zinc phytoextraction potential of seven clones of *Salix spp.* planted on heavy metal contaminated soils. **Plant Soil and Environment**, v. 49, n. 12, p. 542-547, 2003.

XIA, W.; GAO, H.; WANG, X.; CHUNHUA, Z.; LIU, Y.; FAN, T.; WANG, X. Application of EDTA decontamination on soils affected by mining activities and impact of treatment on the geochemical partition of metal contaminants. **Journal of Hazardous Materials**, v. 164, n. 2, p. 936-940, 2009.

2

Multivariate technique for determination of soil pedoenvironmental indicators in Southern Amazonas

Ivanildo Amorim de Oliveira[1*], José Marques Júnior[1], Milton César Costa Campos[2], Renato Eleotério de Aquino[1], Ludmila de Freitas[1] and Antônio Sérgio Ferraudo[3]

*[1]Departamento de Solos e Adubos, Faculdade de Ciências Agrárias e Veterinárias, Universidade Estadual Paulista"Julio de Mesquita Filho", Via de Acesso Prof. Paulo Donato Castellane, s/nº, 14884-900, Jaboticabal, São Paulo, Brazil. [2]Departamento de Agronomia, Universidade Federal do Amazonas, Humaitá, Amazonas, Brazil. [3]Departamento de Ciências Exatas, Faculdade de Ciências Agrárias e Veterinárias, Universidade Estadual Paulista "Júlio de Mesquita Filho", Jaboticabal, São Paulo, Brazil. *Author for correspondence. E-mail: ivanildoufam@gmail.com*

ABSTRACT. The use of statistical techniques to evaluate soil attribute behaviour is an important tool for choosing the most adequate form of soil management. Thus, the aim of this study was to jointly assess the physical and chemical attributes and the magnetic susceptibility characteristics of three environments and to use multivariate statistics to define which attributes have the greatest potential as environmental change indicators. The study was conducted in three areas: one with archaeological black earth (and planted maize); one with pasture; and one of the forest. Sixty-four (64) deformed and non-deformed soil samples each were collected within these areas, with a 70 x 70 m grid spaced every 10 meters at depths of 0.0-0.2 and 0.4-0.6 m, to determine the chemical attributes (fertility), as well as the physical attributes (texture) and magnetic susceptibility. The data were analyzed through multivariate statistics. The archaeological dark earth areas have different behaviours between pastures and forests: chemical attributes have greater interaction in archaeological dark earth due to the high fertility of the anthropic horizons. The attributes that may indicate environmental changes are calcium, potential acidity and clay for archaeological dark earth, forest and pasture areas, respectively.

Keywords: main component, Amazonian, archaeological dark earth.

Introduction

Soil behaviour can be studied by assessing several factors, such as the nutrient levels, which correspond to soil fertility (Silva, Lima, Xavier, & Teixeira, 2010a), and physical attributes that represent the soil structure's organization and the microorganisms that are responsible for mineralization processes and thus represent a potential nutrient supply source for plants.

The evaluation of these soil factors, when performed using univariate methods, increases the scale of the problem and becomes an expensive process (Silva et al., 2010a). An alternative is the use of multivariate analysis, which is an important tool for the exploratory data analysis of soil because it allows the grouping of samples according to similarity while still enabling the selection of the most important variables to discriminate between pre-selected groups (Benites, Moutta, Coutinho, & Balieiro, 2010). However, current research has increasingly reported on the need to develop indicators of global impacts for land use and occupation (Rockström et al., 2009). Such studies are even more necessary for areas with high environmental sensitivity, such as those found in the landscape of the Amazonian deforestation arc located in Southern Amazonas State.

Some multivariate statistical techniques can be applied to soil properties to establish the use of pedoenvironmental indicators; these techniques include principal component analysis (PCA), which aims to reduce the problem dimensionally and ease data interpretation. According to Barroso and Artes (2003), this analysis of factors is a statistical technique that aims to describe the dependence structure of a set of changeable variables and measure commonalities.

The goal of discriminant analysis (DA) is to identify the variables that discriminate groups and choose the most appropriate group to which they should belong based on their characteristics (Hair, Anderson, Tatham, & Black, 2005). Some researchers have used Quality Control Charts to evaluate the behaviour of attributes. This chart is one of the main tools used in statistical quality control and may be indicative of how variables behave in each studied management (Kume, 1993). The combined use of these techniques based on soil attribute behaviours can aid in decision-making, ideal land use and handling and may further indicate possible attributes that are undergoing major changes due to human interference.

The aim of this study was to jointly assess the soil's physical and chemical attributes and its magnetic susceptibility in three different environments and to use these data, by means of multivariate statistics, to define which attributes have great potential as indicators of environmental changes in Southern Amazonas.

Material and methods

The study was performed in farms located around road BR 230, in Manicoré City, Transamazônica, Amazonas State, Brazil. Soil sampling was conducted in areas with three different uses. One was forest, located at 7° 54' 44.5" S and 61° 31' 44.7" W, at an average altitude of 140 m (Figure 1); it is characterized as a Tropical Dense Rainforest that contains by 20- to 50-m-tall trees.

Another was on soil with an anthropic horizon, i.e., archaeological dark earth (ADE) sited at 7° 55' 02.1" S and 61° 31' 45.2" W, which is 102 m in altitude and had been under corn cultivation for 120 days. The last one was a pasture area at 07° 54' 42" S, 61° 31' 50" W and a 135 m mean altitude that had been under intensive pasturing with brachiaria (*Brachiaria brizanta*) at 1 animal unit per hectare capacity. Moreover, the soils were classified as Alitic Plinthic Red Argisols (Ultisols), except for the area with ADE, which was rated as Dystrophic Abrupt Red-Yellow Argisol (Ultisol) (Empresa Brasileira de Pesquisa Agropecuária [Embrapa], 2013).

These soils were derived from Rondonian granites from the Upper Precambrian. The Köppen classification for the region's climate is a rainy tropical type with a short dry period (*Am*), temperatures ranging between 25 and 27°C, and rainfall between 2,250 and 2,750 mm per year, with the greatest concentration of rainfall from October to June (Oliveira, Campos, Freitas, & Soares, 2015).

Seventy-square-meter grids were established in these areas surrounded by 0.49 ha. Soil samples were collected at the cross-points, which were at 10-metre intervals, for a total of 64 sampling points within each grid (Figure 1). Then, at each point, the collection of soil samples from 0.0 to 0.2 m and 0.4 to 0.6 m depths was performed.

Disturbed soil samples were collected for particle size analysis by the pipette melthod, which used a 0.1 mol L^{-1} NaOH solution as a chemical dispersant and mechanical stirring in a high-rotation apparatus for 15 minutes, according to the methodology proposed by Empresa Brasileira de Pesquisa Agropecuária (Embrapa, 2011). Using the ion exchange resin method, the exchangeable calcium (Ca^{2+}), magnesium (Mg^{2+}), potassium (K^{+}), available phosphorus (P), and potential acidity (H+Al) were determined (Raij, Andrade, Cantarella, & Quaggio, 2001). The pH was measured potentiometrically using a 1:2.5 ratio of soil and water (Empresa Brasileira de Pesquisa Agropecuária [Embrapa], 2011). The total carbon was quantified by Walkley-Black's method, and organic matter was estimated from the organic carbon content.

A Bartington MS2 coupled to a Bartington MS2B sensor was used to determine the magnetic susceptibility (MS) in fine air-dried soil samples (FADS). The evaluation was made at a low frequency (0.47 kHz) to obtain more accurate results because it was a single reading (Bartington Instruments [Bartington], 2013).

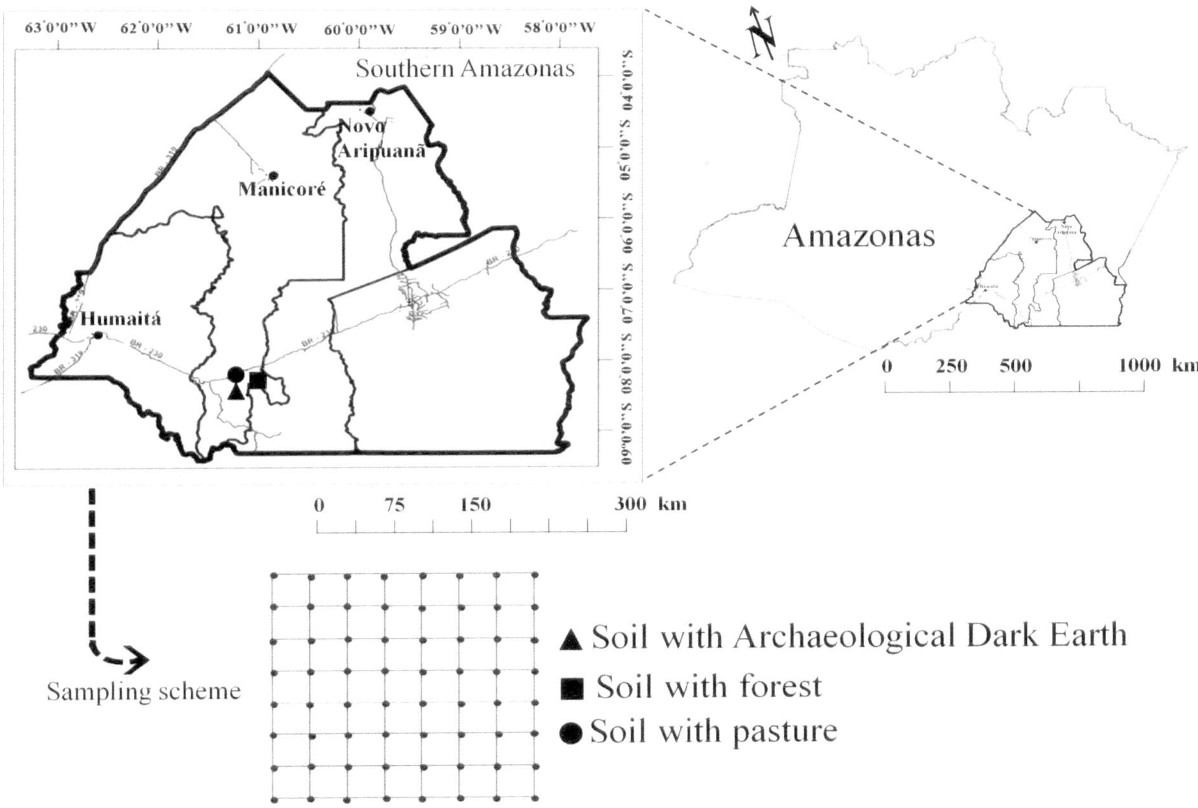

Figure 1. Location map and soil sampling scheme for the studied areas in Southern Amazonas.

To consider the multivariate structure of the data, statistical techniques were used to verify similarities among the land management practices in an attempt to group them in terms of physical and chemical attributes. A "scree-plot" graph of eigenvalues was created to determine the number of components that should be excluded. Such a graph orders the eigenvalues according to the main components, ultimately plotting the variance percentage by each attribute. Notably, this analysis of components should explain over 70% of the total variance (Hair et al., 2005), and these constitute the chosen response variables for principal component analysis (PCA).

Knowing the components to be used, principal component analysis was performed to obtain a smaller set of variable linear combinations that would preserve most of the information provided by the original ones (Silva et al., 2010b). Such analysis accounts for the simultaneous qualitative interactions among those attributes whose original values had been normalized to a mean of 0 and a variance equal to 1. The criteria for choosing the number of components were an eigenvalue greater than 1.00 and a synthesized cumulative variance greater than 70% (Hair et al., 2005).

Subsequently, factor analysis was performed, which allowed the relationships between variables to be explained as a limited number of new variables by extracting the principal components calculated from the correlation matrix among variables. The Orthogonal Varimax Rotation method was used to facilitate interpretation. This method transforms the factor coefficient such that the correlation values among factors and *n* original variables come closer to zero or 1 (Hair et al., 2005).

Recently, Control Chart analysis has been applied to better understand how the soil attribute variability is at all collected points. According to Kume (1993), a control chart is one of the main tools used for statistical quality control and may be indicative of how the variables behave in each area studied.

To compare the area profiles by using a combination of all variables, discriminant composed by a data classification matrix was performed for further analysis. From this analysis, it was possible to define how different the areas are because they are represented in a graph where the boundaries can be observed. The graph axis (x and y) represent the canonical variables, which are new multivariate variables created from the original set of variables and attributes. All statistical analyses were performed using the Statistica software version 7.0.

Results and discussion

The scree plot (Figure 2) shows that it is possible to use only the first two principal components because they explain over 70% of the total variance (Hair et al., 2005) and can graphically represent the discriminant power of soil attributes and the contribution of each variable in the total variance.

In the first and second retained components, each environmental attribute's importance shows the most significance, with 83.07 and 74.88% of the explained variability at depths of 0.0-0.2 and 0.4-0.6 m, respectively. The attributes had the following order of significance for their components: K^+ (9.96%), Ca^{2+} (9.80%), pH in H_2O (9.58%), MS (9.28%), Mg^{2+} (9.26 %) and P (9.20%) at 0.0 to 0.2 m depth (Figure 2A); and Ca^{2+} (9.60%), pH in H_2O (9.24%), Mg^{2+} (9.12%), P (8.64%) and OM (8.57%) at 0.4 to 0.6 m (Figure 2B). For these attributes, greater impacts or alterations may occur when the area is subjected to some form of anthropogenic management.

The eigenvalues confirm that the first and second main components were needed to explain the total variance because they presented high eigenvalues (7.30 and 1.00 at a depth of 0.0 to 0.2 m; 6.12 and 1.35 at a depth of 0.4-0.6 m), thus justifying the use of principal component analysis 1 (PC1) and principal component analysis 2 (PC2). According Kaiser (1958), the PCs with eigenvalues greater than 1 (one) can be used for a two-dimensional ordering of access and variables, thereby enabling biplot graph construction.

Variable correlations and principal component graphic representation (Table 1 and Figure 3) allowed the characterization of the variables that were most favoured in the environmental formation and differentiation.

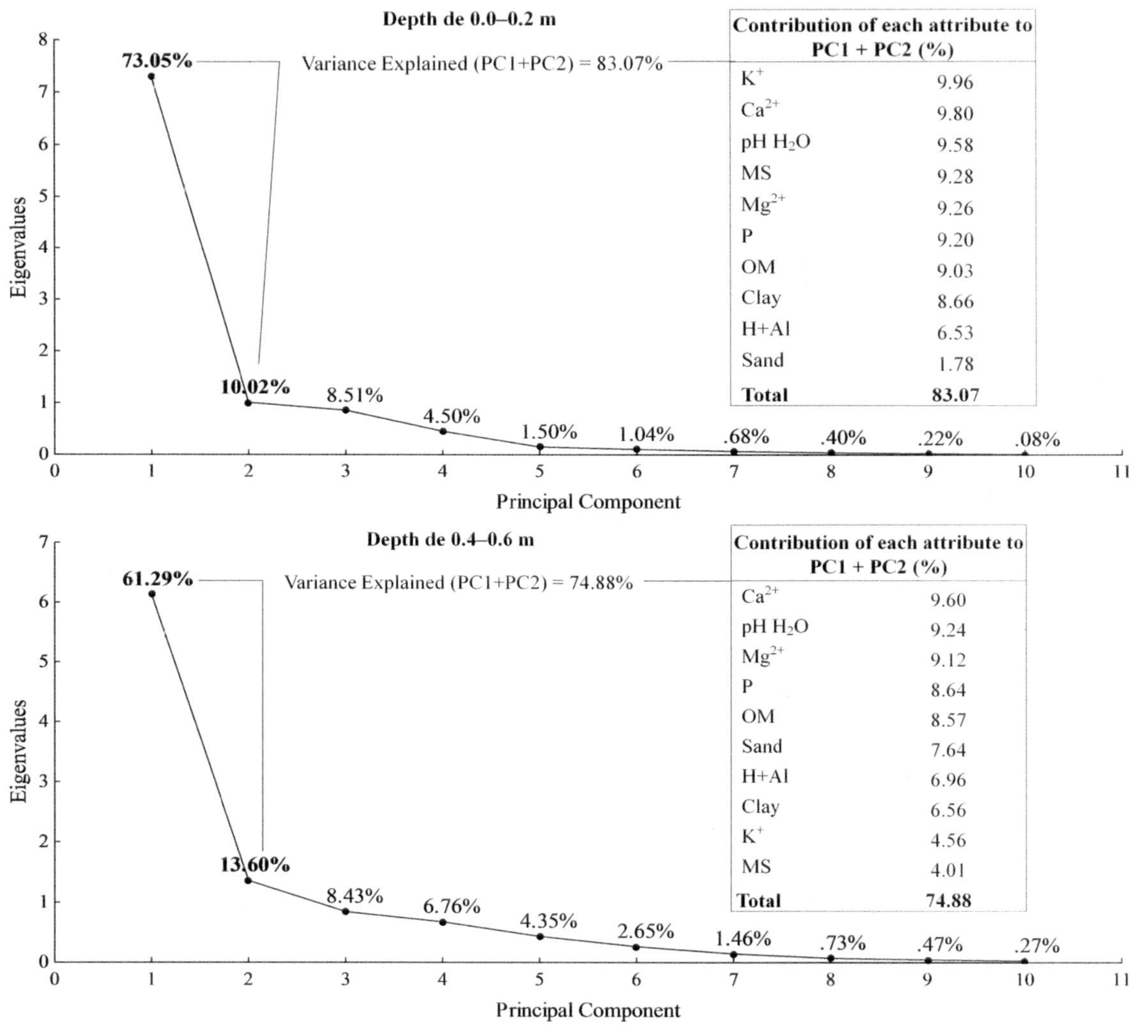

Figure 2. Data set variation ratio explained by principal component (PC) and contribution of each variable in the total variance by "scree plot" method.

Table 1. Correlation among each principal component and the studied variables.

Variables	Principal component --Depth 0.0 – 0.2 m---		--Depth 0.4 – 0.6 m--	
	PC1	PC2	PC1	PC2
Sand	-0.420577	0.034579	-0.850816*	0.199056
Clay	0.930401*	-0.016233	0.747539*	0.311334
pH H_2O	-0.978538*	0.019005	-0.960924*	-0.015644
Potential acidity	0.804035*	-0.079144	0.798272*	0.241429
Organic matter	-0.948272*	-0.057533	-0.329314	0.865057*
Phosphorus	-0.958909*	0.012626	-0.928323*	-0.047597
Potassium	-0.080809	-0.994790*	0.293114	0.608593
Calcium	-0.989511*	0.026008	-0.977922*	0.058434
Magnesium	-0.962230*	-0.016517	-0.934860*	0.194632
MS	-0.963315*	0.001161	-0.631510*	-0.047930

*Most significant values; PC1: principal component 1; PC2: principal component 2; PC3: principal component 3. MS: magnetic susceptibility

The variables that were most strongly correlated with the ADE area at a depth of 0.0 to 0.2 m were pH in H_2O, OM, Ca^{2+}, Mg^{2+}, P and MS (Table 1 and Figure 3A), which appeared in the third and fourth quadrants with a small angle to the abscissa. The ADE soils have higher OM contents, which gives them greater fertility. High levels of P are also a striking feature of the ADEs. Cunha et al. (2009) reported P contents of 116 mg kg^{-1} in ADE areas under the forests of archaeological sites in Amazonas State; 290.5 mg kg^{-1} under ADE cropped areas; and 4.7 mg kg^{-1} in non-anthropogenic areas. The incorporation of organic waste, especially fish bones, turtle shells, animal or even human bones, which are rich in P, contributed to the high content in ADEs (Lima, Schaefer, Mello, Gilkes, & Ker, 2002).

The Ca^{2+} and Mg^{2+} values are also high in ADEs as observed by Barros, Lima, Canellas, and Kern (2012), which justifies their connection to this environment. This behaviour is due to either the incorporation of calcium-rich bones from the Pre-Columbian period or a higher Ca^{2+} from exchange surfaces, which results in less leaching. Campos, Santos, Mantovanelli, and Soares (2012a) found Ca^{2+} and Mg^{2+} higher values in ADEs for non-anthropogenic soils, thus affirming a strong correlation of these attributes to ADE (Table 1).

MS is also highly related to ADE area at 0.0 to 0.2 m of depth (Table 1 and Figure 3A). The OM presence favours the conditions that are required for iron reduction, which promotes increased susceptibility.

Pasture and forest areas are featured by greater bond to clay and potential acidity (H+Al) (Table 1 and Figure 3A). The potential acidity values of forest areas were greater across the environments, probably due to the higher leaching from intense rainfall with drainage systems (Campos, Ribeiro, Souza Júnior, Ribeiro Filho, & Oliveira, 2010; Campos, Ribeiro, Souza Júnior, Ribeiro Filho, & Almeida, 2012b).

In the subsurface layer (0.40-0.6 m), a behaviour similar to the surface layer (0.0 to 0.2 m) can be observed. The subsurface variables that were strongly correlated with ADE were sand pH in H_2O, OM, Ca^{2+}, Mg^{2+}, P and MS (Table 1 and Figure 3B). In contrast, the pasture area was characterized by a closer link to clay (Figure 5B), in accordance with the smallest angle relative to the abscissa's axis, probably due to the presence of a clay-like soil. The forest has a strong connection with the H+Al (Figure 3B), as stated by Campos et al. (2012b); this condition is due to the increased leaching promoted by the intense rainfall associated with better drainage conditions in the Amazonian environment.

Based on these results, it can be stated that the forest and pasture environments are more similar or that they have more physical, chemical and MS homogeneous behaviours (Table 1 and Figure 3). However, the ADE presents as a unique environment that has peculiar chemical, physical and MSs for this type of environment (Figure 3A and B).

The factor analysis (Table 2) indicates that 73.69% of the data variability is explained by the first factor and that 10.24% is explained by the second factor; thus, 83.93% of the data variability is explained by the first two factors. This shows that of the 10 variables, two factors needed to be used, with 64 observations in the original set, resulting in a dimensional reduction of the original variables with an explanation loss lower than 20% (16.07%).

Table 2. Factorial analysis of soil attributes at the evaluated depths with factors (Factors 1 and 2) corresponding to the studied environments in Southern Amazonas.

Attributes	---Depth 0.0-0.2 m---		--Depth 0.4-0.6 m--	
	Factor 1	Factor 2	Factor 1	Factor 2
Sand	0.526836*	0.205837	0.830345*	0.267486
Clay	-0.927258*	0.031663	-0.764111*	0.266531
pH H_2O	0.978019*	-0.021800	0.959871*	0.048330
Potential acidity	-0.798974*	0.084820	-0.811210*	0.200701
Organic matter	0.947510*	0.094369	0.248974	0.918742*
Phosphorus	0.955780*	-0.002876	0.927550*	0.036452
Potassium	-0.027139	0.981456*	-0.478394	0.514503*
Calcium	0.989056*	-0.013465	0.969478*	0.139332
Magnesium	0.961184*	-0.020344	0.915273*	0.265287
MS	0.960728*	0.008044	0.632019*	0.046771
Relative variance	73.69%	10.24%	61.81%	13.87%

*Most discriminant values; MS: Magnetic susceptibility.

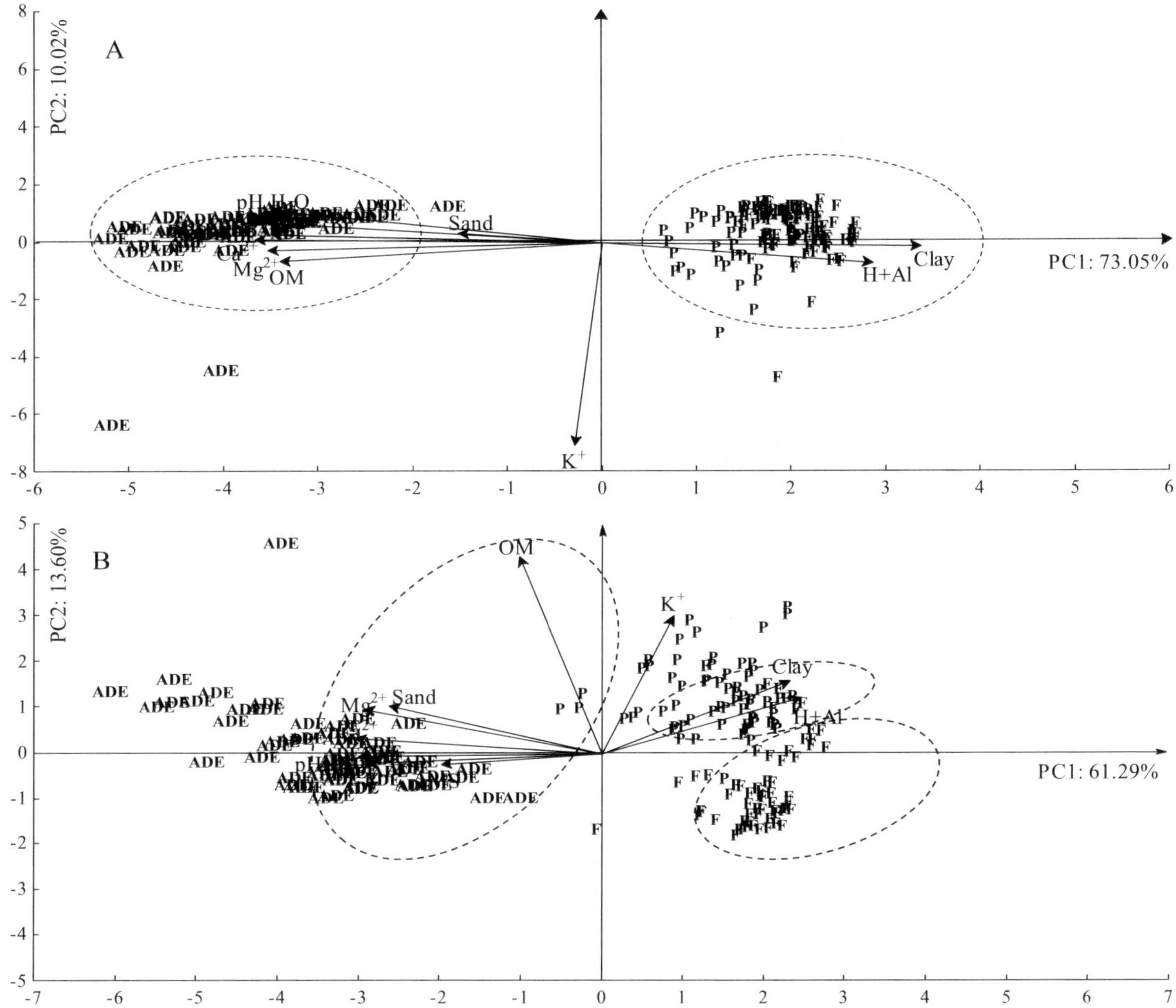

Figure 3. Principal component analysis of soil attributes in the studied environments. A = 0.0 – 0.2 m depth; B = 0.4 – 0.6 m depth; ADE= archaeological dark earth; P= pasture; F= forest.

In factor analysis, the weighting that defines which variables are the most important indicate that the variables that form factor 1 have high factor loadings (Table 2). These variables explain the higher percentage of variation and are the highest contributors to changes in soil characteristics as a function of land use. However, only potassium has a low discriminatory power among the studied attributes, which indicates that there is no difference in this attribute among the three environments in the first layer, as shown for Factor 2.

Clay and H+Al are directly proportional to the other attributes, which means that when the clay and H+Al contents increase, the other attributes decrease and vice-versa. Barreto, Lima, Freire, Araújo, and Freire (2006) observed a relation between potential acidity and soil pH. These authors observed that the higher the pH is, the lower the potential acidity is in the forest. At the 0.4–0.6 m depth, it can be noted that there is a very similar behaviour among the attributes, in which the OM had a lower discriminating power and was expressed in Factor 2 together with K^+. Thus, the higher the OM content is, the lower the K^+ discriminant power is.

A Control Chart of attribute variability was designed for each environment to evaluate each attribute's performance in the variability of all attributes together (Figure 4). The analysis enables clearly visualizing the boundaries between environments, showing their specific variability. However, ADE differs from pasture and forest at depths in the subsurface layer, where the variability in MS is smaller or where the attributes are more homogeneous. This analysis facilitates the interpretation of human interference on surface horizons, which differs from that observed for adjacent soils.

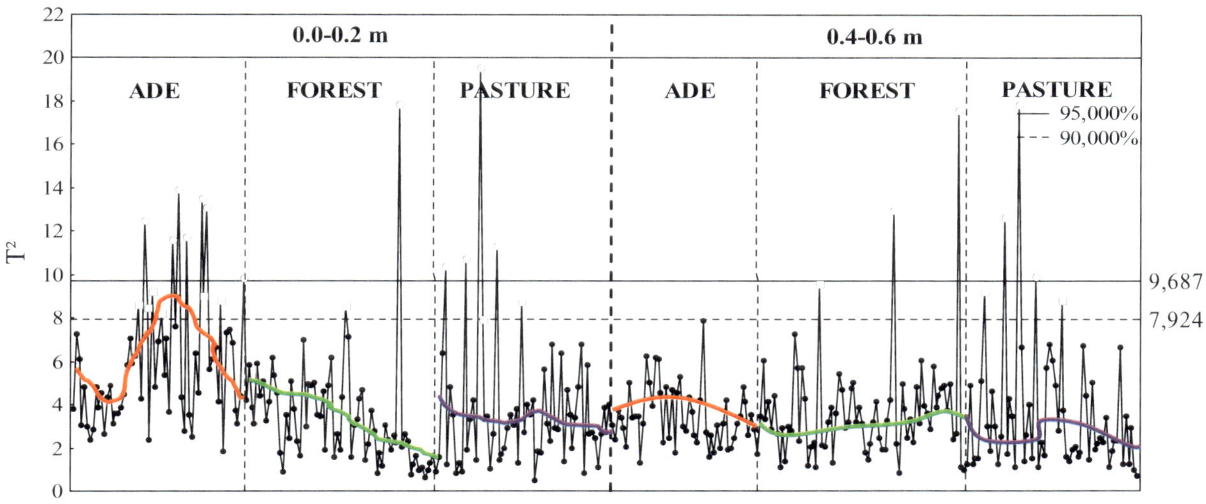

Figure 4. Control Chart of attribute variability in each studied environment in Southern Amazonas.

It should be noted that ADE has the highest data variability, followed by forest and pasture. This finding demonstrates the influence of anthropogenic transformation in soils with anthropic horizons (ADEs). Archaeological evidence indicates that ancient human activities in Amazonian habitats significantly transformed the landscape within neighbouring settlements, especially in the late prehistoric period. A remarkable record of this is the soil areas altered by prehistoric men, which feature a dark colour, archaeological material remains (ceramic fragments and lithic artefacts) and high Ca^{2+}, Mg^{2+}, Zn, Mn, P and C contents (Kern et al., 2009).

As mentioned above, ADE differs in that it is highly fertile. This behaviour is directly linked to greater data variability. In this sense, to verify the relationship between the existing variability among environments and their chemical, physical and MS, we applied a complementary technique using discriminant analysis, which has efficiently matched the results for all areas (Figure 5). By discriminant analysis, it was found that 51.62% of the studied relationship variability was explained in the first two roots (Figure 5) at the 0.0-0.20 m depth. The first canonical root (1 Can) explained 49.33%, and the second (Can 2) explained 2.29%. At the depth of 0.40-0.60 m, the first two roots explained 32.74% of the variance, with 29.09% in the first canonical root (Can 1) and 3.64% in the second (Can 2), totalling 32.73% of the variability among studied relationships. The discriminant analysis results show that Can 1 is responsible for separating ADE from the other areas because ADE has soil with better physical and chemical qualities and has more distant characteristics of other environments. This separation is in accordance with the results of both the principal component analysis (Figure 5) and clustering analysis (Figures 3 and 4). Because the soil with anthropic horizon presents a higher fertility and higher OM content in relation to forest and pasture, this may have been the factor that most influenced the attribute characterization and MS of each area.

Through discriminant analysis, the most powerful attributes to discriminate between environments are pH in H_2O, H+Al, OM and Ca^{2+} at both depths (Table 3). These results demonstrate that these chemical properties may be the most important and contribute the most to area discrimination.

Table 3. Canonical Discriminant Analysis values, highlighting each studied variable weight in Southern Amazonas.

Attributes	--Depth of 0.0-0.2 m--		-Depth of 0.4-0.6 m-	
	Can 1	Can 2	Can 1	Can 2
Sand	0.06287	-0.49534	-0.34401	-0.64447
Clay	-0.19757	-0.50629	-0.22099	-0.46587
pH H_2O	0.80518*	-0.85966*	-1.14584*	-0.21075
Potential acidity	0.92197*	0.55627	-1.15091*	0.61098
Organic matter	-1.01512*	-1.89896*	0.96410*	-1.32450*
Phosphorus	0.01017	-0.46156	0.01965	-0.16534
Potassium	-0.23310	0.37203	0.19983	0.23770
Calcium	1.02901*	1.86653*	-1.02969*	0.83193
Magnesium	0.39721	0.09631	0.11768	0.06386
MS	0.25995	-0.04610	-0.05913	0.00575

* Main characterization of canonical relationships; MS: Magnetic susceptibility.

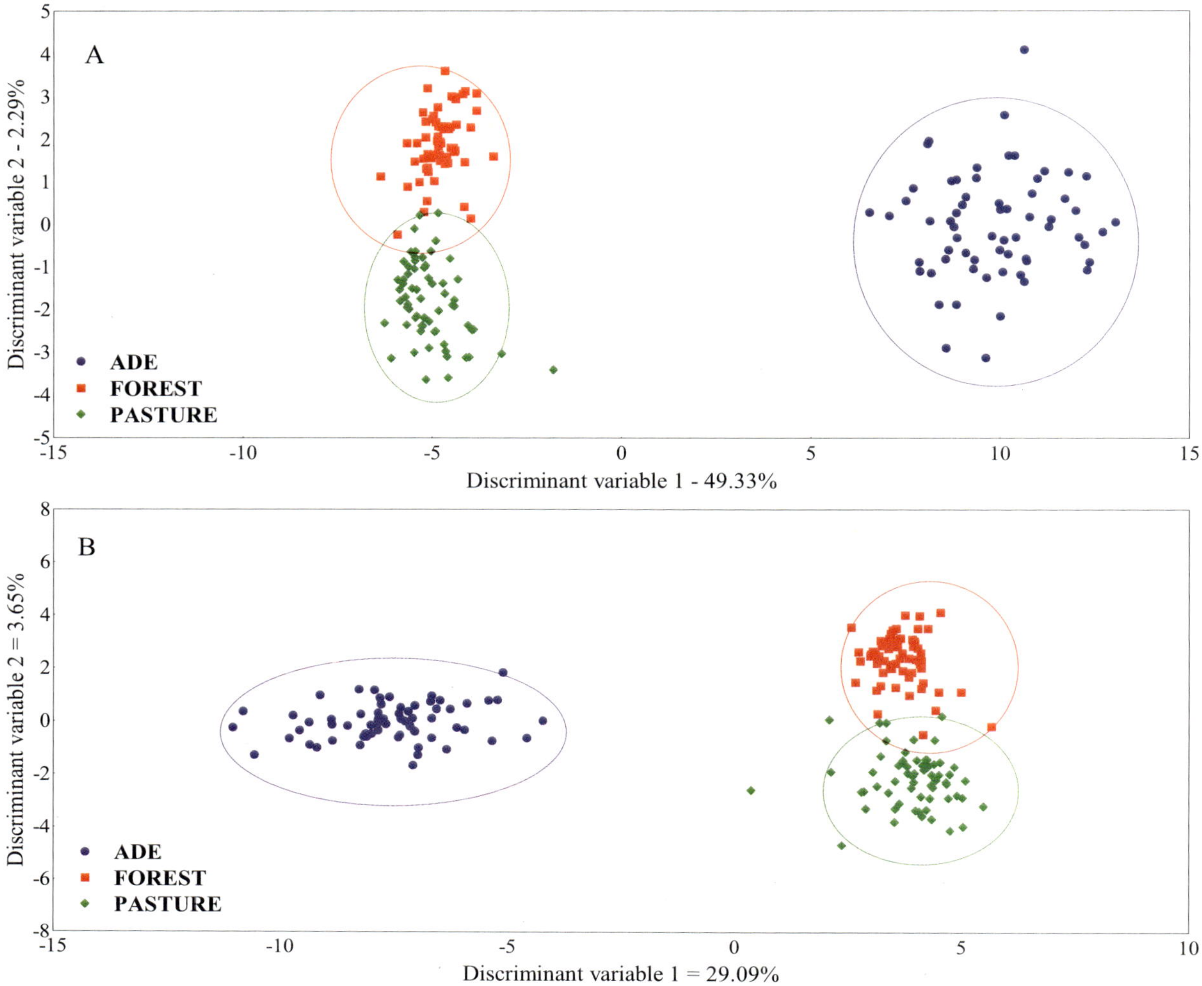

Figure 5. Area clustering by Canonical Discriminant Analysis based on studied attributes in Southern Amazonas. A = 0.0–0.2 m depth; B = 0.4–0.6 m depth. ADE= archaeological dark earth.

In contrasting environments, Table 4 shows that the highest F values are found in ADE area compared to other areas, especially pasture, which demonstrates that ADE differs significantly from the other soils. In contrast, pasture areas and forest have similar characteristics, as observed by their smaller F values, which had no significant difference for both depths.

Table 4. Multivariate analysis contrasting the studied areas in Southern Amazonas.

Managements	Depth of 0.0 – 0.2 m	Depth 0.4 – 0.6 m
	Value of F	Value of F
ADE x Pasture	687.8156***	417.1080***
ADE x Forest	664.7170***	401.0062***
Pasture x Forest	41.5622	65.8633

(***) All values are significant at $p < 0.001$.

This behaviour can be explained by the higher soil fertility and the anthropic horizon presence, as shown by Campos et al. (2012a), who conducted a physical and chemical characterization of soils with anthropic and non-anthropogenic horizons. The authors found that soils with anthropic horizons have better physical and chemical qualities than do soils without the presence of an anthropic horizon, which is common in this region.

In this regard, while assessing Figures 3, 4 and 5 in terms of the main component, it can be seen that the ADE area lies in the third and fourth quadrants and has a greater distance from the forest and pasture areas. The forest is in the same quadrant as the pasture area, demonstrating that these environments have similar features. Within this context, evaluating these areas by discriminant analysis, a similar behaviour is noted; additionally, it shows which connections exist between the areas. Pasture overlaps the forest area, thus demonstrating a greater homogeneity between these environments.

Therefore, in this sense, we can highlight that Ca^{2+}, clay and H+Al are closely linked to ADE, pasture and forest areas, respectively. These attributes may undergo major alterations when the areas are subjected to human actions; thus, the attributes may be possible indicators of environmental changes. An attribute that has been related to a particular environment can provide a measure of environmental change or impact if the components or attributes in the soil are found to have been switched. The use of soil attribute identifiers from different environments is a key tool for driving the practices that reduce depletion to acceptable levels because the pH in H_2O, H+Al, OM and Ca^{2+} contribute greatly to distinguishing the environments, which demonstrates the chemical attributes that are responsible for differentiation among the three environment.

Conclusion

Forest and pasture areas present similar behaviours and attributes, which differ from those of archaeological dark earth areas (ADE) that have greater fertility from anthropic horizons in Amazonian soils.

The soil attributes that are most suitable to use as pedoenvironmental indicators are calcium, potential acidity and clay for dark archaeological, forest and pasture soils, respectively.

Acknowledgements

We would like to thank the Foundation of Research Support of the Amazonas State – FAPEAM (Proc.1270/2012-FAPEAM) and Foundation of Research Support of the São Paulo State - FAPESP and the Amazonian State Registrar's Office of Science, Technology and Innovation – SECT/AM by financial support to the research.

References

Barreto, A. C., Lima, F. H. S., Freire, M. B. G. S., Araújo, Q. R., & Freire, F. J. (2006). Características químicas e físicas de um solo sob floresta, sistema agroflorestal e pastagem no Sul da Bahia. *Revista Caatinga*, *19*(4), 415-425.

Barros, K. R. M., Lima, H. V., Canellas, L. P., & Kern, D. C. (2012). Fracionamento químico da matéria orgânica e caracterização física de Terra Preta de Índio. *Revista de Ciências Agrárias*, *55*(1), 44-51.

Barroso, L. P., & Artes, R. (2003). *Análise multivariada*. Lavras, MG: UFLA.

Bartington Instruments [Bartington]. (2013). *Operation Manual for MS2 Magnetic Susceptibility System*. Oxford, England: Bartington Instruments. Retrieved from www.bartington.com/Literaturepdf/Operation%20 Manuals/om0408%20MS2.pdf

Benites, V. M., Moutta, R. O., Coutinho, H. L. C., & Balieiro, F. C. (2010). Análise discriminante de solos sob diferentes usos em área de mata atlântica a partir de atributos da matéria orgânica. *Revista Árvore*, *34*(4), 685-690.

Campos, M. C. C., Ribeiro, M. R., Souza Junior, V. S., Ribeiro Filho, M. R., & Almeida, M. C. (2012b). Topossequência de solos na transição Campos Naturais-Floresta na região de Humaitá, Amazonas. *Acta Amazonica*, *42*(3), 387-397.

Campos, M. C. C., Ribeiro, M. R., Souza Júnior, V. S., Ribeiro Filho, M. R., & Oliveira, I. A. (2010). Interferências dos pedoambientes nos atributos do solo em uma topossequência de transição Campos/Floresta. *Revista Ciência Agronômica*, *41*(4), 527-535.

Campos, M. C. C., Santos, L. A. C., Silva, D. M. P., Mantovanelli, B. C., & Soares, M. D. R. (2012a). Caracterização física e química de terras pretas arqueológicas e de solos não antropogênicos na região de Manicoré, Amazonas. *Revista Agro@mbiente On-line*, *6*(2), 102-109.

Cunha, T. J. F., Madari, B. E., Canellas, L. P., Ribeiro, L. P., Benites, V. M., & Santos, G. A. (2009). Soil organic matter and fertility of anthropogenic dark earths (terra preta de índio) in the brazilian amazon basin. *Revista Brasileira de Ciência do Solo*, *33*(1), 85-93.

Empresa Brasileira de Pesquisa Agropecuária [Embrapa]. (2011). *Manual de métodos de análises de solo* (2a. ed.). Rio de Janeiro, RJ: Embrapa Solos.

Empresa Brasileira de Pesquisa Agropecuária [Embrapa]. (2013). *Sistema Brasileiro de Classificação de Solos* (3a. ed., revisada e ampliada). Rio de Janeiro, RJ: Embrapa Solos.

Hair, J. R., Anderson, R. E., Tatham, R. L., & Black, W. C. (2005). *Análise multivariada de dados*. Porto Alegre, RS: Bookman.

Kaiser, H. F. (1958). The varimax criterion for analytic rotation in factor analysis. *Psychometrika*, *23*(3), 187-200.

Kern, D. C., Kämpf, N., Woods, W. I., Denevan, W. M., Costa, M. L., Frazão, F. J. L., & Sombroek, W. (2009). As Terras Pretas de Índio na Amazônia: Evolução do Conhecimento em Terra Preta de Índio. In W. G., Teixeira, D. C., Kern, B. E., Madari, & H. N. Lima, (Eds.). *As Terras Pretas de Índio da Amazônia: sua caracterização e uso deste conhecimento na criação de novas áreas* (p. 72-81), Manaus, AM: Embrapa Amazônica Ocidental.

Kume, H. (1993). *Métodos estatísticos para melhoria da qualidade* (4a. ed.). São Paulo, SP: Gente.

Lima, H. N., Schaefer, C. E. R., Mello, J. W. V., Gilkes, R. J., & Ker, J. C. (2002). Pedogenesis and pre-Columbian land use of "Terra Preta Anthrosols" (Indian black earth") of Western Amazonia. *Geoderma*, *110*(1-2), 1-17.

Oliveira, I. A., Campos, M. C. C., Freitas, L., & Soares, M. D. R. (2015). Caracterização de solos sob diferentes usos na região sul do Amazonas. *Acta Amazonica*, *45*(1), 1-12.

Raij, B. V., Andrade, J. C, Cantarella, H., & Quaggio, J. A. (2001). *Análise química para avaliação da fertilidade do solo*. Campinas, SP: Instituto Agronômico.

Rockström, J., Steffen, W., Noone, K., Persson, A., Chapin, F. S., Lambin, E. F., ... Foley, J. A. (2009). A safe operating space for humanity. *Nature*, *461*(7263), 472-475.

Silva, M. O., Freire, M. B. G. S., Mendes, M. A. S., Freire, F. J., Campos, M. C. C., & Amorim, L. B. (2010b). Discriminação de diferentes classes de solos irrigados com águas salinas, na região de Mossoró, RN, com o uso de análise multivariada. *Ambiência*, *6*(2), 261-270.

Silva, A. S., Lima, J. S. S., Xavier, A. C., & Teixeira, M. M. (2010a). Variabilidade espacial de atributos químicos de um Latossolo Vermelho-Amarelo húmico cultivado com café. *Revista Brasileira de Ciência do Solo*, *34*(1), 15-22.

3

Soil enzyme activities under pig slurry addition and different tillage systems

Elcio Liborio Balota*, **Oswaldo Machineski and Priscila Viviane Truber**

*Laboratório de Microbiologia do Solo, Área de Solos, Instituto Agronômico do Paraná, Cx. Postal 481, 86047-902, Londrina, Paraná, Brazil. *Author for correspondence. E-mail: balota@iapar.br*

ABSTRACT. The objective of this work was to evaluate the changes in soil enzyme activities due to application of pig slurry under different soil tillage systems. The experiment was conducted in a clayey Oxisol in Palotina, PR, using different quantities of pig slurry (0, 30, 60 and 120 m^3 ha^{-1} $year^{-1}$) applied to the soil prior to the summer and winter crop season under conventional tillage (CT) and no tillage (NT), with three replicates. The areas were cultivated with soybean (*Glycine max* L.) and maize (*Zea mays* L.) in the summers of 1998 and 1999, respectively, and with wheat (*Triticum sativum* Lam.) in the winters of both years. The soil samples were collected in March and October of 1998 and 1999 at depths of 0-5 , 5-10 and 10-20 cm. The pig slurry application and the soil tillage systems influenced enzyme activities in the soil. The increase of pig slurry application decreased the ratio of soil enzyme activities to microbial biomass carbon. The phosphatase activity had a negative relationship with soil-available phosphorus. The acid phosphatase activity decreased both under CT and NT systems at all depth studied due to pig slurry application.

Keywords: arylsulfatase, acid phosphatase, swine manure, no-tillage systems, nutrient cycling.

Introduction

More than 37 million heads of swine are produced in Brazil annually (MIRANDA, 2007). This level of swine production produces a large amount of waste: about 300 million liters of liquid dejections (pig slurry) a day. This volume of pig slurry is so large because of the large volumes of water used for the removal of swine dejections (feces, urine and ration remains) from the production units to a storage tank, where it is stored for at least 120 days to reduce its pollutant potential, before its application to the soil (OLIVEIRA; PARIZOTTO, 1993; SEGANFREDO, 2007).

Almost 50% of Brazilian swine production occurs in the South States (Paraná, Santa Catarina and Rio Grande do Sul), which are characterized by a high density of small farms (MIRANDA, 2007). This leads to the production of a huge amount of pig slurry, which often exceeds the quantity than can be safely applied to the soil. The slurry may also eventually reach the water supply. For these reasons, swine production is considered an activity accompanied by great potential for environmental damage.

The utilization of pig slurry as fertilizer is the most common practice, mainly because it is an easy and inexpensive solution for disposal of swine dejections. Although it is a valuable fertilizer, the successive application of high amounts of pig slurry to soil can cause soil nutritional disorders and environmental damage (SEGANFREDO, 2007).

Different soil management systems can significantly affect the soil's properties, especially the microbiological properties that play important roles in determining nutrient cycling and soil quality. Microbiological properties can be used as soil quality indicators because soil microorganisms are the second most important (after plants) biological factor in the agricultural ecosystem (YAKOVCHENKO et al., 1996). Therefore, biological and biochemical mediated processes in soils are fundamental to terrestrial ecosystem function (DICK, 1994).

The decomposition of soil organic matter is mediated by microorganisms, which produce enzymes that catalyze the innumerable reactions necessary for their own metabolic processes, decomposition of organic residues, nutrient cycling, and formation of organic matter and soil structure (DICK, 1994). Most soil enzymes are produced by microorganisms and can accumulate in the soil to have a considerable impact on nutrient recycling (DICK, 1997; TABATABAI, 1994). This allows soil enzyme activity to be used to indicate the intensity of certain biochemical processes because the enzyme activity is considered to be a major contributor to the overall microbial activity in soil (DICK, 1992). As enzyme levels increase, the rate of the relevant chemical conversion also increases.

The ecological significance of soil enzyme activities has been discussed in several papers (NANNIPIERI et al., 1990; DICK, 1997). In general, increases in enzyme activity are associated with increased biological activity in the soil, which is considered to be an important indicator of soil quality (DICK, 1992, 1994). Soils managed with organic amendments, such as animal manure, generally have a larger and more active microbial community than those managed with mineral fertilizers (McGILL et al., 1986; FAUCI; DICK, 1994). However, in studies on the effects of soil management, the microbiological properties of the soil have received less emphasis than its chemical and physical properties. For example, there are no studies to determine the effects of pig slurry application on soil enzymes under Brazilian conditions.

The objective of this study was to evaluate the effects that crescent and successive pig slurry application under different soil tillage systems have on soil enzyme activities.

Material and methods

Experimental characteristics and soil sampling

The experiment was established in 1996 at the Experimental Station of IAPAR in Palotina, west of Paraná State (24° 17' S, 53° 50' W). The soil was a clayey Oxisol classified as a Latossolo Vermelho eutrófico (Rhodic Eutrudox), according to Brazilian system of soil classification (BHERING; SANTOS, 2008), having 60% clay, 16% silt and 24% sandy. The soil had a pH of 5.2 ($CaCl_2$), 14.8 mg kg^{-1} of P (Mehlich) and 20.0 g kg^{-1} of organic carbon (Walkley-Black) in the surface layer (0-20 cm).

The experiment was installed in a split-plot block design with tillage as the main plot (100 x 5 m) and pig slurry addition as the subplot (20 x 5 m), which was separated by a 2.0 m buffer with three replicates. Tillage treatments were NT planting into undisturbed soil by opening a narrow trench and CT with one disc plowing at a 20 cm depth and two light harrowings for seedbed preparation. Liquid pig manure was added in four doses (0, 30, 60 and 120 m^3 ha^{-1} $year^{-1}$). Half was applied during the summer crop season, and the other half was applied during the winter crop season. Both applications occurred before the soil preparation. The area was cultivated with soybean (*Glycine max* L.) in summer of 1998 and maize (*Zea mays* L.) in 1999 and wheat (*Triticum sativum* Lam.) in the winter of both years. For both the fall and spring operations each year, crop stubbles were retained on the surface in the NT system and they were tilled conventionally (plowed to 20 cm depth) following harvest.

Five sub-samples of soil were taken randomly within each replicate (0-5, 5-10 and 10-20 cm depths) in March and October of 1998 and 1999, at the end of the summer and winter crops. The fresh soil samples were sieved through a 4 mm mesh to remove large plant material and were stored at 4°C until analysis.

Enzyme activity analyses

Amylase (EC 3.2.1) and cellulase (EC 3.2.1.4) activities were determined according to a modified methodology of Pancholy and Rice (1973) and Deng and Tabatabai (1994a). The activities of arylsulfatase and acid phosphatase were measured according to the methods of Tabatabai (1994). Arylsulfatase activity (arylsulfate sulfohydrolase, EC 3.1.6.1) was determined by using a colorimetric determination of *p*-nitrophenol released when soil samples were incubated with *p*-nitrophenyl sulfate. Acid phosphatase activity (EC 3.1.3) was analyzed with a modified universal buffer (MUB) (pH 6.5) by using a colorimetric determination of *p*-nitrophenol released when soil samples were incubated with *p*-nitrophenyl phosphate. Activities of arylsulfatase and phosphatase were expressed as μg *p*-nitrophenol (PNP) g^{-1} h^{-1}. All enzyme activities were measured using a one-hour incubation period. All determinations were made in triplicate and were expressed on a dry-weight basis. Within each treatment, data were averaged over the four seasons and two years and were then subjected to ANOVA statistical analysis using the SAS statistical software package. Soil enzyme activities for each tillage system and depth were submitted to regression analysis. It was chosen the significant equation by F test ($p \leq 0.05$) that presented higher correlation coefficient (R^2).

Results and discussion

The average measurements obtained from twelve samples in four seasons' samplings indicate that crescent pig slurry application and soil tillage systems did influence the soil enzyme activities. The different enzymes presented diverse tendencies due to the pig slurry doses, the soil tillage and the depth. In general, enzyme activities in the NT system were higher than those observed in the CT system (Figures 1, 2, 3 and 4).

Amylase activity varied from 176 to 647 μg g^{-1} h^{-1} in the soil under CT and from 529 to 995 μg g^{-1} h^{-1} in the NT system (Figure 1). For CT at the 0-5-cm depth and for both CT and NT at the 5-10-cm depth, the amylase activity increased linearly due to an increase in pig slurry doses. Under NT at the 0-5-cm depth and under CT and NT at the 10-20-cm depth, the amylase activity was significantly increased due to the addition of 30 m^3 ha^{-1} of pig slurry. The activity decreased with the addition of 60 m^3 ha^{-1} of pig slurry, and application of 120 m^3 ha^{-1} of pig slurry led to amylase activity values similar to those observed with the addition of 30 m^3 ha^{-1}. However, regression analysis for these data resulted in a non-significant model.

Cellulase activity varied from 18 to 176 μg g^{-1} h^{-1} in the soil under CT and from 39 to 189 μg g^{-1} h^{-1} in the NT system (Figure 2). At the 0-5-cm depth, the cellulase activity fit a linear model and was inversely proportional to the addition of pig slurry under the NT system. While under CT system the cellulase activity increased with increases of pig slurry addition. At 5-10-cm depth, there was a slight increase in cellulase activity with pig slurry application, and the data fit a quadratic model for both the CT and the NT systems. For the 10-20-cm depth, there was a slight decrease (which fit a quadratic model) for the NT system, and there was a significant linear decrease for the CT system.

The amylase and cellulase activities were within the ranges reported in the literature; additionally, they were similar to those observed in high fertility soils in Brazil (from 350 to 830 μg g^{-1} for amylase and from 67 to 220 μg g^{-1} for cellulose) (BALOTA et al., 2004). These enzymes play important roles in nutrient mineralization because they participate actively in residue decomposition. This decomposition provides readily available carbon for the growth of microorganisms (DENG; TABATABAI, 1994b). Cellulose is the most abundant compound in the biosphere and makes up almost half of the biomass synthesized by photosynthetic fixation of CO_2. Addition of organic material generally increases the enzyme activities in soils, because the added organic compounds represent a supply of readily available substrates for microorganisms. The additional organic compounds lead to increasing glycosidase activity, a chemical activity that plays a major role in the degradation of carbohydrates in soils (DENG; TABATABAI, 1996).

Arylsulfatase activity in the soil varied from 10.1 to 20.1 μg PNP g^{-1} h^{-1} under the CT system and from 14.9 to 33.4 μg PNP g^{-1} h^{-1} under the NT system (Figure 3). At the 0-5-cm depth, the arylsulfatase activity fit a linear model with an inverse relationship to the addition of pig slurry under the NT system. While under CT system, the cellulase activity increased due to an increase in pig slurry addition. The arylsulfatase activity under NT was about 204 and 100% higher than the CT system with application of 0 and 30 m^3 ha^{-1} of pig slurry, respectively. At 120 m^3 ha^{-1} of pig slurry, the values of arylsulfatase are practically the same for both the CT and the NT systems. At the 5-10-cm depth with pig slurry application, there was a slight increase in arylsulfatase activity that fit a quadratic model for the NT system, and there was a significant increase that fit a linear model due to pig slurry addition under the CT system. At the 10-20-cm depth, there was a significant increase in activity that a fit linear model under both the NT and the CT systems.

Arylsulfatase activity levels reported in the literature vary widely (from 4 to 770 μg g^{-1} h^{-1}) in many regions of the world (BALIGAR et al., 1999; BALOTA et al., 2004; GUPTA et al., 1993). In Brazil, the reported levels of arylsulfatase activity vary from 4.0 to 104 μg g^{-1} h^{-1} (BALIGAR et al., 1999; BALOTA et al., 2004). However, the level of arylsulfatase activity obtained in this study was similar to that observed in another area in Paraná

State under clayey soil with high natural fertility and subjected to different soil tillage systems (BALOTA et al., 2004).

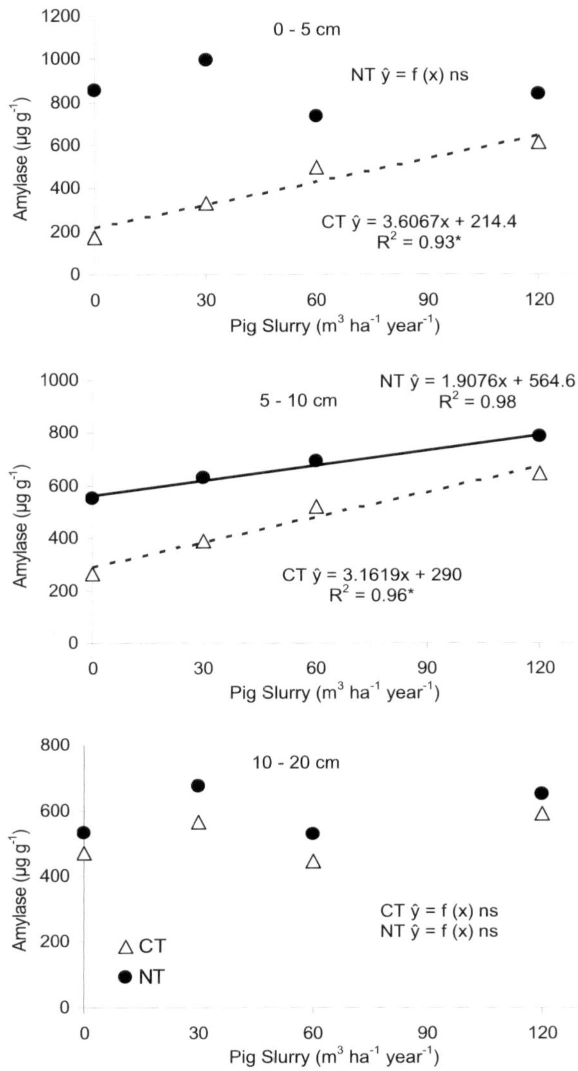

Figure 1. Amylase activity as affected by the application of pig slurry under different soil tillage systems. All values represent the mean of twelve replicates (four sampling seasons and three replicates).
*Significant at 5% of probability; ns: non-significant.

It has been suggested that the addition of organic compounds can significantly increase arylsulfatase activity (DENG; TABATABAI, 1997). The organic carbon from plant residues constitutes a principal reservoir of sulfate esters, which are the substrates for this group of enzymes. Arylsulfatases are one of many types of sulfatases involved in the mineralization of sulfate esters. Most arylsulfatases are not constitutive enzymes, and their synthesis by microorganisms may be controlled by the carbon and sulfur content of the system (DICK, 1997; TABATABAI, 1994).

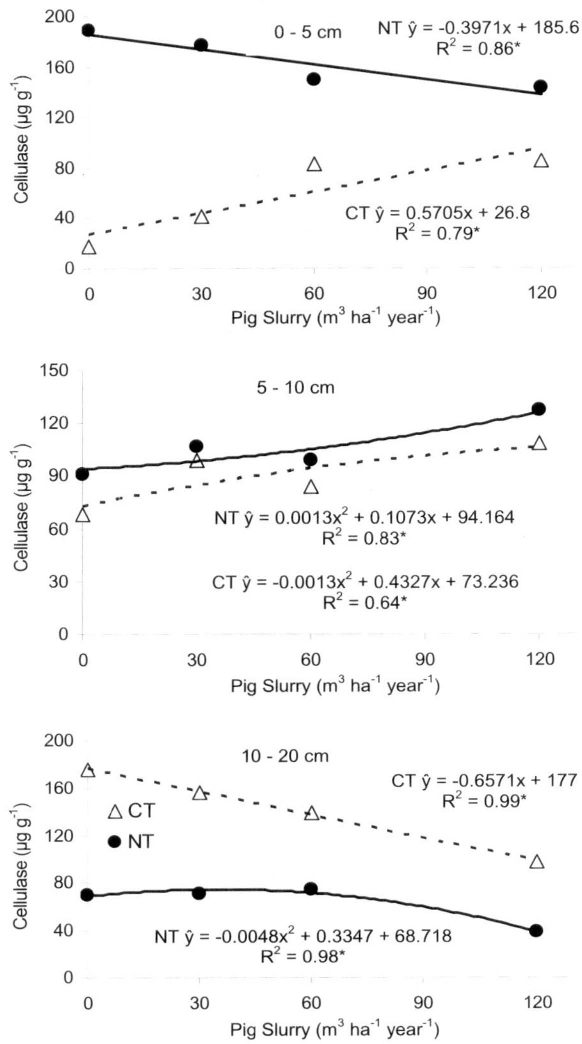

Figure 2. Cellulase activity as affected by the application of pig slurry under different soil tillage systems. All values represent the mean of twelve replicates (four sampling seasons and three replicates).
*Significant at 5% of probability.

Acid phosphatase activity varied from 83 to 300 μg PNP g^{-1} h^{-1} in the soil under CT and from 171 to 387 μg PNP g^{-1} h^{-1} under the NT system (Figure 4). At the 0-5-cm depth, the phosphatase activity decreased linearly due to the addition of pig slurry when using NT system, and the change in activity fit a quadratic model under the CT system. At the 5-10 cm and the 10-20-cm depth, the phosphatase decreased slightly under NT and significantly decreased under CT due to increase of pig slurry addition.

Our results for acid phosphatase activity are consistent with those reported in the literature, which vary widely from 55 to 1165 μg g^{-1} h^{-1} in several studies in Brazil (BALIGAR et al., 1999; CARNEIRO et al., 2004). However, the phosphatase activity level obtained was about two times lower than that observed in another region of Paraná State under clayey soil with high natural fertility and subjected to different soil tillage (BALOTA et al., 2004).

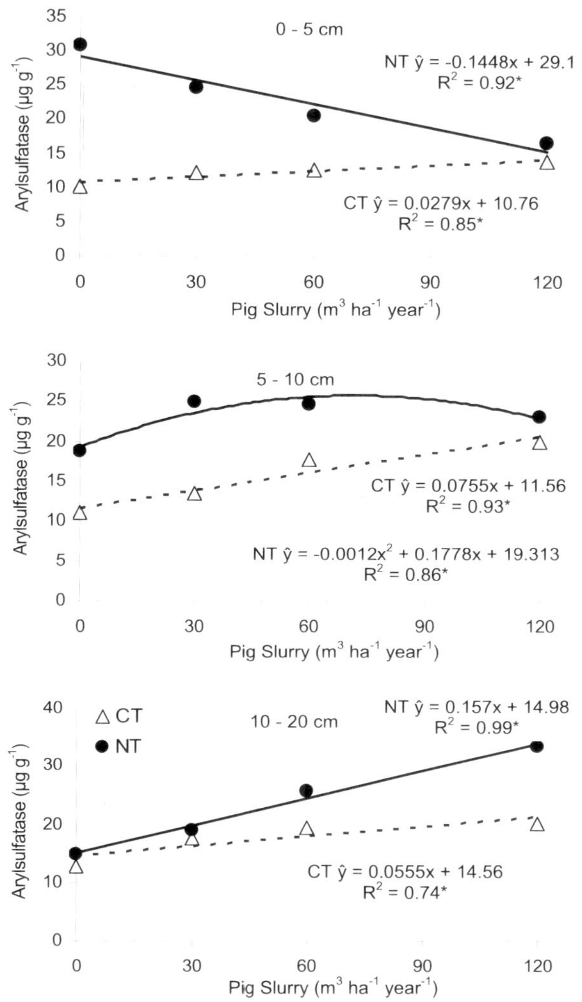

Figure 3. Arylsulfatase activity as affected by the application of pig slurry under different soil tillage systems. All values represent the mean of twelve replicates (four sampling seasons and three replicates).
*Significant at 5% of probability.

Acid phosphatases have been studied extensively because of their importance in soil organic phosphorus mineralization and plant nutrition and because of their optimum activity under acidic conditions (DICK, 1997; TABATABAI, 1994). Phosphatase is the general name of a broad group of enzymes that catalyze the hydrolysis of both esters and anhydrides of H_3PO_4. Because of their high metabolic activity and short lifespan, microorganisms are the most important sources of phosphatases in the soil. It has been suggested that phosphatases are produced when the available phosphorus content reaches critical levels for plant and microorganism growth (SPIERS; McGILL, 1979). In natural systems, such as forests, there is a large amount of microorganism growth without the addition of phosphate fertilizers, even when the level of available phosphorus is low. In these systems, available phosphorus is controlled by the cycling of organic phosphorus, and microbial biomass is an essential component of this process. On the other hand, in agricultural systems, added phosphorus fertilizers may depress phosphatase activity (SPIERS; McGILL, 1979).

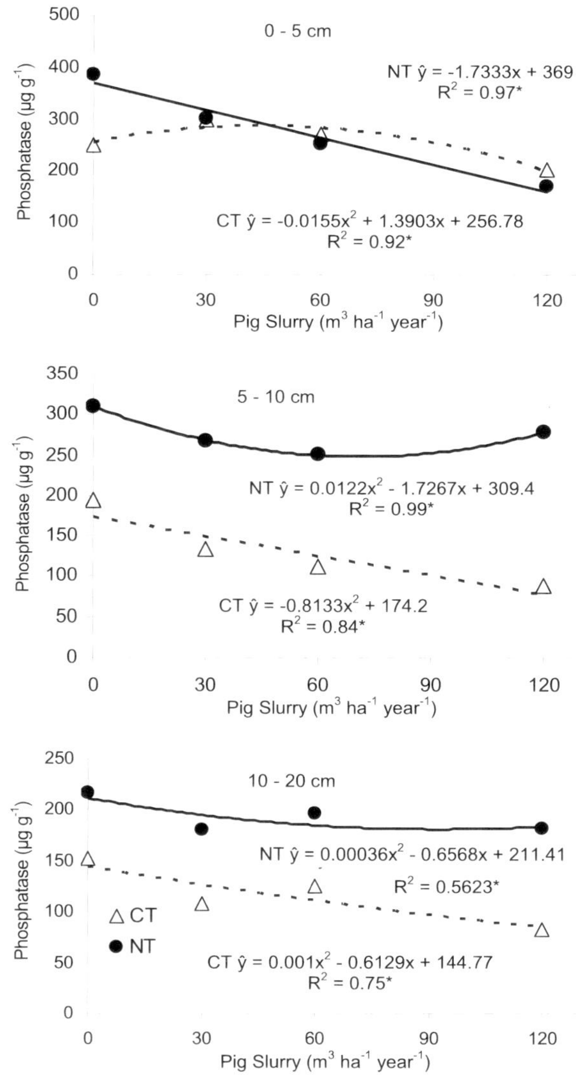

Figure 4. Acid phosphatase activity as affected by the application of pig slurry under different soil tillage systems. All values represent the mean of twelve replicates (four sampling seasons and three replicates).
*Significant at 5% of probability.

Our measurements of the change in soil enzyme activities due to pig slurry application are in agreement of those observed previously by Lalande et al. (2000) and by Rochette et al. (2000).

The application of 60 and 120 Mg ha^{-1} $year^{-1}$ for 19 consecutive years increased the denitrifying enzyme activity by up to 200% for up to three months following the pig slurry application (ROCHETTE et al., 2000). In another long-term field experiment, the application of 90 m^3 ha^{-1} of pig slurry for 18 consecutive years increased the acid phosphatase and arylsulfatase activities at the 0-15 cm depth by 80% when compared to the control 28 days after pig slurry application. At the 15-30 cm depth, there were no observed differences (LALANDE et al., 2000).

The alteration of soil enzyme activities due to pig slurry application may be in response to a higher nutrient immobilization capacity of the microbial community. An increasing supply of organic compounds increases the supply of readily available substrates, such as carbohydrates, for the microorganisms that produce the majority of soil enzymes. Pig slurry is a rich source of both organic and inorganic nutrients. It contains, on a dry matter basis, approximately 3.4% N, 4.2% P and 5.6% K (OLIVEIRA; PARIZOTTO, 1993) as well as several micronutrients needed for crop production. The inorganic nutrients are readily available, while the organic nutrients gradually become available over time. The application of 30, 60 and 120 m^3 ha^{-1} $year^{-1}$ of pig slurry is equal to the addition of 220, 440, and 860 kg ha^{-1} of ammonium sulfate; 570, 1140 and 2280 kg ha^{-1} of simple superphosphate; and 134, 268 and 536 kg ha^{-1} of potassium chloride, respectively. Although pig slurry only provides low levels of available carbon (approximately 1%) for soil microorganisms, it is a source of large amount of nutrients and organic compounds that can stimulate the activity of the microbial biomass.

This great variation in soil enzyme activities due to crescent pig slurry application and soil tillage systems shows that these enzymes are sensitive to soil disturbance. In general, amylase, cellulase and arylsulfatase activity increased with depth under the CT system and decreased with depth under the NT system. The acid phosphatase activity decreased with depth under both the CT and the NT systems. The decrease in soil enzymes with depth may be related to the lower labile C- and O_2 content, which may result in a smaller microbial biomass and lower activity than that in the surface layer. These results indicate that the addition of pig slurry under different soil tillage systems causes diverse effects on the microbial population composition (responsible for producing different enzymes) at different depths.

The decrease in soil enzyme activities due to pig slurry application may be due to a decrease in microorganism growth or due to the stimulation of O_2 demand that consequently creates anaerobic zones (LALANDE et al., 2000). Alternatively, the high rate of pig slurry application may simply create an anaerobic zone in the soil. Also, the application of high amounts of pig slurry was associated with the rainy season and may have caused temporary anaerobic conditions in some soil zones, and effect that could have affected microorganism growth.

The application of high amounts of pig slurry (120 m^3 ha^{-1}) has been associated with the decline of the microbial population and its activity, and it has also been associated with decreased levels of acid phosphatase (LALANDE et al., 2000). However, this decrease in acid phosphatase was also attributed to high soil phosphorus content due to consecutive pig slurry applications. Another important point is that, normally, long-term experiments result in a selection pressure on the microbial community due to the soil management practiced. This selection may promote the development of microorganisms specific for that soil's conditions.

Another effect of manure additions on soil is that the alteration of the organic carbon pool can also protect soil enzymes via association with organic and inorganic colloids that contribute to enzyme stabilization in soils (DICK, 1994; NANNIPIERI et al., 1990). Plant roots also stimulate enzyme activity by creating favorable microhabitats (e.g., increased porosity, water and a diversity of compounds) for microorganisms (DICK, 1997). In addition to the input of carbon contributed to the soil by pig slurry by above- and below-ground biomass, soil carbon is also contributed by root exudates. In general, from 7.0 to 64% of the net carbon fixed by plants is lost through respiration and rhizodeposition in the soil (WHIPPS, 1990). Root exudates can contain more than 200 carbon compounds (KUMAR et al., 2006) including sugars, amino acids, organic acids, hormones, vitamins, polymeric carbohydrates, enzymes and gases (ethylene and CO_2) (WHIPPS, 1990). These compounds, released by roots in the rhizosphere, can be utilized immediately by microorganisms to increasing microbial diversity, number and activity. These increases consequently lead to greater production and release of soil enzymes.

The increase in enzyme activities in the NT system relative to the CT system was 54% for amylase, 16% for cellulase, 53% for arylsulfatase and 48% for acid phosphatase. These results suggest that the decrease in soil disturbance (i.e., no-tillage) can improve microbial activity. Conventional tillage

(CT), which disturbs the soil to prepare the land, reduces soil structure by degrading soil aggregates. On the other hand, the lack of soil disturbance in the NT system increases microbial activity because it favors the formation and stabilization of macro aggregates, which provide habitats for microbiota (BEARE et al., 1994; POWLSON; JENKINSON, 1981). These habitats then allow for the expansion of the microbial community and the production of soil enzymes (DICK, 1992, 1997).

Important ecological information from soil enzymes can be obtained by calculating the ratio of soil enzyme activity to microbial biomass (KANDELER; EDER, 1993). Our results indicate that the increase in pig slurry application decreased the rate of soil enzyme activities to microbial biomass carbon (Figure 5). This is suggests that although the pig slurry application increased the soil enzyme activity, the amount of soil enzyme produced per unit of microbial biomass decreased. According to Kandeler and Eder (1993), this result also suggests that the amount of enzyme immobilized on clay or humic colloids in the soil was also decreased. It has been observed that enzymes produced in excess after the addition of organic compounds are rapidly decomposed (NANNIPIERI et al., 1983). However, it is important to study the mechanisms that increase and maintain a stable biological activity that affects enzyme activities.

Correlations

The enzyme activities were correlated with each other (Table 1), as observed in previous studies (BALOTA et al., 2004; DENG; TABATABAI, 1996, 1997; FRANKENBERGER; DICK, 1983). Amylase and arylsulfatase activity showed a strong relationship with microbial biomass carbon and organic carbon (Table 1), an effect that was not observed for cellulase and phosphatase activity. These close correlations between enzyme activities and organic carbon are likely due to higher carbon levels that support greater microbial biomass and thus have more enzymatic activity (DENG; TABATABAI, 1996, 1997). Furthermore, more organic matter provides a better environmental condition for stabilizing and protecting extracellular enzymes. There was a negative relationship between acid phosphatase activity and extractable phosphorus (Figure 6), confirming the observation that phosphatases are stimulated when soil phosphate levels are low (SPIERS; McGILL, 1979). However, this negative correlation was not found in several other studies (BALIGAR et al., 1999; BALOTA et al., 2004).

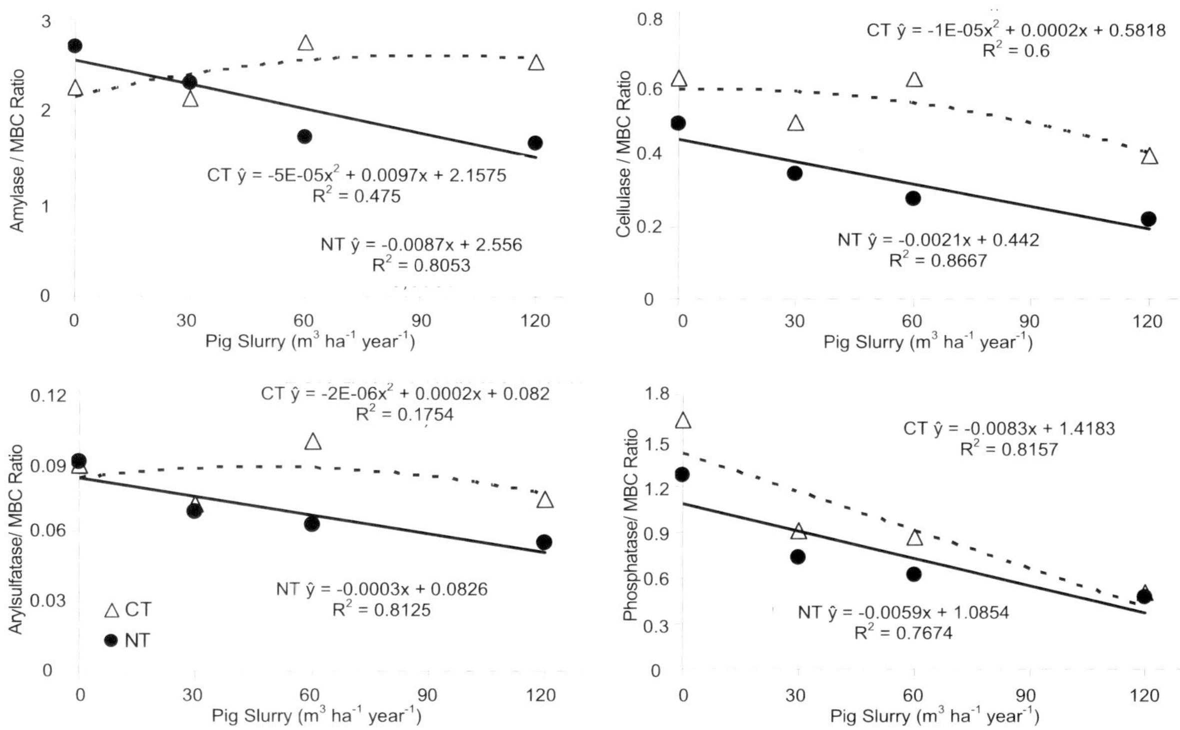

Figure 5. Ratio of soil enzyme activities to microbial biomass carbon (MBC) at the 0-20 depth, as affected by application of pig slurry doses under different soil tillage systems. All values represent the mean of 36 samples (four sampling seasons, three depths and three replicates).

Table 1. Simple correlations (r) between soil enzyme activities and microbial biomass across all treatments, depth and sampling season.

Variables*	Amylase	Cellulase	Arylsulfatase	Acid Phosphatase
Cellulase	0.62*	-	-	-
Arylsulfatase	0.67*	0.27	-	-
Acid Phosphatase	0.25	0.068	0.23	-
MBC	0.78*	0.21	0.56*	0.23
MBN	0.65*	0.35	0.24	0.18
MBP	0.53*	0.40	0.0073	0.033
Organic C	0.78*	0.50	0.40	0.34

*MBC: microbial biomass carbon; MBN: microbial biomass nitrogen; MBP: microbial biomass phosphorus.

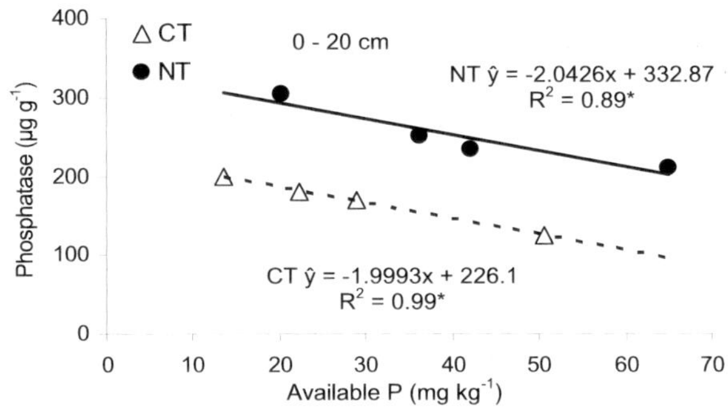

Figure 6. Relationship between acid phosphatase and available phosphorus at the 0-20 depth, as affected by application pig slurry doses under different soil tillage systems. All values represent the mean of 36 samples (four sampling seasons, three depths, and three replicates).
*Significant at 5% of probability.

Our results confirm previous observations that demonstrate the responsiveness of soil enzyme activities to changes in soil management practices. The results demonstrate that the measurement of soil enzyme activities presented a good sensitivity for detecting soil alterations due to the application of pig slurry and changes in soil tillage systems. Although swine dejections have a high potential for causing environmental damage, they can serve as an excellent source of nutrients for crop production and can increase the microbial activity and enzyme activities in the soil. These increases can consequently enhance soil fertility and soil quality.

Conclusion

Pig slurry application and changes in soil tillage systems influenced the soil enzyme activities. The soil enzyme activity was a sensitive measure for detecting changes in the soil due to the application of pig slurry and to changes in soil tillage systems. The increase of pig slurry application decreased the ratio of soil enzyme activities per unit of microbial biomass carbon. The phosphatase activity was inversely related to the available phosphorus in soil. The NT system provides for higher enzyme activities than the CT system.

References

BALIGAR, V. C.; WRIGHT, R. J.; FAGERIA, N. K.; PITTA, G. V. E. Enzyme activities in cerrado soils of Brazil. **Communications in Soil Science and Plant Analysis**, v. 30, n. 9, p. 1551-1560, 1999.

BALOTA, E. L.; KANASHIRO, M.; COLOZZI-FILHO, A.; ANDRADE, D. A.; DICK, R. P. Soil enzyme activities under long-term tillage and crop rotations systems in subtropical agro-ecosystems. **Brazilian Journal of Microbiology**, v. 35, n. 4, p. 300-306, 2004.

BEARE, M. H.; CABRERA, M. L.; HENDRIX, P. F.; COLEMAN, D. C. Aggregate-protected and unprotected organic matter pools in conventional and no-tillage soils. **Soil Science Society of American Journal**, v. 58, n. 3, p. 787-795, 1994.

BHERING, S. B.; SANTOS, H. G. **Mapa de solos do estado do Paraná**: legenda atualizada. Rio de Janeiro: Embrapa Florestas/Embrapa Solos/IAP, 2008.

CARNEIRO, R. G.; MENDES, I. C.; LOVATO, P. E.; CARVALHO, A. M.; VIVALDI, L. J. Indicadores biológicos associados ao ciclo do fósforo em solos de Cerrado sob plantio direto e convencional. **Pesquisa Agropecuária Brasileira**, v. 39, n. 7, p. 661-669, 2004.

DENG, S. P.; TABATABAI, M. A. Cellulase activity of soils. **Soil Biology and Biochemistry**, v. 26, n. 10, p. 1347-1354, 1994a.

DENG, S. P.; TABATABAI, M. A. Colorimetric determination of reducing sugars in soils. **Soil Biology and Biochemistry**, v. 26, n. 4, p. 473-477, 1994b.

DENG, S. P.; TABATABAI, M. A. Effect of tillage and residue management on enzyme activities in soils: II. Glycosidases. **Biology and Fertility of Soils**, v. 22, n. 3, p. 208-213, 1996.

DENG, S. P.; TABATABAI, M. A. Effect of tillage and residue management on enzyme activities in soils: III. Phosphatases and arylsulfatase. **Biology and Fertility of Soils**, v. 24, n. 2, p. 141-146, 1997.

DICK, R. P. A review: long-term effects of agricultural systems on soil biochemical and microbial parameters. **Agriculture Ecosystems and Environment**, v. 40, n. 1, p. 25-36, 1992.

DICK, R. P. Soil enzyme activities as indicators of soil quality. In: DORAN, J. W.; COLEMAN, D. C.; BEZDICEK, D. F.; STEWART, B. A. (Ed.). **Defining soil quality for a sustainable environment**. Madison: SSSA, 1994. p.107-124.

DICK, R. P. Soil enzyme activities as integrative indicators of soil health. In: PANKHURST, C. E.; DOUBE, B. M.; GUPTA, V. V. S. R. (Ed.). **Biological Indicators of Soil Health**. New York: CAB International, 1997. p. 121-156.

FAUCI, M. F.; DICK, R. P. Soil microbial dynamics: short- and long-term effects of inorganic and organic nitrogen. **Soil Science Society of American Journal**, v. 58, n. 3, p. 801-806, 1994.

FRANKENBERGER, W. T.; DICK, W. A. Relationships between enzyme activities and microbial growth and activity indices in soil. **Soil Science Society of American Journal**, v. 47, n. 5, p. 945-951, 1983.

GUPTA, V. V. S. R.; FARREL, R. E.; GERMIDA, J. J. Activity of arylsulfatase in Saskatchewan soils. **Canadian Journal of Soil Science**, v. 73, n. 3, p. 341-347, 1993.

KANDELER, E.; EDER, G. Effect of cattle slurry in grassland on microbial biomass and on activities of various enzymes. **Biology and Fertility of Soils**, v. 16, n. 4, p. 249-254, 1993.

KUMAR, R.; PANDEY, S.; PANDEY, A. Plant roots and carbon sequestration. **Current Science**, v. 91, n. 7, p. 885-890, 2006.

LALANDE, R.; GAGNON, B.; SIMARD, R. R.; CÔTÉ, D. Soil microbial biomass and enzyme activity following liquid hog manure application in a long-term field trial. **Canadian Journal of Soil Science**, v. 80, n. 2, p. 263-269, 2000.

McGILL, W. B.; CANNON, K. R.; ROBERTSON, J. A.; COOK, F. D. Dynamics of soil microbial biomass and water-soluble organic C in Breton L after 50 years of cropping to two rotations. **Canadian Journal of Soil Science**, v. 66, n. 1, p. 1-19, 1986.

MIRANDA, C. R. Aspectos ambientais da suinocultura brasileira. In: SEGANFREDO, M. A. (Ed.). **Gestão ambiental na suinocultura**. Brasília: Embrapa, 2007. p. 14-36.

NANNIPIERI, P.; GREGO, S.; CECCANTI, B. Ecological significance of biological activity in soil. In: BOLLAG, J. M.; STOZKY, G. (Ed.). **Soil Biochemistry**. New York: Marcel Dekker, 1990. v. 6, p. 293-357.

NANNIPIERI, P.; MUCCINI, L.; CIARDI, C. Microbial biomass and enzyme activities: production and persistence. **Soil Biology and Biochemistry**, v. 15, n. 6, p. 679-685, 1983.

OLIVEIRA, E.; PARIZOTTO, M. L. V. **Características e uso fertilizante do esterco de suíno**. Londrina: Iapar, 1993. (Circular, 83).

PANCHOLY, S. K.; RICE, E. L. Soils enzymes in relation to old field succession: amylase, cellulase, invertase, dehydrogenase, and urease. **Soil Science Society of American Journal**, v. 37, n. 1, p. 47-50, 1973.

POWLSON, D. S.; JENKINSON, D. S. A comparison of the organic matter, biomass, adenosine triphosphate and mineralizable nitrogen contents of ploughed and direct-drilled soils. **The Journal of Agricultural Science**, v. 97, n. 3, p. 713-721, 1981.

ROCHETTE, P.; BOCHOVE, E.; PRÉVOST, D.; ANGERS, D. A.; CÔTÉ, D.; BERTRAND, N. Soil carbon and nitrogen dynamics following application of pig slurry for the 19th consecutive year: II-Nitrous oxide and mineral nitrogen. **Soil Science Society of American Journal**, v. 64, n. 4, p. 1396-1403, 2000.

SEGANFREDO, M. A. Uso de dejetos suínos como fertilizante e seus riscos ambientais. In: SEGANFREDO, M. A. (Ed.). **Gestão ambiental na suinocultura**. Brasília: Embrapa, 2007. p. 150-175.

SPIERS, G. A.; McGILL, W. B. Effects of phosphorus addition and energy supply on acid phosphatase activity in soils. **Soil Biology and Biochemistry**, v. 11, n. 1, p. 3-8, 1979.

TABATABAI, M. A. Enzymes. In: WEAVER, R. W.; AUGLE, S.; BOTTOMLY, P. J.; BEZDICEK, D.; SMITH, S.; TABATABAI, A.; WOLLUM, A. (Ed.). **Methods of soil analysis**. Part 2. Microbial and biochemical properties. Madison: SSSA, 1994. p. 775-833.

WHIPPS, J. M. Carbon economy. In: LYNCH, J. M. (Ed.). **The Rhizosphere**. New York: John Wiley and Sons, 1990. p. 59-97.

YAKOVCHENKO, V.; SIKORA, L. J.; KAUFMAN, D. D. A biologically based indicator of soil quality. **Biology and Fertility of Soils**, v. 21, n. 4, p. 245-251, 1996.

4

Micromorphometric analysis of porosity changes in the surface crusts of three soils in the Piracicaba region, São Paulo State, Brazil

Selene Cristina de Pierri Castilho[1], Miguel Cooper[2*] and Laura Fernandes Simões da Silva[1]

*[1]Programa de Pós-graduação em Solos e Nutrição de Plantas, Escola Superior de Agricultura "Luiz de Queiroz", Universidade de São Paulo, Piracicaba, São Paulo, Brazil. [2]Escola Superior de Agricultura "Luiz de Queiroz", Universidade de São Paulo, Avenida Pádua Dias, 11, 13418-900, Piracicaba, São Paulo, Brazil. *Author for correspondence. E-mail: mcooper@usp.br*

ABSTRACT. The aim of this study was to micromorphologically analyze modifications in soil porosity due to the occurrence of erosive rainfall events. Three types of soils, Typic Hapludox, Typic Hapludult and Rhodic Kandiudalf, were analyzed to determine their total porosity and the distribution and number of pores. The soils were tilled with a heavy disk and leveling harrow to ensure homogeneity of surface conditions at the beginning of the trial period and to simulate cropping conditions. The soils were kept uncovered during the trial period, and rainfall events with intensities greater than 25 mm h^{-1} were recorded. The results showed that all of the soils presented crust formation, influenced by the soil texture. NVef showed significant changes in total porosity, pore numbers and pore morphology, while LVAd and PVAd did not display marked modifications in these attributes.

Keywords: pore morphology, erosive rainfall, structural changes.

Introduction

The formation of surface soil crusts has direct consequences for crop development and environmental quality because it reduces water infiltration into the soil and water availability for plant roots, prevents the emergence of seedlings, increases runoff and erosion (ASSOULINE, 2004), reduces agricultural yields and causes environmental impacts (TOUMA et al., 2011). Crusts are cohesive structures of variable thickness, formed by the reorganization and consolidation of soil particles, thereby changing the physical-hydraulic properties of the soil (BRANDÃO et al., 2006).

The susceptibility of aggregates to destruction caused by the impac of raindrops depends on soil properties such as texture, organic matter content, mineralogical composition and moisture. Other aspects, such as topographic factors, vegetation and rainfall features, also affect aggregate stability (FOX et al., 2004; HILLEL, 2004). The destruction of aggregates can be caused by natural conditions, such as the impact of raindrops on the soil, or by anthropogenic influences, such as the intensive tilling techniques used for the cultivation of agricultural lands (PAGLIAI et al., 2004). The disruption caused by agricultural machinery prior to planting leaves the soil in an unstable condition, breaking aggregates, releasing particles and making the soil susceptible to crusting (HAO et al., 2011). The formation of surface crusts reduces the number and continuity of pores in the soil, affecting water infiltration and its redistribution along the profile, thus promoting water erosion and increasing erosion rates (CARMI; BERLINER, 2008).

Evaluation of structural changes caused by management practices and soil crusting and quantification of attributes such as the pore size distribution can be achieved using micromorphometric methods (FOX et al., 2009), which are currently applied in a number of fields as a basis for soil morphology analyses (DALLA ROSA et al., 2012; PAGLIAI et al., 2004).

Based on the knowledge that rain and the intensive tillage practices applied in agricultural areas are the main factors leading to the degradation of soil structure and that soil structure is an important indicator of soil quality, this study aimed to monitor and evaluate porosity modifications in surface crusts in three types of soil under erosive rainfall events. Micromorphometric methods and image analysis were performed to better understand how rainfall events affect the reorientation and redistribution of soil pores.

Material and methods

The experiment was conducted in three soils in the Piracicaba region of São Paulo state, Brazil, in 5 x 5 m plots. The soils were a Typic Hapludox (LVAd) with a loamy texture and a 3% average slope; a Typic Hapludult (PVAd) with a loamy/sandy texture and a 4% average slope; and a Rhodic Kandiudalf (NVef) with a clayey texture and a 10% average slope.

Homogeneity of the initial conditions of each soil surface was achieved by preparing the plots using a leveling harrow, simulating planting conditions. The changes in the formation of surface crusts and soil porosity were monitored during the rainy season between November 2008 and May 2009. To ensure that rain would act directly on the soil surface, the plots remained uncovered during the entire trial period. Samplings were performed in four stages based on the sequence of erosive rainfall events. In this study, events with an intensity exceeding 25 mm h^{-1} were considered erosive (GONÇALVES et al., 2006). Thus, four sampling periods were defined: after tillage (T0), after the occurrence of three erosive rainfall events (T3), after the occurrence of five erosive rainfall events (T5) and after the occurrence of ten erosive rainfall events (T10). The samplings were conducted in three replicates at 24 hours after a rainfall event to allow drainage of any excess water.

Rainfall data were provided by the automatic meteorological station of the Department of Biosystems Engineering, ESALQ/USP, College of Agriculture "Luiz de Queiroz", municipality of Piracicaba, São Paulo state, Brazil.

To characterize the soil surface layer, a particle size distribution analysis was performed on disturbed samples collected at depths of 0 to 0.05 m and 0.1 to 0.15 m according to the methodology proposed by Camargo et al. (1986).

Micromorphometric analyses were conducted to monitor the formation and evolution of surface crusts and changes in soil porosity caused by crusting and raindrops. Undisturbed soil samples were used for the micromorphometric analyses, which were collected in three replicates using cardboard boxes of 0.12 x 0.07 x 0.05 m. The samples were air dried for 15 days and then dried in an oven at 40°C with ventilation for 48 hours. Impregnation was carried out using polyester resin and the fluorescent pigment Tynopal OB (BASF® - The Chemical Company, Germany) (MURPHY et al., 1977). Polished blocks were employed to obtain digital images for the analysis and quantification of soil porosity.

The blocks from each treatment (T0, T3, T5, T10) were illuminated with ultraviolet light and photographed using a charged couple device (CCD) digital camera (Sony®, model DFW-X700) attached to a petrographic microscope for porosity quantification. Images were acquired and processed using Noesis Visilog 5.4 image analysis software (MURPHY et al., 1977). The obtained images were treated via image binarization using the threshold function of the Noesis Visilog 5.4 program, establishing the limits of black and white in each image and transforming it from an RGB (red, green and blue) image to a black and white image. After binarization, the images were filtered using the open and close function, in which the image is opened and closed by 1 pixel, and any object smaller than 4 pixels is erased. This function ensures that all possible noise caused by the threshold is eliminated. After determining the basic and derived variables for the pores, they were classified by size and shape using the macro-functions developed by Juhász et al. (2007) in Visual Basic language in Microsoft Excel®. The applied quantification of total porosity evaluates the quantity of macro- and mesopores according to the classification presented by Bullock et al. (1985). Pore shape was classified according to the method of Cooper et al. (2010), which classifies pores as rounded (channels and isolated vughs), elongated (fissures) or complex (packing voids). Calculation of the quantity, size and type of pores as well as the porosity of the samples were performed through digital image analysis (COOPER et al., 2005; JUHÁSZ et al., 2007).

From the total porosity data, it is possible to classify the soil according to Pagliai et al. (2004), who categorizes soil as medium dense when it presents a total porosity below 10%; as moderately porous when it shows a total porosity of 10-25%; as porous when the total porosity is 25-40%; and as extremely porous when it displays a total porosity above 40%.

To investigate the differences between the obtained data, a mean comparison analysis was carried out. The experimental design consisted of random blocks, considering the soil types as blocks, and the treatments were represented by the rainfall events. The Duncan test was used at a 5% probability, and statistical analyses were performed using SAS® statistical software.

Results and discussion

In the descriptive analysis of the total porosity data (Table 1), the variability in the data observed for both LVAd and PVAd was considered low (CV < 12%), while the variability for NVef was considered intermediate (12% < CV < 60%) (WARRICK; NIELSEN, 1980). This analysis also showed that the distribution of the total porosity data was symmetrical, with a variability of 1% being observed between the mean and the median, resulting in a mean that adequately represented the dataset.

The results of the variance analysis of the total porosity data from the three analyzed soils showed that the total porosity of LVAd did not change significantly with the occurrence of rainfall, except at T3, when the average total porosity was reduced from approximately 60% (T0, T5 and T10) to approximately 47% (T3) (Figure 1). For PVAd, at T0, T3 and T5, no statistically significant changes were observed in the total porosity, which remained at approximately 60%. A significant reduction of the total porosity was observed only after T10, when it decreased to 55% (Figure 2). NVef showed the lowest mean total porosity, with a statistically significant difference being observed at T5, when the total porosity was reduced by half, from approximately 40% (T0) to approximately 20% (T5) (Figure 3).

Although some differences in total porosity were observed with the occurrence of the rainfall events, it is worth noting that throughout the trial period, both LVAd and PVAd were classified as extremely porous, while NVef changed from a porous to a moderately porous condition (PAGLIAI et al., 2004).

When the total soil porosity during the trial period, it is clear that both the LVAd and PVAd have similar porosity evolutions undergoing seemingly random changes that can be explained by the low clay content in the soils. At the beginning of the experiment (T0), the total porosity of LVAd was considered high (60%) reducing after the first three rains, reaching the value of 47% (T3).

Then, after the occurrence of five rainfall events, an increase in porosity to 66% is observed (T5) and a following decrease to approximately 56% after the occurrence of ten rainfall events (T10) (Table 1).

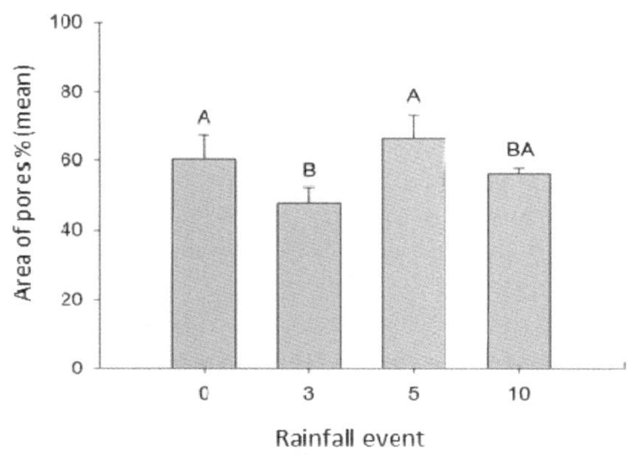

Figure 1. Total porosity of LVAd. Means followed by the same capital letter do not differ at 5% of probability.

Table 1. Descriptive statistical analysis on data of total soil porosity.

			Total Porosity (%)			
LVAd	Mean	Median	S	Max	Min	CV (%)
0 RE	60.24	56.68	7.3	68.64	55.38	12.13
3 RE	47.59	47.06	4.55	52.39	43.33	9.56
5 RE	66.56	65.51	6.7	73.72	60.44	10.06
10 RE	56.24	56.44	1.77	57.92	54.37	3.16
			Total Porosity (%)			
PVAd	Mean	Median	s	Max	Min	CV (%)
0 RE	57.21	58.55	3.35	59.7	53.4	5.86
3 RE	59.2	57.85	5.84	65.6	54.14	9.88
5 RE	63.56	62.5	3.44	67.41	60.78	5.41
10 RE	55.37	54.45	2.8	58.52	53.14	5.06
			Total Porosity (%)			
NVef	Mean	Median	s	Max	Min	CV (%)
0 RE	40.41	40.41	8.94	47.62	30.41	22.12
3 RE	29.02	39.28	9.23	38.65	20.24	31.81
5 RE	20.12	29.02	4.64	23.24	14.79	23.05
10 RE	39.28	20.12	8.62	47.44	30.26	21.95

RE – rainfall events; s – standard deviation; CV – coefficient of variation. LVAd - Typic Hapludox, PVAd - Typic Hapludult; NVef - Rhodic Kandiudalf.

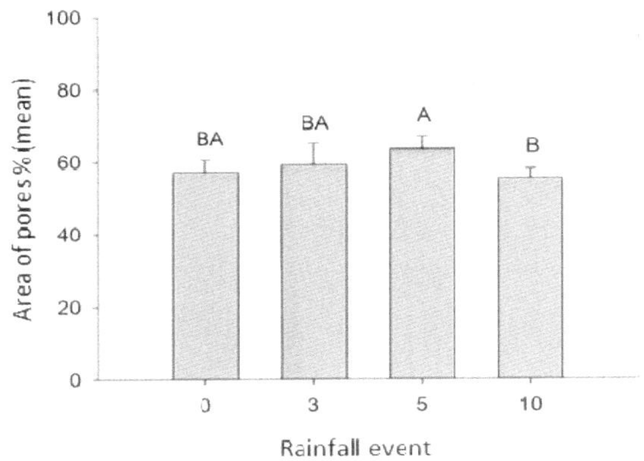

Figure 2. Total porosity of PVAd. Means followed by the same capital letter do not differ at 5% of probability.

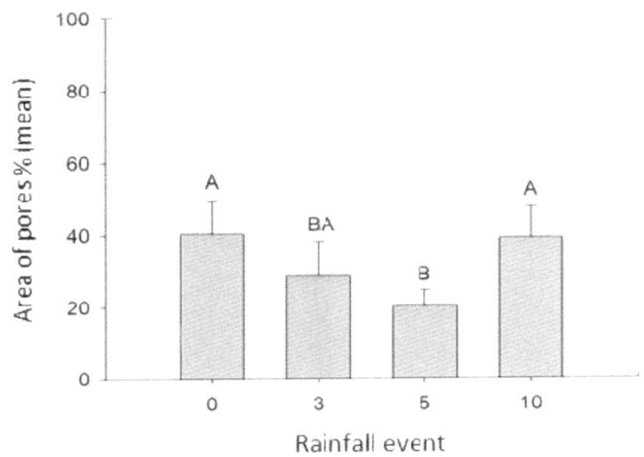

Figure 3. Total porosity of NVef. Means followed by the same capital letter do not differ at 5% of probability.

Following the same trend of random changes in total porosity in LVAd , the pore number also changed with the occurrence of rainfall, ranging from 390 pores at T0 to 545 at T3, 253 at T5 and 364 at T10 (Table 2).

In order to better understand the random behavior of soil porosity during the rainfall events, focus was placed on the particle size analysis of the soils, given that the amount of clay, silt and sand found in the soil can make it more or less susceptible to the action of erosive events (LE BISSONNAIS, 1996). The particle size analysis performed for the topsoil (0 – 0.05 m) and indicated similar clay, silt and sand contents in LVAd and PVAd of 200, 70 and 700 g kg^{-1}, respectively. For NVef, there was an increase in the clay content to 500 g kg^{-1} and a reduction of the sand content to approximately 400 g kg^{-1}.

The soil tillage performed to prepare the plots at the beginning of the trial period, together with the low clay content in the surface layers of LVAd and PVAd (200 g kg^{-1}), favored soil pulverization in the surface layer, altering its structure and facilitating the detachment and mobilization of particles during the occurrence of rainfall events (BERTOL et al., 2010). Thus, the behavior of soil porosity observed in the micromorphological analysis could be explained by the remobilization of less-structured material due to the impact of raindrops on the soil, which directly affects its structure, sometimes thickening or unpacking the soil matrix.

The difference in the clay content of NVef in relation to the other two soils can explain the differences in total soil porosity behavior associated with the occurrence of rainfall events. The higher clay content found in the surface horizon (500 g kg^{-1}) was responsible for the observed changes in porosity, as the soil surface undergoes compression and compaction through the impact of raindrops, increasing soil density and reducing its total porosity (FOX et al., 2004). The increase in total porosity observed at T10 was attributed to the appearance of fissures on the surface of NVef (Figure 4), generated by the compression and expansion of the soil material caused by the repeated wetting and drying cycles in the soil (COOPER et al., 2010).

Table 2. Descriptive statistical analysis for data of total number of pores of the soils.

	Total Number of Pores					
LVAd	Mean	Median	s	Max	Min	CV (%)
0 RE	390	406	126.75	508	256	32.5
3 RE	545	527	64.89	617	491	11.9
5 RE	253	213	102.05	369	177	40.33
10 RE	364	374	29.14	388	332	7.99
	Total Number of Pores					
PVAd	Mean	Median	s	Max	Min	CV (%)
0 RE	363.66	371	79.25	439	281	21.79
3 RE	371.33	357	178.93	557	200	48.18
5 RE	266	267	4.58	270	261	1.72
10 RE	390	427	72.91	437	306	18.69
	Total Number of Pores					
NVef	Mean	Median	s	Max	Min	CV (%)
0 RE	275	284	88.84	359	182	32.3
3 RE	606	585	91.32	706	527	15.07
5 RE	695.66	722	167.06	848	517	24.01
10 RE	467.66	430	215.97	700	273	46.18

RE – rainfall event; s – standard deviation; CV – coefficient of variation. LVAd - Typic Hapludox, PVAd - Typic Hapludult; NVef - Rhodic Kandiudalf.

Figure 4. Fissures on the surface of NVef.

The observed changes in total soil porosity contribute to understanding how the evolution of rainfall events affects soil porosity, especially in soils with a higher clay content. However, analysis of total porosity alone is not sufficient to reveal the effect of rainfall events on the physical quality of the soil, as rain may affect not only total porosity (block area occupied by pores) but also the types of pores that occur in the soil, thus modifying the water dynamics in the soil profile. According to Sun et al. (2008), the greatest loss of porosity during rainfall events corresponds to the loss of elongated pores, which was not shown in the analysis of total porosity. Therefore, a descriptive analysis of pore size and shape was performed in the studied soils (Figures 5, 6 and 7).

LVAd showed a predominance of large complex pores, with pores larger than 0.156 mm^2 and an absence of elongated pores being observed. The total porosity in this soil was mainly contributed by large complex, medium rounded and small pores. With the occurrence of rainfall events, the complex porosity varied in terms of the occupied area, but without altering the sizes of the complex pores, which generally continued to be large pores (greater than 0.156 mm^2).

For PVAd, a predominance of large complex pores was observed, with very few effective changes in the area occupied by these pores being recorded. At T10, an increase in the proportion of rounded porosity was detected. The increase in rounded pores associated with ten erosive rainfall events was related to structural degradation and soil compaction (RINGROSE-VOASE; BULLOCK, 1984), which affects water movement in the profile due to the low connectivity between pores of this type.

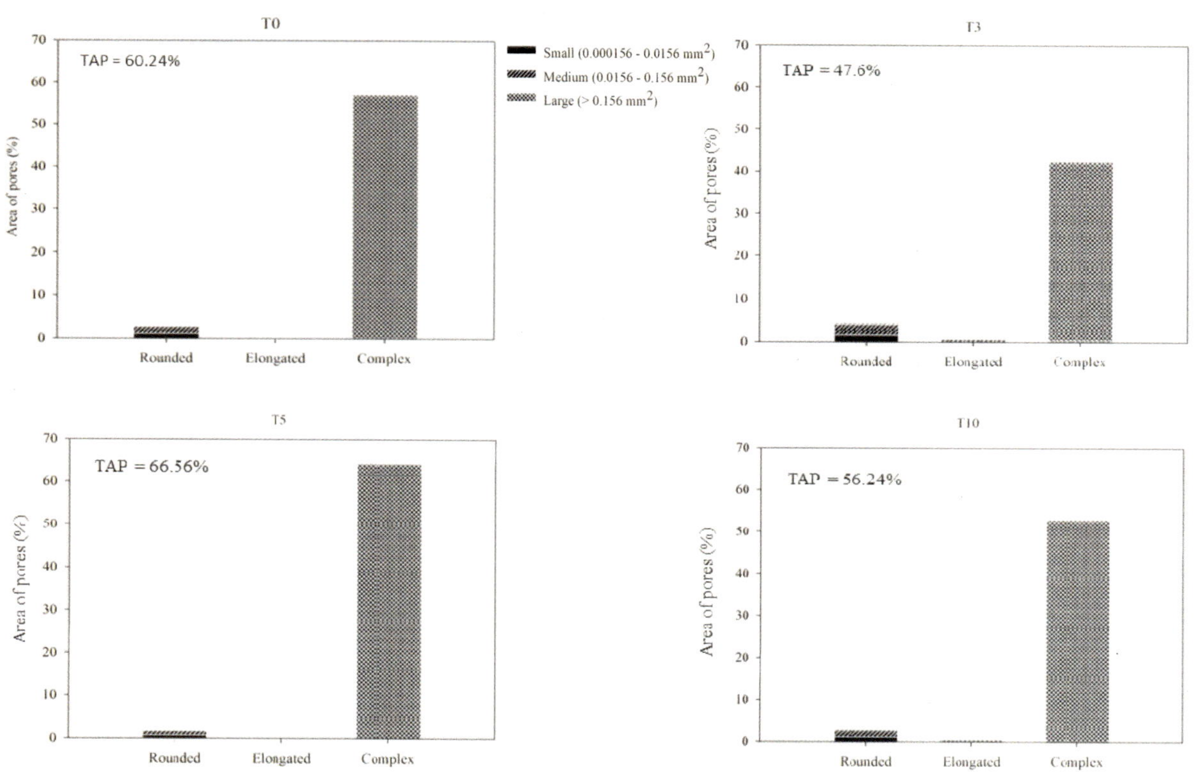

Figure 5. Evolution of the size and type of pores with the occurrence of rainfall events (T0, T3, T5 and T10) for LVAd.

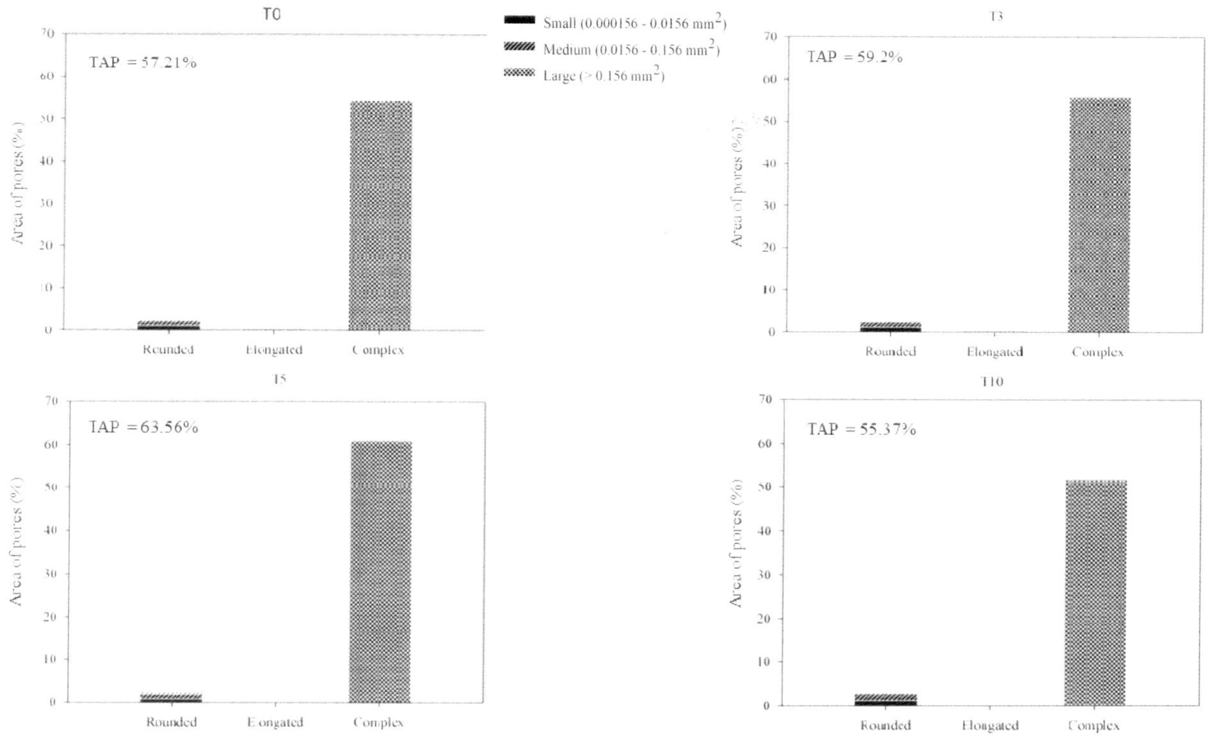

Figure 6. Evolution of the size and type of pores with the occurrence of rainfall events (T0, T3, T5 and T10) for PVAd.

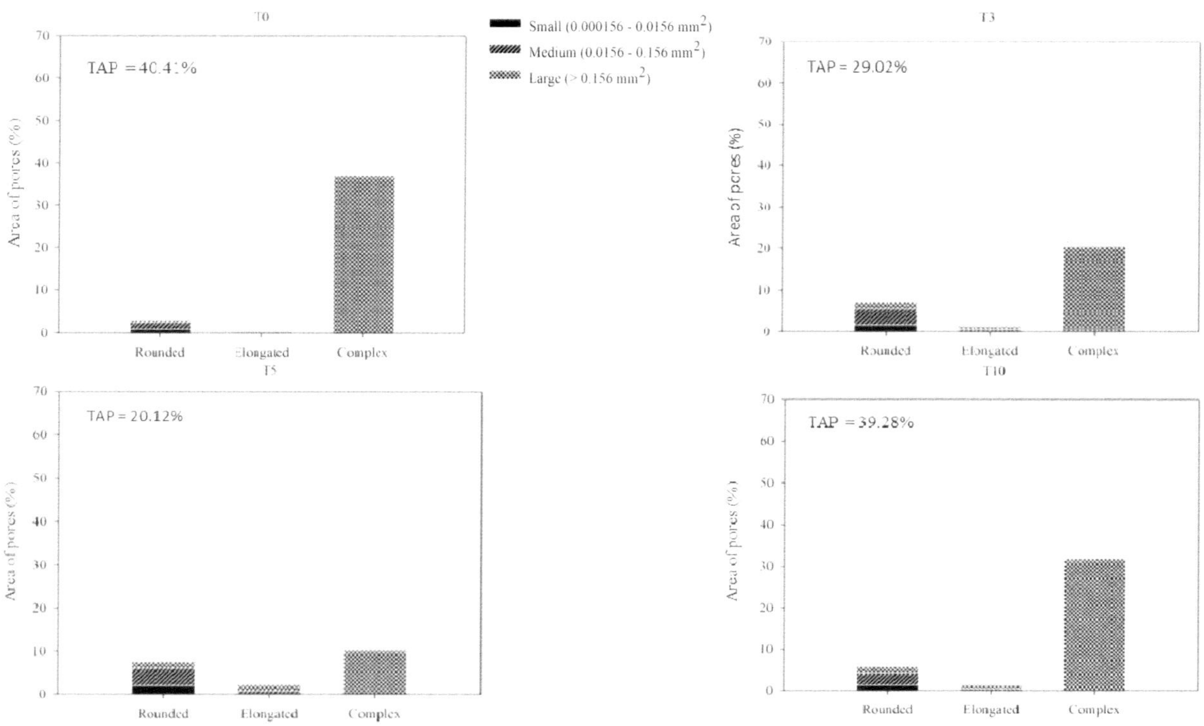

Figure 7. Evolution of the size and type of pores with the occurrence of rainfall events (T0, T3, T5 and T10) for NVef.

NVef displayed a predominance of large complex pores at T0, which suffered a decrease in area as rainfall events occurred. In addition to the loss of large complex pores, increases in the number of small and medium rounded pores and in medium and large elongated pores were observed. These changes in porosity indicate a loss of water circulation in the soil profile, as the rounded pores became smaller and less interconnected. At T10, an increase in large complex pores was noted due to the occurrence of fissures caused by the contraction and expansion processes associated with the wetting and drying cycles in the soil; as these pores coalesced, the soil classification shifted to complex porosity.

The changes in pore type and morphology due to the impact of erosive rainfall events were best understood based on the pore distribution analysis, which provided data on the area (Figures 8, 10 and 12) and the number of pores (Figures 9, 11 and 13) based on the observed pore shapes (round, elongated or complex).

LVAd showed a high total porosity and large number of pores at T0 (Figure 8 and 9). For all rainfall events, a small number of complex pores was observed to occupy a large area of the blocks of this soil, indicating that these pores are interconnected. After the rainfall events (T3, T5 and T10), slight variations were observed in large complex pores (> 1000 μm) in relation to the total block area. However, this small variation did not represent a significant change in total soil porosity.

The redistribution and remobilization of the soil particles caused by the impact of raindrops on the surface of LVAd were not sufficient to cause significant changes in total soil porosity (Table 2) or in the shape and arrangement of pores (Figures 8 and 9).

PVAd presented high total porosity at T0 (Figure 10), with interconnected pores being observed. Similar to LVAd, the sequence of rainfall events caused minor changes in the morphology, distribution and surface area occupied by pores in this soil (Table 2, Figures 10 and 11).

Among the three analyzed soils, major changes in porosity were observed for NVef, with a reduction of the area occupied by pores and changes in the morphology and distribution of pores being recorded for each sequence of rainfall events, with the exception of T10 (Figures 12 and 13). A reduction in porosity can be explained by the redistribution of detached particles and soil compression due to the impact of raindrops on the surface (FOX et al., 2004), thus reducing the number of complex and elongated pores. At T10, an increase of complex porosity was observed, which could be attributed to the coalescence of elongated pores formed by intense fissuration of the soil matrix due to the wetting and drying cycles that occur during rainfall events (Figure 4). Although the changes in porosity recorded in NVef were greater than in the other soils, no significant differences were detected for this parameter (Table 2).

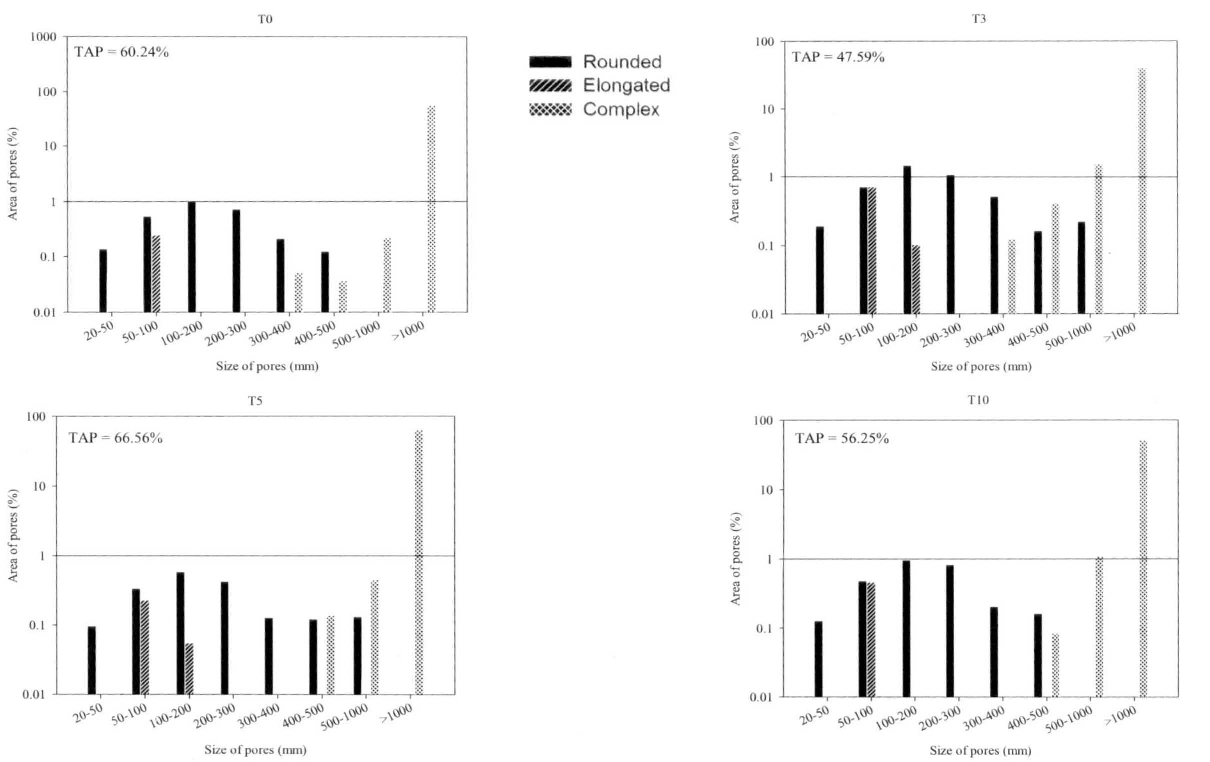

Figure 8. Evolution of the porosity distribution for LVAd.

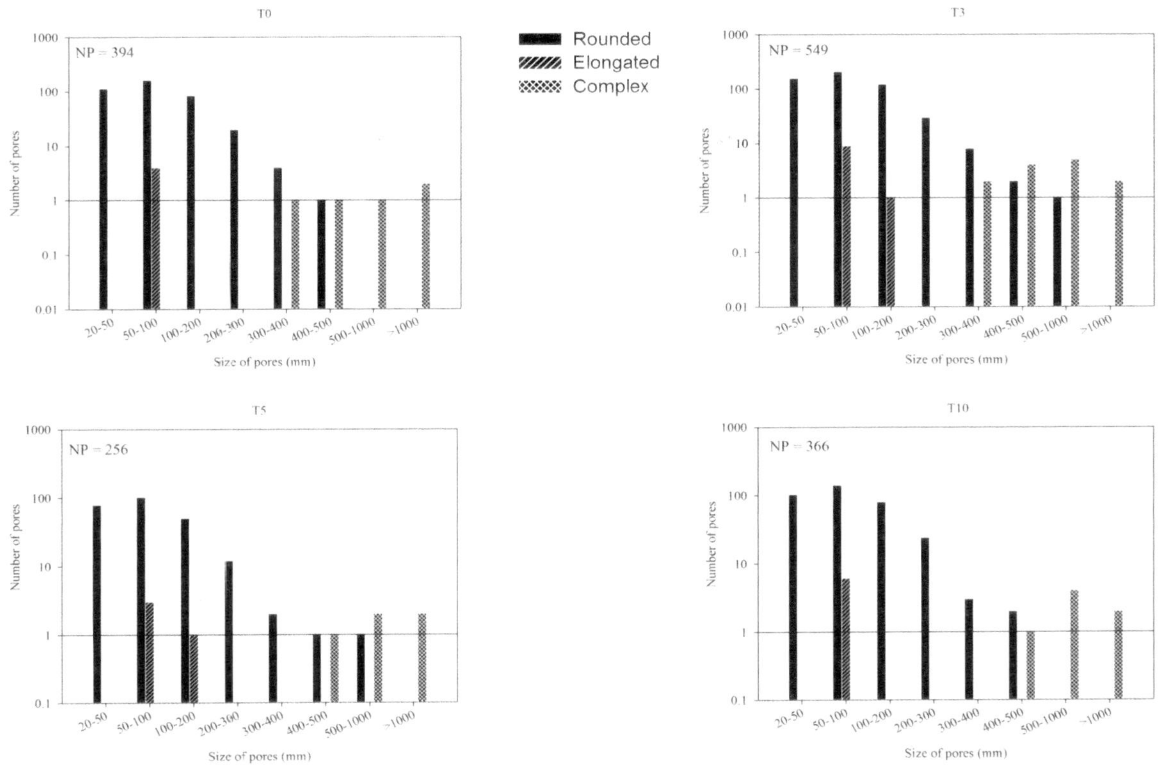

Figure 9. Evolution of the number of pores for LVAd.

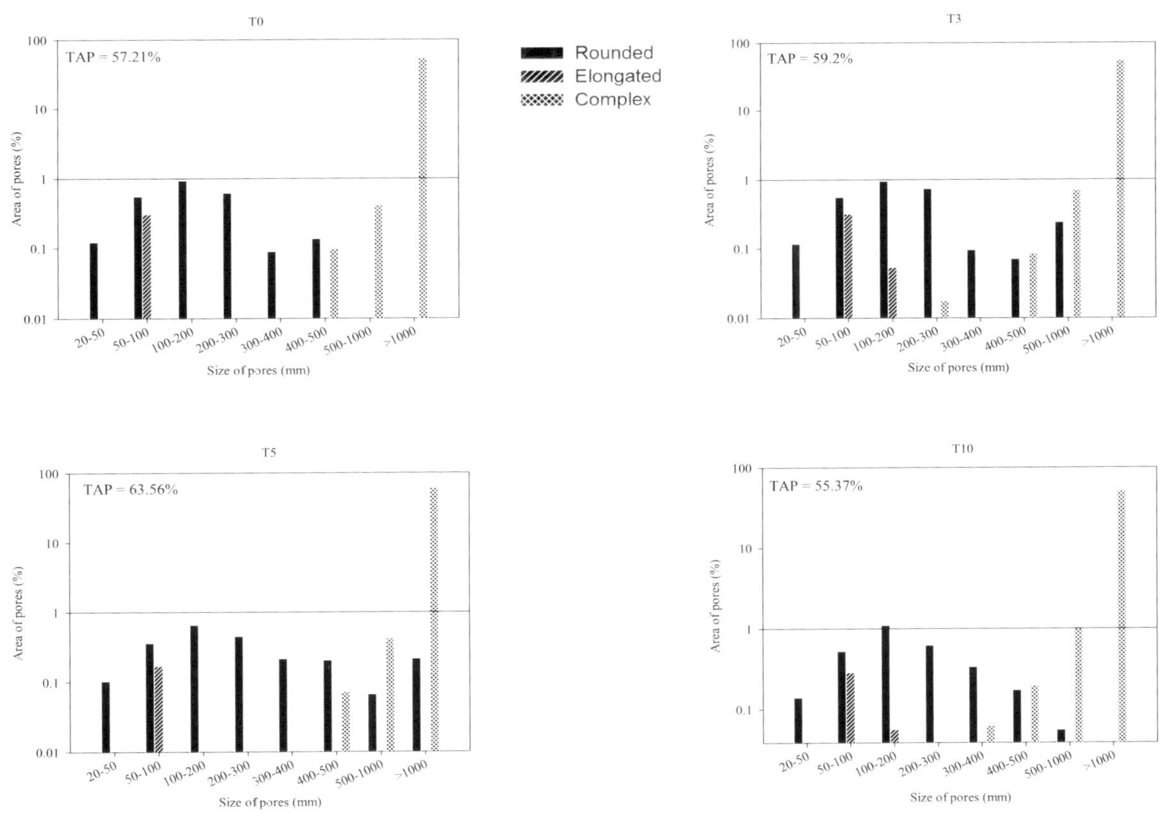

Figure 10. Evolution of the porosity distribution for PVAd.

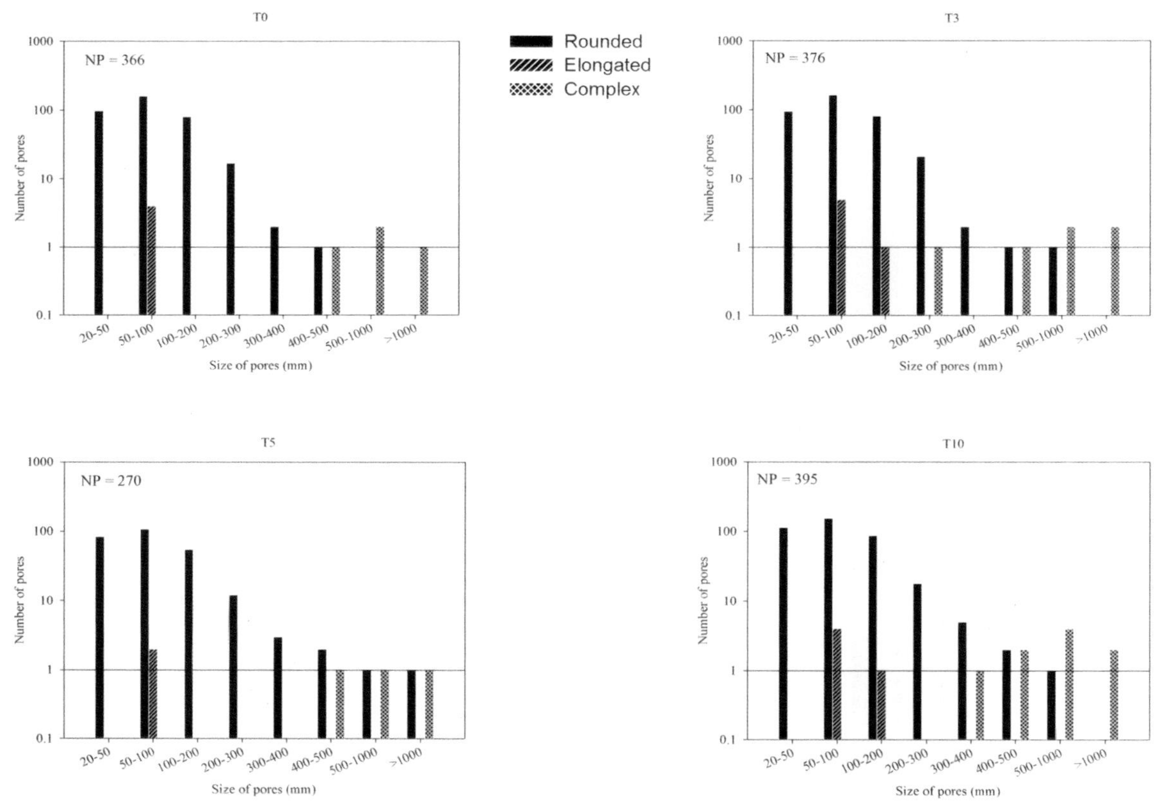

Figure 11. Evolution of the number of pores for PVAd.

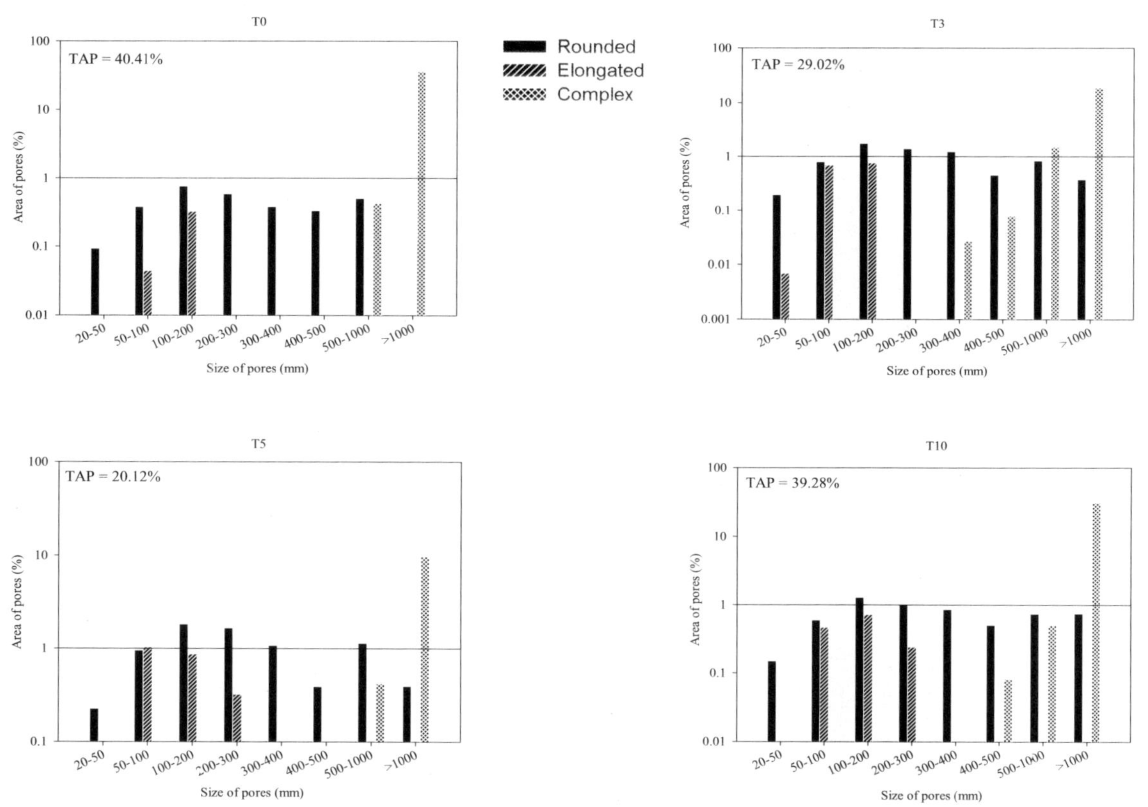

Figure 12. Evolution of the porosity distribution for NVef.

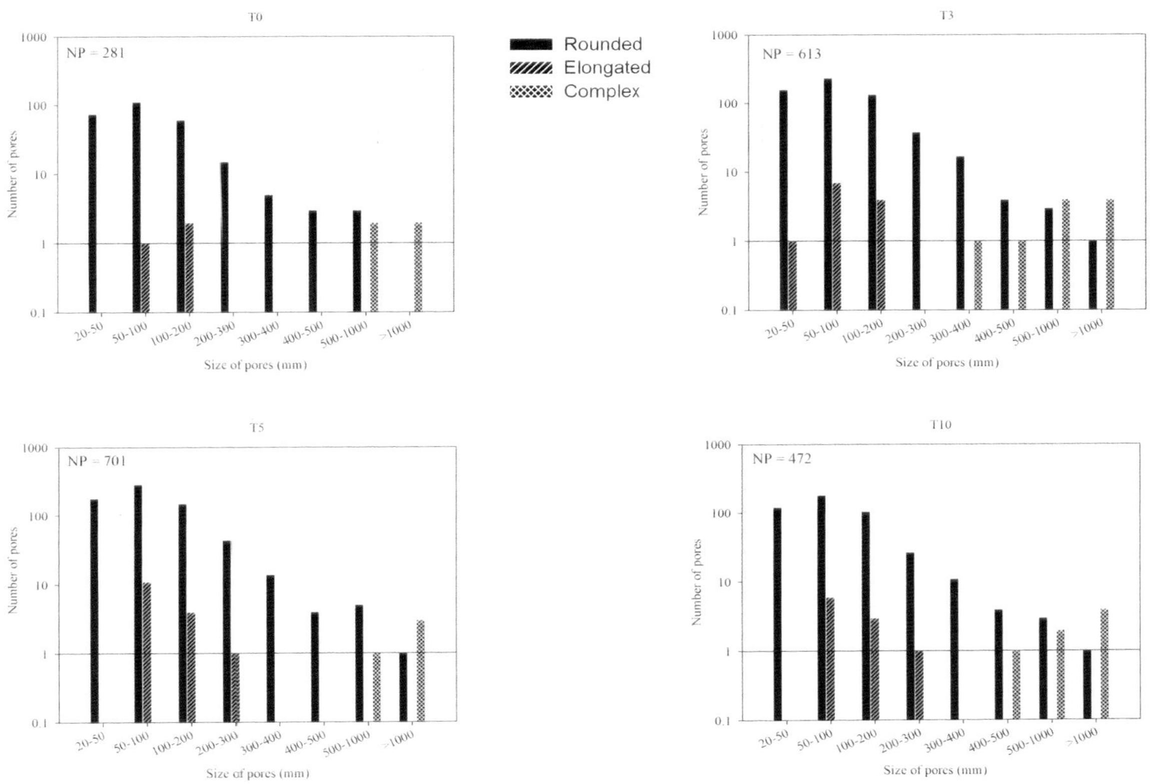

Figure 13. Evolution of the number of pores for NVef.

Conclusion

The three analyzed soils displayed surface crust formation. However, the soil texture was an important factor in differentiating the crust morphology.

NVef showed the most significant changes in total porosity and pore numbers and morphology due to the crust formation caused by erosive rainfall events.

The crust formation observed on the surface of LVAd and PVAd did not significantly alter the total porosity or the number and morphology of pores.

Acknowledgements

The authors wish to thank FAPESP (Foundation for Research Support of São Paulo state) for financing this project (Project 2006/57233-3) and CNPq for providing scholarships. We thank Carlos Eduardo Pinto Juhász (*in memoriam*), Pedro Lian D. B. Castanho, Lucas P. R. Conte, Raul S. Toma, Jaqueline D. Rosa, João Medeiros, Tatiana Rittl, Ariane Teato, Renata C. Bovi, Gabriela Rabello and Renata S. Momoli for their support of this research.

References

ASSOULINE, S. Rainfall-induced soil surface sealing: a critical review of observations, conceptual models and solutions. **Vadose Zone Journal**, v. 3, n. 3, p. 570-591, 2004.

BERTOL, I.; VÁZQUEZ, E. V.; GONZÁLEZ A. P.; COGO, N. P.; LUCIANO, R. V.; FABIAN, E. L. Sedimentos transportados pela enxurrada em eventos de erosão hídrica em um Nitossolo háplico. **Revista Brasileira de Ciência do Solo**, v. 34, n. 1, p. 245-252, 2010.

BRANDÃO, V. S.; SILVA, D. D.; RUIZ, H. A.; PRUSKI, F. F.; SCHAEFER, C. E. G. R.; MARTINEZ, M. A.; MENEZES, S. J. M. C. Resistência hidráulica da crosta formada em solos submetidos a chuvas simuladas. **Revista Brasileira de Ciência do Solo**, v. 30, n. 1, p. 13-21, 2006.

BULLOCK, P.; FEDOROFF, N.; JONGERIUS, A.; STOOPS, G.; TURSINA, T. **Handbook for soil thin section description**. Albrington: Waine Research, 1985.

CAMARGO, O. A.; MONIZ, A. C.; JORGE, J. A.; VALADARES, J. M. A. S. **Métodos de análise química, mineralógica e física de solos do Instituto Agronômico de Campinas**. Campinas: IAC, 1986.

CARMI, G.; BERLINER, P. The effect of soil crust on the generation of runoff on small plots in an arid environment. **Catena**, v. 74, n. 1, p. 37-42, 2008.

COOPER, M.; VIDAL TORRADO, P.; CHAPLOT, V. Origin of microaggregates in soils with ferralic horizons. **Scientia Agricola**, v. 62, n. 3, p. 256-263, 2005.

COOPER, M.; VIDAL-TORRADO, P.; GRIMALDI, M. Soil structure transformations from ferralic to nitic horizons on a toposequence in southeastern Brazil. **Revista Brasileira de Ciência do Solo**, v. 34, n. 5, p. 1685-1669, 2010.

DALLA ROSA, J.; COOER, M.; DARBOUX, F.; MEDEIROS, J. C. Soil roughness evolution in different tillage systems under simulated rainfall using a semivariogram-based index. **Soil & Tillage Research,** v.124, p. 226–232, 2012.

FOX, D. M.; BRYAN, R. B.; FOX, C. A. Changes in pore characteristics with depth for structural crusts. **Geoderma**, v. 120, n. 1-2, p. 109-120, 2004.

FOX, S. C.; MILLS, A. J.; POCH, R. M. Micromorphology of surface crusts in the Knersvlakte, South Africa. **Journal of Mountain Science**, v. 6, n. 2, p. 189-196, 2009.

GONÇALVES, F. A.; DEMETRIUS, D. S.; PRUSKI, F. F.; CARVALHO, D. F.; CRUS, E. F. Índices e especialização da erosividade das chuvas para o Estado do Rio de Janeiro. **Revista Brasileira de Engenharia Agrícola e Ambiental**, v. 10, n. 2, p. 169-276, 2006.

HAO, H.; HARTMANN, C.; APICHART, J.; SIWAPORN, S.; PROMSAKHA, S.; RICHARD, G.; BRUAND, A.; DEXTER, A. R. Slumping dynamics in tilled sandy soils under natural rainfall and experimental flooding. **Soil and Tillage Research**, v. 114, n. 1, p. 9-17, 2011.

HILLEL, D. **Introduction to environmental soil physics**. New York: Elsevier Academic Press, 2004.

JUHÁSZ, C. E. P.; COOPER, M.; CURSI, P. R.; KETZER, A. O.; TOMA, R. S. Savanna woodland soil micromorphology related to water retention. **Scientia Agricola**, v. 64, n. 4, p. 344-354, 2007.

LE BISSONNAIS, Y. Agregate stability and assessment of soils crustability and erodibility: I. Theory and methodology. **European Journal of Soil Science**, v. 47, n. 47, p. 425-437, 1996.

MURPHY, C. P.; BULLOCK, P.; TURNER, R. H. The measurement and characterization of voids in soil thin sections by image analysis. Part I. Principles and techniques. **Journal of Soil Science**, v. 28, n. 3, p. 498-508, 1977.

PAGLIAI, M.; VIGNOZZI, N.; PELLEGRINI, S. Soil structure and the effect of management practices. **Soil and Tillage Research**, v. 79, n. 1, p. 131-143, 2004.

RINGROSE-VOASE, A. J.; BULLOCK, P. The automatic recognition and measurement of soil types by image analysis and computer programs. **Journal of Soil Science**, v. 35, n. 4, p. 673-684, 1984.

SUN, Z.; KANG, Y.; JIANG, S. Effects of water application intensity, drop size and water application amount on the characteristics of topsoil pores under sprinkler irrigation. **Agricultural Water Management**, v. 95, n. 7, p. 869-876, 2008.

TOUMA, J.; RACLOT, D.; AL-ALI, Y.; ZANTE, P.; HAMROUNI, H.; DRIDI, B. In situ determination of the soil surface crust hydraulic resistance. **Journal of Hidrology**, v. 403, n. 3, p. 253-260, 2011.

WARRICK, A. W.; NIELSEN, D. R. Spatial variability of soil physical properties in the field. In: HILLEL, D. (Ed.). **Applications of soil physics**. Part 2. New York: Academic Press, 1980.

Soil water balance in different densities of *Pinus taeda* in Southern Brazil

Jorge Luiz Moretti de Souza[1], Kharyn Freitas Fezer[1], Bruno César Gurski[1*], Daniela Jerszurki[1] and Paulo Eugênio Pachechenik[2]

*[1]Departamento de Solos e Engenharia Agrícola, Universidade Federal do Paraná, Rua dos Funcionários, 1540, 80035-050, Curitiba, Paraná, Brazil. [2]Escola de Ciências Agrárias e Medicina Veterinária, Pontifícia Universidade Católica, Curitiba, Paraná, Brazil. *Author for correspondence. E-mail: brunogurski@ufpr.br*

ABSTRACT. This study aimed to quantify and compare water balance components over the course of a year for different *Pinus taeda* planting densities in an oxisol in southern Brazil. This experiment was conducted on 6-year-old trees in a clay oxisol at the Monte Alegre Farm, a property of the Klabin Company. The experimental design was a randomized block with four replicates and five treatments with different amounts of soil coverage: T100 (100% coverage - standard planting coverage); (T75, 75; T50, 50; and T25, 25%), and; T0 (without cover - clearcutting). The soil water storage and actual evapotranspiration under non-standard conditions were determined in a weekly estimated soil water balance (SWB) with measured components. By the end of the year, the treatments had not reached field capacity or wilting point storage. The average value of total downward drainage was 100.2 mm, and the highest values occurred in the T75 and T100 treatments. The lowest population density (T25) had the highest actual evapotranspiration (*ETr*), due to the growth of the remaining *Pinus taeda* trees. The highest evapotranspiration occurred in September, due to the resumption of *Pinus taeda* growth.

Keywords: exotic forest, spacing, hydric relations, soil water storage, hydric consumption.

Introduction

Forest cultivation depends on local water availability and has been identified as a practice that is capable of reducing soil water resources. However, this is contradicted by numerous authors who report a resulting improvement in soil water storage, a reduction in water losses by evaporation, a decrease in runoff, and soil aggregation that prevents erosion (Chow, 1964; Tundizi, 2000; Serengil et al., 2007; Van Dijk & Keenan, 2007). *Pinus taeda* has high transpiration rates when the soil moisture is close to field capacity, but under conditions of water stress, the crop reduces transpiration rates, suggesting an adaptive mechanism (Trinidad, Hernández, Orozco, & Upton, 2002). Also, Leite, Nogueira and Moreira (2006) conclude that spacing is one of main factors affecting water availability in reforestation, because it alters the physiological activity of the plant, the consumption of soil water, and rainfall interception.

The soil water balance (SWB) is defined as the inputs and outputs of water in a given control volume and time interval (Reichardt & Timm, 2004; Libardi, 2005). Monitoring of the soil water storage dynamics associated with crop water requirements is considered an essential tool for numerous agricultural activities and the efficient use of soil water resources (Frizzone, Andrade Júnior, Souza, & Zocoler, 2005; Souza & Gomes, 2007). Studies of the water requirements of exotic plantation forest plants, such as *Pinus taeda*, can assist in developing alternative approaches that use water efficiently without compromising environmental balance, and promote the expansion of forest cultivation. This study aimed to quantify and compare the components of the water balance for different *Pinus taeda* planting densities in an oxisol, over the course of a year in Telêmaco Borba, Paraná State, Southern Brazil.

Material and methods

This experiment was conducted at the Monte Alegre Farm, which belongs to the Klabin Company and is located in Telêmaco Borba, Paraná State, Brazil. The site has the coordinates 24° 13' 19" S, 50° 32' 33" W and an altitude of 700 m. Data were collected in 2009 in an area of 12.5 ha of 6-year-old *Pinus taeda* trees in a clay oxisol with undulated relief (Empresa Brasileira de Pesquisa Agropecuária [Embrapa], 2006). According to Köppen, the climate classification of the region is Cfa Cfb^{-1}, transitional humid subtropical to temperate, with the coldest average temperatures below 16°C with frosts, and the warmest average temperatures above 22°C, with hot summers (Caviglione, Kiihl, Caramori, & Oliveira, 2000).

The experimental design was a randomized block with four replicates. Each block was 3.125 ha and each treatment comprised 0.625 ha. The treatments consisted of water balances with different trees coverage: T100 (100% coverage - standard planting coverage; T75, 75; T50, 50; T25, 25% coverage); and T0 (without cover – clear cutting) (Figure 1).

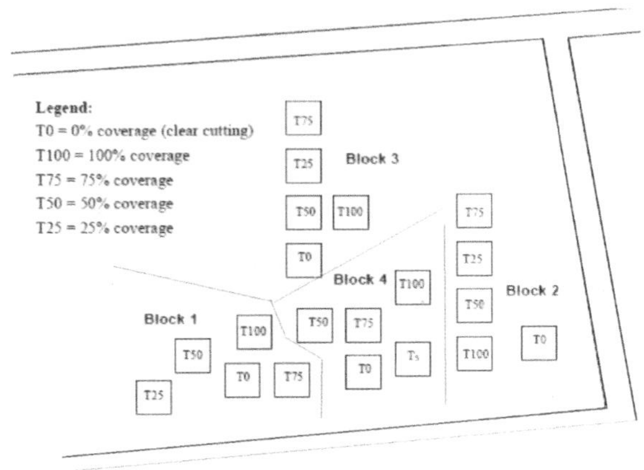

Figure 1. Distribution of treatments in experimental area (12.5 ha) of Monte Alegre Farm in Telêmaco Borba, Paraná State.

Disturbed soil samples were taken weekly throughout 2009 (54 weeks) using a dutch auger. Two samples were collected from each treatment at depths 0-0.1; 0.1-0.2; 0.2-0.4; 0.4-0.6, and 0.6-1.0 m. The gravimetric water content was determined according Empresa Brasileira de Pesquisa Agropecuária (Embrapa, 1997), based on the average of the two moisture samples taken at each depth. During 2009, a total of 10,800 samples were collected to determine gravimetric moisture. Undisturbed samples were collected in October 2009 in two 1.5 x 3.0 x 1.8 m trenches, using soil cores at the same depths as the disturbed samples, with three replicates at each depth. Only two trenches were sampled because the area was homogeneous and commercial and because there were several other studies in the same experimental area (Rigatto, Dedecek, & Matos, 2004). The physical soil characteristics (bulk density, saturated soil hydraulic conductivity, and the soil water retention curve) were determined according to Embrapa (1997) (Table 1).

The SWB was estimated at weekly intervals, with a total of 52 in 2009. The calculations were carried out based on vertical flow only. Because the study area is relatively flat, the input and output runoffs were considered equal, and there was no irrigation in the area.

Table 1. Physical and hydraulic soil characteristics at different depths on Monte Alegre Farm, in Telêmaco Borba, Paraná.

Depth	Sand	Silt	Clay	C organic	$\rho^{(1)}$	Macropores	Micropores	$\alpha^{(2)}$	$K_0^{(3)}$
(m)	%				(kg m^{-3})	(m^3 m^{-3})			(mm day^{-1})
0.0-0.1	41	10	49	1.70	1100 a	0.200 a	0.395 c	0.598 a	15607.34 a
0.1-0.2	40	11	49	1.20	1210 a	0.137 ab	0.403 bc	0.541 a	4097.95 b
0.2-0.4	40	11	50	0.89	1210 a	0.140 ab	0.395 c	0.537 a	5651.00 ab
0.4-0.6	36	11	53	0.68	1230 a	0.080 b	0.462 a	0.542 a	957.20 b
0.6-1.0	36	11	53	0.45	1160 a	0.116 ab	0.443 ab	0.561 a	904.80 b
CV (%)$^{(4)}$	-	-	-	5.8	6.0	24.8	3.5	4.0	71.2

(1)Soil bulk density; (2)Total porosity; (3)Soil saturated hydraulic conductivity; (4)Coefficient of variation. *Means followed by the same letter do not differ by Tukey ($p < 0.05$).

Actual evapotranspiration under non-standard conditions (*ETr*) was calculated as follows as equation 1:

$$ETr = -\Delta S + P - D + U \quad (1)$$

where:
ETr: actual evapotranspiration under non-standard conditions (mm $period^{-1}$);
ΔS: soil-moisture storage variation in the root zone (mm $period^{-1}$);
P: precipitation (mm $period^{-1}$);
D: downward drainage out of the root zone (mm $period^{-1}$);
U: upward capillary flow into the root zone (mm $period^{-1}$).

The volumetric water content (θ_i) was obtained by multiplication of gravimetric water content (u_i) by bulk density (ρ_i) of each soil *i*-th layer. Soil water storage (*S*) was calculated using trapezoidal rule, with a ΔS obtained from the difference between previous (S_j) and current storage (S_{j+1}), according equation 2:

$$S_j = \theta_1 \cdot z_1 + \sum_{i=1}^{n} \frac{(\theta_i + \theta_{i+1}) \cdot z_i}{2} \quad (2)$$

where:
S_j: soil water storage in *j*-th week year (mm);
θ_i: volumetric moisture in *i*-th soil depth ($cm^3\ cm^{-3}$);
z_i: soil depth (mm);
j: weeks over year that samples were taken (52 weeks);
i: sample collection depths: 1: 0-0.1; 2: 0.1-0.2; 3: 0.2-0.4; 4: 0.4-0.6; and, 5: 0.6-1.0 mm.

Precipitation rates (*P* in mm day^{-1}) were measured with 60 acrylic gauges with 40 mm of water capacity installed within the experimental area. Two rain gauges were installed in each treatment 0.50 m from the edge of the trees, 1.30 m above the ground. A third rain gauge was placed at the midpoint between the rows of trees. The mean *P* was calculated as the average of the three rain gauges.

Downward drainage (*D*) and upward capillary flow (*U*) into the root zone were estimated using the Darcy-Buckingham water flow equation. Unsaturated soil hydraulic conductivity $K(\theta)$ was estimated according to Mualem (1976) with regression parameters obtained from the model described in Van Genuchten (1980). Soil water retention values, for each depth, were obtained between -0.006 and -1.500 MPa with a pressure plate apparatus (Richards, 1965) by desorbing the saturated cores at several pressure steps. The saturated cores were used to obtain water retention values over the entire range studied; i.e., under -0.006 MPa (pressure table) and -0.010 to -1.500 MPa (Richard's pressure chamber). Water content at each pressure was calculated from the volume of outflow between pressure steps, the final water content, and the weight of oven-dried soil. Volumetric water content and soil-water pressure potential, obtained for each depth, were adjusted as proposed by Van Genuchten (1980), using the *Soil Water Retention Curve* program (Dourado Neto, Nielsen, Hopmans, Reichardt, & Bachhi, 2001).

Volumetric water content at field capacity (θ_{fc}) and wilting point (θ_{wp}) were set at the -0.01 MPa and -1.500 MPa pressure points, respectively. The field capacity (θ_{fc}) was considered to be constant when the drainage rate was very slow. Total porosity (α) was considered equal to the volumetric water content at saturation (θ_s). Analysis of variance was carried out with a Tukey's test, at 5% of probability, to compare differences in root-zone soil moisture storage.

Results and discussion

The study period was characterized by the occurrence of well-distributed precipitation, with atypical months compared to the historical averages in Telêmaco Borba, Paraná State, Brazil (Figure 2a). The precipitation (*P*) changed between treatments, but the water collected was not related to tree density.

Although *P* in January was above historical average, four months (July, September, October and December) had greater *P* than January. Historically, winter is the driest season of the year in the study region (Figure 1), but in 2009, it fairly rainy, especially during July. In fact, July's monthly *P* was almost three times the historical average (82 mm). Additionally, the spring had higher *P* than the historical average. Conversely, summer and autumn had rainfalls lower than the historical average, by approximately 35 and 27%, respectively.

The T0 treatment had the highest annual *P* (1,802 mm $year^{-1}$) of all treatments. This result is expected, as the absence of trees meant there was no interception of the rainfall by leaves and trunks. The treatment with the greatest number of trees, the T100 treatment, had a significantly lower annual *P* (1,533 mm $year^{-1}$). The T75, T50 and T25 treatments showed no significant difference ($p = 0.05$) in annual *P*.

The smaller number of trees per area did not result in less *ETr* in the T25 (1,458.8 mm $year^{-1}$)

and T50 (1,408.7 mm year^{-1}) treatments, which had higher water consumption than the T100 (1,225.5 mm year^{-1}) and T75 (1,305.8 mm year^{-1}) treatments (Figure 2b). Similar results were reported by Van Dijk and Keenan (2007), who found greater evapotranspiration consumption in a pine forest after thinning. This finding is due to the increased growth of the remaining trees, especially the leaf area index (LAI), because of the greater light and space availability.

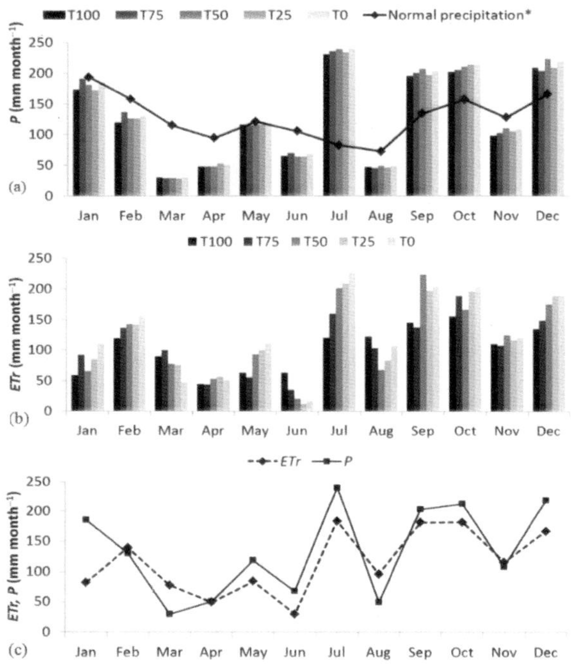

Figure 2. Main component of input and output of soil water for the treatments (T100 – 100; T75 – 75; T50 – 50; T25 – 25% coverage and; T0 – without cover), from January to December 2009, in Monte Alegre Farm, Telêmaco Borba, Paraná State: a) monthly precipitation (*P*); (b) monthly actual evapotranspiration under non-standard conditions (*ETr*), and; c) monthly average comparison between *P* and *ETr*. *Climatological normal observed between the period 1947 to 2005 for the same region (Nunes, Batista, Soares, Fier, & Santos, 2009).

In at least 6 months of 2009, the T0 treatment had the highest *ETr*, which is theoretically impossible because there are no plants in the treatment. Weeding was conducted in the T0 treatment area to prevent plant growth. Therefore, one possible explanation of this result is that the high *ETr* occurred because the trunks and roots were not removed from experimental area. Thus, the addlement roots may have formed infiltration pores and channels, causing runoff out of the experimental control volume. The portion of the water that was precipitated could not be computed as *D* with the Darcy-Buckingham equation because it was not stored in the soil of the experimental volume control, so it was computed as *ETr* at *SWB*. However, this hypothesis needs further study to be verified.

The average *ETr* across all treatments was 1,388 mm year^{-1}, or 3.8 mm day^{-1}. Studies conducted with *Pinus taeda* in United States (Lu, Sun, Amatya, & Mcnulty, 2003; Ford, Hubbard, Kloeppel, & Vose, 2007), in a similar climate, reported values of a similar magnitude to those found in this study.

According Carneiro et al. (2008), stomatal conductance and leaf water potential decrease when the soil water content and transpiration rate are reduced. Under conditions of high soil water availability and liquid radiation, stomatal closure does not occur, and the plant regulates water availability itself. This may explain the increase in monthly *ETr* with higher *P* (Figure 2c). Trinidad, Hernández, Orozco and Upton (2002) reported that *Pinus taeda* species have high transpiration rates when the soil moisture is close to field capacity. Higher water consumption also occurred in September, likely as a result of resumed *Pinus* growth.

In the analysis, the T0 and T25 treatments showed the highest and lowest soil water storage (*S*), respectively. The T100, T75, and T50 treatments were intermediate (Table 2). Interestingly, contrary to expectations, decreasing the number of trees per area did not promote greater *S* (Table 2 and Figure 3a). Similar results were obtained by Van Dijk and Keenan (2007), who found a decrease in soil water storage after thinning a forest. According to the same authors, the growth of the remaining trees, especially with respect to Leaf Area Index (LAI), increased with the increased availability of light and space among the trees.

Table 2. Statistical analysis of average monthly soil water storage with different densities of *Pinus taeda* (T100 – 100; T75 – 75; T50 – 50; T25 – 25% coverage and; T0 – without cover), over 2009, in Monte Alegre Farm, Telêmaco Borba, Paraná State.

Treatments	Average monthly soil water storage (mm)*
T100	336,6 bc
T75	337,0 bc
T50	328,6 bc
T25	326,4 c
T0	349,5 a
F Value	11,23**

*Means followed by same letter do not differ by Tukey test ($p < 0.05$); **$p < 0.01$.

From June to December, the T0 treatment had an average soil water storage that was 18.3; 16.4; 17.1 and 18.3% higher than the T100, T75, T50, and T25 treatments, respectively (Figure 3a). In January and February, the average soil water storage for the T0 treatment was only 2.5; 3.6; 3.8 and 5.3% higher than the T100, T75, T50, and T25 treatments, respectively. However, from March to May, the T0 treatment had an average soil water storage that was 3.2; 3.7; 2.9, and 1.7% lower than T100, T75, T50, and T25 treatments, respectively (Figure 3a).

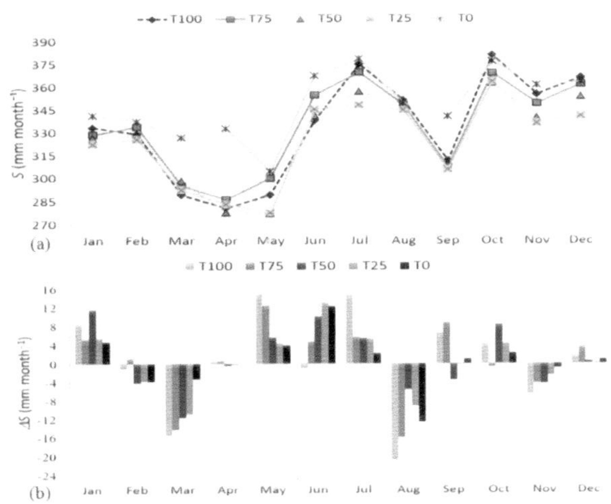

Figure 3. Soil water storage (S) from January to December of 2009, in Monte Alegre Farm, Telêmaco Borba, Paraná State: a) Average monthly; and b) soil water storage variation (ΔS) with different densities of *Pinus taeda* (T100 – 100; T75 – 75; T50 – 50; T25 – 25% coverage and; T0 – without cover).

The average depth of stored water for the T100, T75, T50, T25, and T0 treatments was approximately 425.3 and 205.0 mm, respectively, considering the volumetric water content at field capacity and wilting point. In samples collected weekly over the year, none of treatments reached this S (Figure 3a). Based on S values during the monitored period, the *Pinus taeda* did not use the water that was retained in soil at tension lower than -1.5 MPa. However, Myers and Talsma (1992) concluded that *Pinus radiata* managed to remove water from soil retained at even higher tension. Thus, the available soil water did not limit *Pinus* development in the study period.

Among treatments, the average S in the T0 treatment (in July) was 13.51% lower than the S at field capacity. With absence of vegetation cover (clearcut - T0), it is clear that the S varied in same proportion as P (contrast between Figures 1 and 2a). The highest water storage value in year occurred in April for other treatments, on average 32.7% below field capacity storage (Figure 3a). The S in the T100, T75, T50, and T25 treatments showed the same trends over the year; S decreased gradually from June until December. From September to December, plants had higher water consumption, resulted in higher ETr.

The average ΔS values in January, May, June, July, September, October, and December were 7.1; 8.3; 7.9; 6.7; 2.6; 3.8, and 1.4 mm, respectively; and in February, March, April, August, and November were -2.4; -11.1; -0.1; -12.8, and -3.7 mm, respectively (Figure 3b). The ΔS in the soil profile for all treatments followed the same trends as P, which agrees with Antonino, Sampaio, Dall'olio and Salcedo (2000) and Lima, Antonino, Soares, Souza and Lira (2006).

In January, May, July, and December, water was entering the soil in all treatments. However, because S remained high during this period, only the T75 treatment reflected the water entering the soil in February. Over the course of the year, the T50 (12 mm year^{-1}) and T25 (5 mm year^{-1}) treatments had the highest and lowest recorded amount of water entering the soil, respectively. The average drainage (D) was highest in July (16.8 mm month^{-1}), followed by October (16.7 mm month^{-1}), and December (14.3 mm month^{-1}). The total annual drainage was 100.2 mm year^{-1}, on average (Figure 4). Cruz, Libardi, Carvalho and Rocha (2005) found an average D of 72.4 mm year^{-1} for citrus in red oxisol in Piracicaba, São Paulo State, Brazil. Azevedo, Silva and Silva (2003) found an average D of 99.5 mm year^{-1} in oxisol cultivated with mango in the Petrolina, Pernambuco State, Brazil.

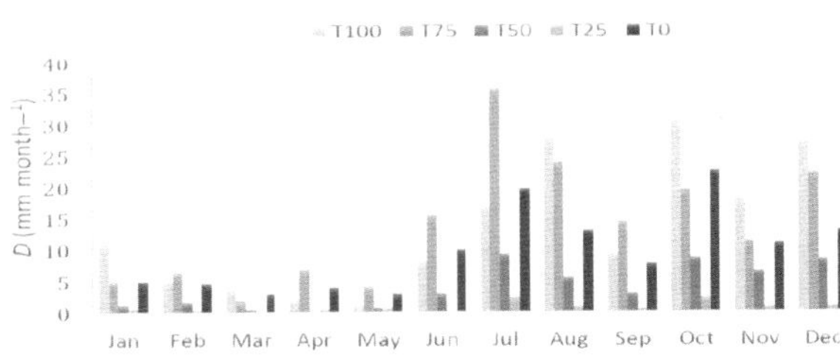

Figure 4. Downward drainage (D) estimated using the Darcy-Buckingham equation, with different densities of *Pinus taeda* (T100 – 100; T75 – 75; T50 – 50; T25 – 25% coverage and; T0 – without cover), from January to December of 2009, in Monte Alegre Farm, Telêmaco Borba, Paraná State.

The highest P values (Table 2), with D of 16.8, 16.7, 14.3 and 6.7 mm month^{-1}, occurred in July, September, October, and December, respectively (Figure 3). These values are considered high for to soil type in the region; oxisols are characterized by high infiltration capacity and are naturally well drained (Embrapa, 2006). The T75 treatment had the highest annual D (166.8 mm year^{-1}), followed by the T100 (160.3 mm year^{-1}) and T0 (116.3 mm year^{-1}) treatments. The T25 and T50 treatments had the lowest annual D: 9.6 and 47.9 mm year^{-1}, respectively. The T50 and T25 treatments had lower D because they had higher water consumption. With increasing light availability and space between trees, there was greater tree growth and an increase in tree canopy diameter. The T0 treatment (without cover - clearcutting) had the highest *D* value, possibly because it had greater amounts of water entering the soil (15% more than other treatments). In the absence of trees, there was no *P* interception by leaves and trunks. Consequently, the T0 treatment had the highest annual *ETr* (1,542.4 mm year^{-1}) (Figure 2c).

Conclusion

In 2009, the treatments did not reached field capacity and wilting point storages. The average total downward drainage was 100.2 mm, and the highest values occurred in the T75 and T100 treatments. The lowest population density treatment (T25) had the greatest *ETr*, due to greater *Pinus taeda* growth. The highest evapotranspiration occurred in September, as a result of resumed *Pinus taeda* growth.

References

Antonino, A. C. D., Sampaio, E. V. S. B., Dall'olio, A., & Salcedo, I. H. (2000). Balanço hídrico em solo com cultivos de subsistência no semiárido do nordeste do Brasil. *Revista Brasileira de Engenharia Agrícola e Ambiental, 4*(1), 29-34. doi: 10.1590/S1415-43662000000100006

Azevedo, P. V., Silva, B. B., & Silva, V. P. R. (2003). Water requirements of irrigated mango orchards in northeast Brazil. *Agricultural Water Management, 58*(3), 241-254. doi: 10.1016/S0378-3774(02)00083-5

Carneiro, R. L. C., Ribeiro, A., Huaman, C. A. M. Y., Leite, F. P., Sediyama, G. C., & Neves, J. C. L. (2008). Consumo de água em plantios de eucalipto: parte 2 modelagem da resistência estomática e estimativa da transpiração em tratamentos irrigados e não-irrigados. *Árvore, 32*(1), 11-18. doi: 10.1590/S0100-67622008000100002

Caviglione, J. H., Kiihl, L. R. B., Caramori, P. H., & Oliveira, D. (2000). *Cartas climáticas do Paraná*. Londrina, PR: Iapar. Retrevied from http://www.iapar.br/modules/conteudo/conteudo.php?conteudo=677

Chow, V. T. (1964). *Handbook of applied hydrology: a compendium of water-resources technology*. New York City, NY: McGraw-Hill Book Company.

Cruz, A. C. R., Libardi, P. L., Carvalho, L. A., & Rocha, G. C. (2005). Balanço de água no volume de solo explorado pelo sistema radicular de uma planta de citros. *Revista Brasileira de Ciência do Solo, 29*(1), 1-10. doi: 10.1590/S0100-06832005000100001

Dourado Neto, D., Nielsen, D. R., Hopmans, J. W., Reichardt, K., & Bachhi, O. O. S. (2001). Software to model soil water retention curves. SWRC, version 2.00. *Scientia Agricola, 57*(1), 191-192.

Empresa Brasileira de Pesquisa Agropecuária (1997). *Manual de métodos de análise de solo*. Rio de Janeiro, RJ: SNLCS.

Empresa Brasileira de Pesquisa Agropecuária (2006). *Sistema brasileiro de classificação de solos*. Rio de Janeiro, RJ: Embrapa.

Ford, C. R., Hubbard, R. M., Kloeppel, B. D., & Vose, J. M. (2007). A comparison of sap flux based evapotranspiration estimates with catchment-scale water balance. *Agricultural and Forest Meteorology, 145* (3-4), 176-185. doi: 10.1016/j.agrformet.2007.04.010

Frizzone, J. A., Andrade Júnior, A. S., Souza, J. L. M., & Zocoler, J. L. (2005). *Planejamento da irrigação: análise de decisão de investimento*. Brasília, DF: Embrapa.

Leite, H. G., Nogueira, G. S., & Moreira, A. M. (2006). Efeito do espaçamento e da idade sobre variáveis de povoamentos de *Pinus taeda* L. *Árvore, 30*(4), 603-612.

Libardi, P. L. (2005). *Dinâmica da água no solo*. São Paulo, SP: USP.

Lima, J. R. S., Antonino, A. C. D., Soares, W. A., Souza, E. S., & Lira, C. A. B. (2006). Balanço hídrico no solo cultivado com feijão caupi. *Revista Brasileira de Ciências Agrárias, 1*(1), 89-95.

Lu, J., Sun, G., Amatya, D. M., & Mcnulty, S. G. (2003). Modeling actual evapotranspiration from forested watersheds across the Southeastern United States. *Journal of American Water Resources Association, 39*(4), 887-896. doi: 10.1111/j.1752-1688.2003.tb04413.x

Mualem, Y. (1976). A new model for predicting the hydraulic conductivity of unsaturated porous media. *Water Resources Research, 12*(3), 513-522. doi: 10.1029/WR012i003p00513

Myers, B. J., & Talsma, T. (1992). Site water balance and tree water status in irrigated and fertilized stands of *Pinus radiata*. *Forest Ecology and Management, 52*(1-4), 17-42. doi: 10.1016/0378-1127(92)90494-T

Nunes, J. S. N., Batista, A. C., Soares, R. V., Fier, I. S. N., & Santos, C. C. (2009). Climatologia do comportamento da precipitação no distrito florestal de Monte Alegre, PR, Brasil. *Floresta, 39*(4), 783-792. doi: 10.5380/rf.v39i4.16313

Reichardt, K., & Timm, L. C. (2004). *Solo, planta e atmosfera: conceitos, processos e aplicações*. Barueri, SP: Manole.

Richards, L. A. (1965). Physical conditions of water in soil. In Black, C. A., Evans, D. D., White, J. L., Ensminge, L. E., & Clark, F. E. (Orgs.), *Methods of soil analysis: physical and mineralogical properties, including statistics of measurements and sampling* (p. 128-152). Madison, WI: ASA-SSSA.

Rigatto, P. A., Dedecek, R. A., & Matos, J. L. M. (2004). Influência dos atributos do solo sobre a qualidade da madeira de pinus taeda para produção de celulose kraft. *Árvore, 28*(2), 267-273. doi: 10.1590/S0100-676220040 00200013

Serengil, Y., Gokbulak, F., Ozhan, S., Hyzal, A., Sengonul, K., Balc, A., & Ozyuvaci, N. (2007).

Hydrological impacts of a slight thinning treatment in a deciduous forest ecosystem in Turkey. *Journal of Hidrology, 333*(2-4), 569-577. doi:10.1016/j.jhydrol.2006.10.017

Souza, J. L. M., & Gomes, S. (2007). Avaliação e desempenho das equações de estimativa do armazenamento de água no solo em um balanço hídrico climatológico decendial irrigacionista. *Acta Scientiarum. Agronomy, 29*(4), 433-446. doi: 10.4025/actasciagron. v29i4.383

Trinidad, T. M., Hernández, J. V., Orozco, A. M., & Upton, J. L. (2002). Respuesta al déficit hídrico en *Pinus leiophylla*: consumo de agua y crecimento en plântulas de diferentes poblaciones. *Agrociencia, 36*(3), 365-376.

Tundisi, J. G. (2000). Limnologia e gerenciamento integrado de recursos hídricos. *Ciência e Ambiente, 21*(1), 9-20.

Van Dijk, A. I. J. M., & Keenan. R. J. (2007). Planted forests and water in perspective. *Forest Ecology and Management, 251*(1-2), 1-9. doi: 10.1016/j.foreco.2007.06.010

Van Genuchten, M. T. (1980). A closed-form equation for predicting the hydraulic conductivity of unsaturated soil. *Soil Science Society of America Journal, 44*(5), 892-898. doi: 10.2136/sssaj1980.03615995004400050002x

6

Mineralogy and phosphorus adsorption in soils of south and central-west Brazil under conventional and no-tillage systems

Jessé Rodrigo Fink[1], Alberto Vasconcellos Inda[1*], Cimélio Bayer[1], José Torrent[2] and Vidal Barrón[2]

*[1]Departamento de Solos, Universidade Federal do Rio Grande do Sul, Av. Bento Gonçalves, 7712, Cx. Postal 15100, 97105-400, Porto Alegre, Rio Grande do Sul, Brazil. [2]Departamento de Agronomía, Universidad de Córdoba, Córdoba, Spain. *Author for correspondence. E-mail: alberto.inda@ufrgs.br*

ABSTRACT. The low phosphorus availability in tropical and subtropical soils, normally related to adsorption of phosphate to the minerals surfaces, can be attenuated when organic matter (OM) accumulates in the soils. Herein, we report the results of long-term experiments (18–32 years) aimed at quantifying the maximum phosphorus adsorption capacity (MPAC) and its determinant mineralogical variables in Brazilian soils and at assessing the effect of no-tillage (NT) in mitigating the phosphorus adsorption of soils. The MPAC of soils ranged from 297 to 4,561 mg kg^{-1} in the 0.00–0.10 m layer and from 285 to 4,961 mg kg^{-1} in the 0.10–0.20 m layer. The MPAC was correlated with the concentrations of iron oxides, goethite and ferrihydrite, gibbsite/(gibbsite+kaolinite) ratio and the specific surface area. The OM increased in the 0.00–0.10 m layer of NT soils, which was not reflected on the decrease of MPAC for the no-tillage soils.

Keywords: iron oxides, gibbsite, organic carbon, phosphate.

Introduction

Phosphorus is an essential plant nutrient that is usually present in forms of low availability in the highly weathered soils of tropical and subtropical regions from Brazil, a phenomenon that has been suggested as one of the main limiting factors for agricultural production (NOVAIS; SMYTH, 1999). This lack of available phosphorus mainly is attributed to its strong adsorption (this term is used for all forms of retention of the element) by mineral surfaces, making a part of the total phosphorus unavailable to plants (ALMEIDA et al., 2003; CESSA et al., 2009; JOHNSON; LOEPPERT, 2006).

The main minerals related to the adsorption of P in soils are iron (Fe) and aluminum (Al) oxides, which together with kaolinite, make up the dominant mineralogy in the clay fraction of most soils in the southern (subtropical) and central-western (tropical) regions of Brazil (NITZSCHE et al., 2008; SCHAEFER et al., 2008). Hematite and goethite are the predominant pedogenic Fe oxides, followed by maghemite and ferrihydrite, and gibbsite is Al oxide. The concentrations of these minerals in the soil and characteristics such as crystallinity and specific surface area vary with the parent material, intensity of weathering, composition of the soil solution and drainage conditions (MOTTA; KÄMPF, 1992; SCHAEFER et al., 2008) and may also be affected by soil management systems (SILVA NETO et al., 2008).

The degree of interaction between P and soil mineral constituents can be evaluated by the maximum phosphorus adsorption capacity (MPAC), which can be calculated by fitting the Langmuir equation, among others (OLSEN; WATANABE, 1957). Several studies have shown significant variations in the MPAC in well-drained soils with advanced weathering that are generally correlated with the crystalline phases of Fe and Al oxides (ALMEIDA et al., 2003; BARBIERI et al., 2009; BROGGI et al., 2010; CESSA et al., 2009; CURI; FRANZMEIER, 1984; FONTES; WEED, 1996; LAIR et al., 2009). Most authors have stressed that soils rich in goethite and gibbsite adsorb more P than those rich in hematite and kaolinite. Torrent et al. (1994), however, reported that the average P adsorption capacity per unit area is similar between goethite and hematite in soils and that the increased P adsorption in goethites is determined by structural features such as a larger specific surface area (SSA). Thus, the high SSA of poorly crystalline Fe oxides such as ferrihydrite, even in small concentrations in well-drained soils, may play a crucial role in P adsorption (JOHNSON; LOEPPERT, 2006; RANNO et al., 2007).

In addition to the mineralogical composition of the soil, attention has been focused on the role of organic matter (OM) in reducing P adsorption capacity and improving the availability of P for plants grown in a soil under a long-term conservation management system such as the no-tillage system (PAVINATO et al., 2009). Organic ligands in the soil solution compete with P for adsorption sites on the surface of minerals, thus are decreasing P adsorption by these minerals (LIU et al., 1999; MIKUTTA et al., 2006; SCHWERTMANN et al., 1986). However, there is no consensus on this issue, and some studies have failed to verify the ability of increased levels of organic C to reduce P adsorption (RANNO et al., 2007; RHEINHEIMER et al., 2003).

The objective of this study was to determine the MPAC in five highly weathered soils and to establish relationships with (i) the pedogenic Fe oxides content, (ii) the concentrations of specific phases of Fe oxides (goethite, hematite, maghemite and ferrihydrite), (iii) the proportions of gibbsite and kaolinite and (iv) the specific surface area of these soils. The ability of no-tillage system to reduce soil phosphorus adsorption in comparison with conventional tillage was also evaluated.

Material and methods

Description of sites and soil sampling

The study was based on five soils from different regions of Brazil (Table 1). Long-term experiments were conducted in these soils with successions and/or rotations of crops under conventional tillage (CT) and no-tillage (NT) systems. The experiments were set up in a randomized block design with three replicates. In each experimental replicate of both tillage systems (CT and NT), the soil was sampled as follows: (i) at depths of 0.00 to 0.10 m and 0.10 to 0.20 m, for a total 60 soil samples, and (ii) in the 0.00–0.20 m layer, for a total of 30 soil samples. Soil samples representing native grassland conditions were collected in areas adjacent to the experimental field at depths of 0.00 to 0.20 m for particle size analysis and the determination of the mineralogical composition of the soil by X-ray diffraction. After air-dried and ground, soil samples were passed through a 2-mm sieve to obtain the air-dried fine soil (ADFS) fractions.

Physical analyses

Particle size analysis was performed using the pipette method after dispersion of the particles by ultrasound. For this purpose, 20 g ADFS samples from the 0.00–0.20 m layer of each soil were added into 100 mL centrifuge tubes containing 80 mL of distilled water, which were subjected to various levels of energy according to the soil (between 200 and 2,100 J mL^{-1}) in a Vibracel ultrasound machine equipped with a 136 mm long × 13 mm diameter probe. To each soil, the energy required for the complete dispersion of soil primary particles was estimated from a previously determined calibration curve (NORTH, 1976). Having determined the soil particle size, the total clay fraction (Ø < 0.002 mm) of each soil sample was collected by sedimentation according to Stokes's Law. The suspended clay was flocculated with 0.1 M HCl, washed with an ethanol/water solution (1:1) and dried at 50°C.

The specific surface area (SSA) of the ADFS samples of the 0.00–0.10 m and 0.10–0.20 m layers were estimated using the water adsorption method at 20% relative humidity (QUIRK, 1955).

X-ray diffraction analysis

The mineralogical composition of the total clay and iron oxide concentrate fractions at a depth of 0.00-0.20 m of the soils under native grassland was determined by X-ray diffraction (XRD) (Siemens D5000 diffractometer with CoKα radiation and a Fe filter). Analyses were conducted on non-oriented slides for the total clay fraction (at a 2θ range of 4−52°) and the iron oxide concentrate fraction (at a 2θ range 20−70°). Iron oxides were concentrated by treating the total clay fraction with hot 5 mol L^{-1} NaOH (KÄMPF; SCHWERTMANN, 1982). The minerals were identified based on the diffraction constant data provided by Brown and Brindley (1980).

Table 1. Soil classes, locations, environmental characteristics, duration of the experiment and cropping system/crop rotation.

Soil[1]	Location	Parent material	Mean annual Temperature °C	Mean annual Precipitation mm	Duration of experiment years	Cropping system/ crop rotation
			Southern region			
Ultic Hapludalf	Santa Maria, RS	Sandstone	18.8	1686	17	Summer: Corn/Soybeans Winter: Ryegrass/clover
Rhodic Paleudult	Eldorado do Sul, RS	Granite	19.2	1446	25	Summer: Corn/Cowpea Winter: Oats/Vetch
Humic Hapludox	Guarapuava, PR	Basalt	17.1	1941	32	Summer: Soybeans/Corn Winter: Wheat/Vetch/Oats
			central-west			
Rhodic Hapludox	Dourados, MS	Basalt	22.2	1409	20	Summer: Soybeans Winter: Wheat
Typic Hapludox	Luziânia, GO	Basalt	21.2	1553	18	Summer: Corn/ Soybean/Cotton

[1]Soil Survey Staff (2010).

Chemical analyses

The concentrations of extractable Ca^{+2}, Mg^{+2}, K^{+}, Al^{+3}, H+Al, pH and P were determined in the ADFS samples for the 0.00–0.20 m layer according to methods described by Embrapa (1997) and served as the basis for calculating the sum of bases (S), cation exchange capacity (CEC_{pH7}), base saturation (V) and aluminum saturation (m). The total organic carbon (organic C) contents in the ADFS samples of the 0.00–0.10 m and 0.10–0.20 m layers were determined by dry combustion in a carbon analyzer (Shimadzu VCSH).

Selective dissolutions of Fe oxides were prepared in the ADFS samples of the 0.00–0.10 m and 0.10–0.20 m layers. The Fe in pedogenic Fe oxides (Fe_d) was extracted with dithionite-citrate-bicarbonate at 80°C (MEHRA; JACKSON, 1960). The Fe in Fe oxides of low crystallinity (basically ferrihydrite, Fe_o) was extracted using 0.2 mol L^{-1} ammonium oxalate at pH 3.0 in darkness (SCHWERTMANN, 1964). The levels of dissolved Fe were determined by atomic absorption spectroscopy (AAS).

The proportions of the different Fe oxides phases identified in the XRD analyses of the concentrated Fe oxides fraction were estimated from the areas of the 110 reflection plane of goethite (Gt-110), the 012 plane of hematite (Hm-012) and the 220 plane of maghemite (Mh-220) that were obtained using the Siemens DIFFRAC plus-EVA computer program. The Gt-110 reflection area was multiplied by 0.30 because the intensities of the Hm-012 and MH-220 reflections are 30% of the Gt-110 reflection intensity (ALMEIDA et al., 2003). The Al^{3+} substitution was estimated according to the regression Al(mol mol^{-1})=17.30-5.72c (Å) in goethite (SCHULZE, 1984), according to the regression Al(mol mol^{-1})=31.09-6.17a_0 (Å) in hematite (SCHWERTMANN et al., 1979) and according to the regression Al(mol mol^{-1})=37.54-4.50a_0 (Å) in the maghemite (SCHWERTMANN; FECHTER, 1984). The concentration of each Fe oxide specific phase was obtained by allocation of the Fe content relative to the crystalline Fe oxides (Fe_d – Fe_o) according to their concentration in each soil and considering their chemical formulas. The gibbsite/(gibbsite + kaolinite) ratio, Gb/(Gb + Kt), was calculated from the diffraction patterns of non-oriented slides prepared from the clay total fraction according to the relationship between the intensity of the 002 reflection plane of gibbsite (I_{Gb-002}) and the 001 plane of kaolinite (I_{Kt-001}) (MOTTA; KÄMPF, 1992).

For the evaluation of P adsorption, 3.00-g ADFS samples were used, in duplicate, and each sample was equilibrated with 30 mL of KCl (0.02 mol L^{-1}) containing doses of 30, 60, 120, 240, 480, 960 or 1,920 mg L^{-1} P for the Oxisols and 15, 30, 60, 120, 240, 480 or 960 mg L^{-1} P for the Alfisol and Ultisol. The suspension was shaken horizontally for 5h (Oxisols) or 2h (Alfisol and Ultisol) [these times were based on the observed time course of adsorption (data not shown)], and then left to rest for 20h before measuring the P content in the supernatant according to the method of Murphy and Riley (1962). The amount of P adsorbed by the sample was estimated using the difference between the added P and the amount in the supernatant. The maximum phosphorus adsorption capacity (MPAC) was obtained from the parameters of the fitted Langmuir equation

$$P_{adsorbed} = K * P_{max} * \frac{P_{solution}}{(1 + K * P_{solution})}$$

where:

$P_{adsorbed}$ is the P adsorbed in the sample, $P_{solution}$ is the concentration in the solution, K is a constant related to the sorption energy, and P_{max} is the maximum P adsorption capacity.

Data analysis

The data analysis was performed with the Statistical Analysis System software (SAS for Windows). The

Kolmogorov-Smirnov test for normality showed that at the 5% level the parameters follow a normal distribution only within the soil classes. In this case, comparison of means were performed with Tukey test ($\alpha = 0.05$). Spearman nonparametric correlations (r_s) were obtained for parameters that were not normal distributed.

Results and discussion

Soil characteristics

The Alfisol and Ultisol developed from sandstone and granite, respectively, showed lower clay contents ($\leq$ 228 g kg^{-1} of soil) than the Oxisols developed from basalt (clay $\geq$ 596 g kg^{-1} of soil) (Table 2).

The soil pH ranged from 4.8 to 5.2, with the exception of the Typic Hapludox, which exhibited pH values near 6.0. The pH reflected the levels of H+Al and Al saturation (m), and the Typic Hapludox had the lowest values for these parameters. The sum of bases (S) ranged from 2.96 to 9.79 $cmol_c$ kg^{-1} in the Ultic Hapludalf and Rhodic Hapludox soils, respectively, and Ca^{+2} was the predominant cation. The CTC_{pH7} values ranged from 7.3 $cmol_c$ kg^{-1} in the CT of Ultic Hapludalf to 16.6 $cmol_c$ kg^{-1} in the NT of Rhodic Hapludox, with base saturations (V) greater than 50% only in the Typic and Rhodic Hapludox soils. The levels of available P were higher in the NT than in the CT samples in three of the five soils, ranging from 9.5 mg kg^{-1} in the CT of Humic Hapludox to 59 mg kg^{-1} in the NT of Typic Hapludox.

The organic C content differed significantly among the soils (Figure 1) as the likely result of differences in clay content (Alfisol and Ultisol < Oxisols), mineralogy and pedoclimate that are known to affect the accumulation and stabilization of OM (Table 1). The significant effect of NT on the increase of organic C relative to CT was observed only in the 0.00–0.10 m layer in the Rhodic Paleudult and Humic Hapludox. With the exception of the Typic Hapludox, the organic C content decreased with increasing depth in all soils subjected to NT; however, this effect was not observed under CT (Figure 1).

The SSA values in the soils ranged from 9.9 to 63.4 m^2 g^{-1} in both the NT and CT samples (Table 3). The SSA tended to be greater in the NT- than in the CT-treated soils, probably because of the increase in organic C in the no-tilled surface soil layer. The positive relationship ($p < 0.0001$) between SSA, estimated by the water-adsorption method, and organic C content was in sharp contrast with the inverse relationships observed in several studies in which the SSA was estimated by N_2 adsorption (AFIF et al., 1995; ALMEIDA et al., 2003). This discrepancy can be attributed to a more effective hindrance of N_2 to enter pores or covering the mineral surface by associated with clay fraction than the polar water molecule, thus difficulting the access and covering of the mineral surfaces (ALMEIDA et al., 2003).

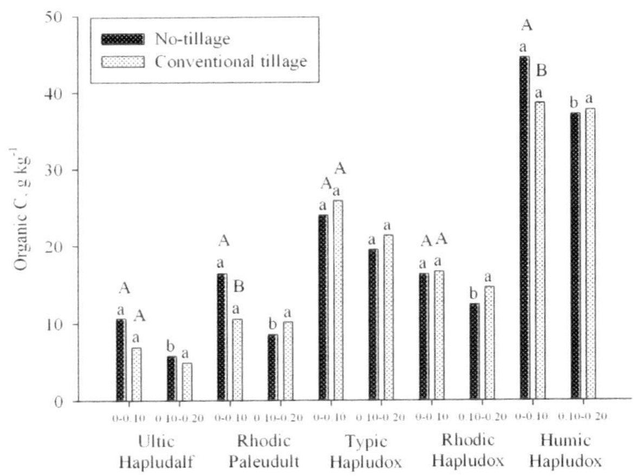

Figure 1. Total organic carbon contents in the layers of each soil under both management systems (uppercase letters refer to the comparison of management systems at the depth of 0.00–0.10 m; lowercase letters refer to comparisons between depths with the same management system – Tukey's test at the 5% level).

Table 2. Clay contents and characteristics of the adsorptive complexes of soils in the 0.00–0.20 m layer.

Soil[1]	Management system	Clay	pH	K	Ca	Mg	Al	H + Al	CTC_{pH7}	S	V	m	P
		g kg^{-1}		-------cmol$_c$ kg^{-1}-------							-----%----		mg kg^{-1}
Ultic Hapludalf	NT	65	4.9	0.21	2.4	0.9	1.0	10.9	14.4	3.51	24	22	22.0
	CT		5.0	0.16	2.0	0.8	0.6	4.4	7.3	2.96	40	17	11.0
Rhodic Paleudult	NT	228	4.9	0.21	3.2	1.8	0.9	7.7	13.0	5.21	40	15	16.0
	CT		4.8	0.28	4.3	1.5	1.0	8.7	14.8	6.98	41	15	16.0
Humic Hapludox	NT	642	5.0	0.42	4.3	1.8	0.5	7.7	14.3	6.52	46	7	10.0
	CT		5.0	0.48	3.3	1.2	0.7	8.7	13.7	4.98	36	12	9.5
Rhodic Hapludox	NT	596	5.0	0.49	6.5	2.7	0.4	6.9	16.6	9.69	58	4	37.0
	CT		5.2	0.49	6.8	2.5	0.3	6.2	15.9	9.79	61	3	19.0
Typic Hapludox	NT	886	5.8	0.22	4.5	2.2	0.0	4.4	11.3	6.02	61	0	59.0
	CT		6.1	0.29	4.6	2.2	0.0	2.8	9.8	7.09	72	0	11.0

[1]Soil Survey Staff (2010). V, base saturation; m, aluminum saturation; S, sum of bases.

Table 3. Fe contents in pedogenic Fe oxides, concentrations of Fe oxides specific phases, mineralogical attributes and specific surface area of the soils.

Soil[1]	Depth	Fe_d		Ferrihydrite		Goethite		Hematite		Maghemite	
		NT	CT	NT	CT	NT	CT	NT	CT	NT	CT
	--m--	g kg^{-1}									
Ultic Hapludalf	0.00–0.10	8.2	7.8	1.2	1.2	7.2	6.8	4.3	4.1	0.0	0.0
	0.10–0.20	8.2	7.6	1.2	1.1	7.2	6.6	4.3	4.0	0.0	0.0
Rhodic Paleudult	0.00–0.10	12.1 bB	13.5 bA	2.1	2.2	4.8 bB	5.4 bA	11.2 bB	12.6 bA	0.0	0.0
	0.10–0.20	13.8 a	14.3 a	2.0	2.2	5.6 a	5.8 a	13.0 a	13.4 a	0.0	0.0
Humic Hapludox	0.00–0.10	101.5	104.7	6.4	5.4	115.0	119.5	36.3	37.8	0.0	0.0
	0.10–0.20	101.6	106.3	6.5	5.7	115.1	121.2	36.4	38.3	0.0	0.0
Rhodic Hapludox	0.00–0.10	115.9	120.3	3.5 B	5.5 A	5.4	5.6	97.7	100.5	60.2	62.0
	0.10–0.20	117.1	120.2	3.7 B	5.2 A	5.5	5.6	98.6	100.5	60.8	62.0
Typic Hapludox	0.00–0.10	34.3	34.4	6.6	6.3	37.8	38.2	9.6	9.7	0.0	0.0
	0.10–0.20	36.1	33.9	6.3	6.1	40.2	37.6	10.2	9.5	0.0	0.0

Soil[1]	Depth	Fe_{ox}/Fe_d		Gt/(Gt+Hm)	Gb/(Gb+Ct)	SSA	
		NT	CT			NT	CT
	--m--					$m^2\ g^{-1}$	
Ultic Hapludalf	0.00–0.10	0.08	0.09	0.77	0.10	13.4	9.9
	0.10–0.20	0.08	0.09			11.3	10.0
Rhodic Paleudult	0.00–0.10	0.10 a	0.09	0.45	0.09	20.1	17.6 b
	0.10–0.20	0.09 b	0.09			18.8	19.2 a
Humic Hapludox	0.00–0.10	0.04	0.03	0.89	0.78	63.4	61.0
	0.10–0.20	0.04	0.03			61.5	60.8
Rhodic Hapludox	0.00–0.10	0.02	0.03	0.05	0.04	43.0	44.8
	0.10–0.20	0.02	0.03			43.2	44.6
Typic Hapludox	0.00–0.10	0.11	0.11	0.90	0.69	35.2	36.2
	0.10–0.20	0.10	0.11			33.8	33.8

[1]Soil Survey Staff (2010). Fe_d: concentration of Fe relative to the total pedogenic Fe oxides; Fe_o, concentration of Fe extracted by ammonium oxalate; SSA, specific surface area; NT, no-tillage system; CT, conventional tillage system; Gb, gibbsite; Kt, kaolinite; Gt, goethite; Hm, hematite. Lowercase letters indicate the differences between the depth measurements, and the uppercase letters indicate the differences between management systems (Tukey's test at the 5% level).

Mineralogical composition and phosphorus adsorption capacity of the soils

The X-ray diffraction patterns in the total clay fraction (Figure 2a) revealed the predominance of kaolinite in the Rhodic Hapludox, Rhodic Paleudult and Ultic Hapludalf ($Gb/(Gb+Kt) \leq 0.10$) (Table 3) associated with 2:1 or 2:1:1 minerals (Rhodic Hapludox and Ultic Hapludalf) and quartz (Rhodic Paleudult and Ultic Hapludalf) that occurred to a less extent. The Typic and Humic Hapludox soils showed predominance of gibbsite [$Gb/(Gb+Kt) \geq 0.69$] with the presence of 2:1 or 2:1:1 minerals in the Humic Hapludox. In all soils, the goethite and hematite Fe oxides were present in the fraction of concentrated Fe oxides (Figure 2b). In the Ultic Hapludalf, Humic and Typic Hapludox soils, $Gt/(Gt+Hm) \geq 0.77$, i.e. goethite predominated over hematite (Table 3). The Rhodic Hapludox showed a predominance of hematite over goethite [$Gt/(Gt+Hm) = 0.05$], whereas the proportions of goethite and hematite were similar in the Rhodic Paleudult (Table 3). The presence of maghemite was detected only in the Rhodic Hapludox (Figure 2b), where it occurs in significant proportion according to the ratio $Mh/(Mh+Gt+Hm) = 0.36$.

The Fe present in pedogenic Fe oxides (Fe_d) and the concentrations of the various Fe oxides identified in each soil are presented in Table 3. The Fe-for-Al substitution in the goethite, hematite and maghemite, as well as the chemical formulas considered in the allocation of Fe levels are listed in Table 4. In each soil, the Fe_d levels and the concentrations of Fe oxide phases were similar between the management systems and at both depths studied, whereas wide ranges were observed among the soils (Table 3). The highest Fe_d concentrations were observed in the Rhodic and Humic Hapludox developed from basalt (105–120 g kg^{-1}), whereas the Ultic Hapludalf and Rhodic Paleudult, formed from sandstone and granite, respectively, exhibited lower Fe_d values (7–15 g kg^{-1}). The ferrihydrite concentrations in the soil, estimated from the Fe extracted by ammonium oxalate (Fe_o), also showed a relationship with the parent material, ranging from 3.5 to 6.5 g kg^{-1} in the Oxisols and from 1.1 to 2.2 g kg^{-1} in the Alfisoil and Ultisoil. A $Fe_o/Fe_d < 0.11$ in the soils indicated a predominance of crystalline phases of Fe oxides (SCHWERTMANN et al., 1982). Concentrations of crystalline Fe oxides showed significant amplitudes, where the contents of goethite ranged from 4.8 g kg^{-1} in Rhodic Paleudult to 121.2 g kg^{-1} in Humic Hapludox; hematite ranged from 4.0 g kg^{-1} in the Ultic Hapludalf to 100.5 g kg^{-1} in the Rhodic Hapludox. In the Rhodic Hapludox, where maghemite was detected, its concentration was approximately 61 g kg^{-1} (Table 3).

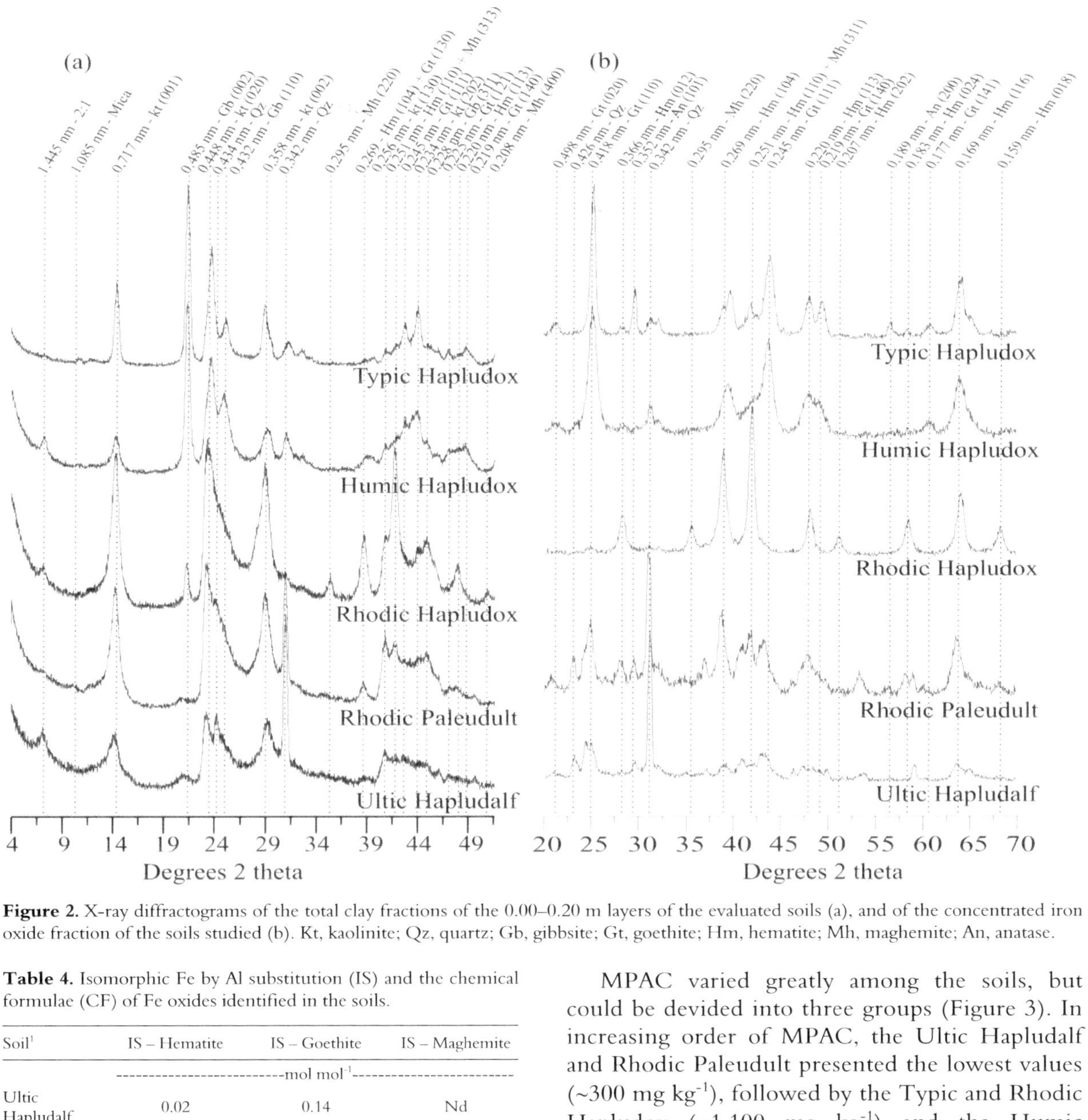

Figure 2. X-ray diffractograms of the total clay fractions of the 0.00–0.20 m layers of the evaluated soils (a), and of the concentrated iron oxide fraction of the soils studied (b). Kt, kaolinite; Qz, quartz; Gb, gibbsite; Gt, goethite; Hm, hematite; Mh, maghemite; An, anatase.

Table 4. Isomorphic Fe by Al substitution (IS) and the chemical formulae (CF) of Fe oxides identified in the soils.

Soil[1]	IS – Hematite	IS – Goethite	IS – Maghemite
	------------------------mol mol^{-1}------------------------		
Ultic Hapludalf	0.02	0.14	Nd
Rhodic Paleudult	0.11	0.16	Nd
Humic Hapludox	nd	0.31	Nd
Rhodic Hapludox	0.16	nd	0.16
Typic Hapludox	0.14	0.33	nd
Soil[1]	CF - Hematite	CF - Goethite	CF - Maghemite
Ultic Hapludalf	α-$Fe_{1.96}Al_{0.04}O_3$	α -$Fe_{0.86}Al_{0.14}OOH$	
Rhodic Paleudult	α -$Fe_{1.78}Al_{0.22}O_3$	α -$Fe_{0.84}Al_{0.16}OOH$	
Humic Hapludox	α -Fe_2O_3	α -$Fe_{0.69}Al_{0.31}OOH$	
Rhodic Hapludox	α -$Fe_{1.68}Al_{0.32}O_3$	α -FeOOH	γ-$Fe_{1.68}Al_{0.32}O_3$
Typic Hapludox	α -$Fe_{1.72}Al_{0.28}O_3$	α -$Fe_{0.67}Al_{0.33}OOH$	

[1]Soil Survey Staff (2010).

MPAC varied greatly among the soils, but could be devided into three groups (Figure 3). In increasing order of MPAC, the Ultic Hapludalf and Rhodic Paleudult presented the lowest values (~300 mg kg^{-1}), followed by the Typic and Rhodic Hapludox (~1,100 mg kg^{-1}) and the Humic Hapludox, which showed the greatest P adsorption capacity (~4,700 mg kg^{-1}). As for the soil mineralogical attributes, the MPAC was positively correlated with Fe_d and Gb/(Gb+Kt (Table 5). The observed lower capacity of kaolinite to adsorb P compared with that of gibbsite (BARBIERI et al., 2009; BROGGI et al., 2010; CESSA et al., 2009; FONTES; WEED, 1996) has been related to the lower concentrations of OH sites on its surface (CELI et al., 2003).

Considering the different Fe oxides, MPAC was positively correlated with the concentration of goethite and ferrihydrite (Table 5). These results reinforce the importance of low crystallinity phases such as ferrihydrite in the adsorption of P (JOHNSON; LOEPPERT, 2006; RANNO et al., 2007), even at low concentrations in the soil, and the greater P adsorption capacity of goethite compared with hematite observed in previous studies (ALMEIDA et al., 2003; CESSA et al., 2009; CURI; FRANZMEIER, 1984; FONTES; WEED, 1996). In addition, the results suggest that the higher amount of P adsorption by goethite is related to its greater specific surface area, as evidenced by the significant relationship (p 0.0027) obtained between the concentration of goethite and the SSA values of the soils (Table 5), thus corroborating the results of Torrent et al. (1990, 1992, 1994). It should be noted that the formation of goethite is favored in environments that promote the accumulation of OM in the soil (BIGHAM et al., 2002), which suggests that the correlation between MPAC and organic C (Table 5) is affected by the covariance between the concentrations of goethite and organic C ($p < 0.0001$).

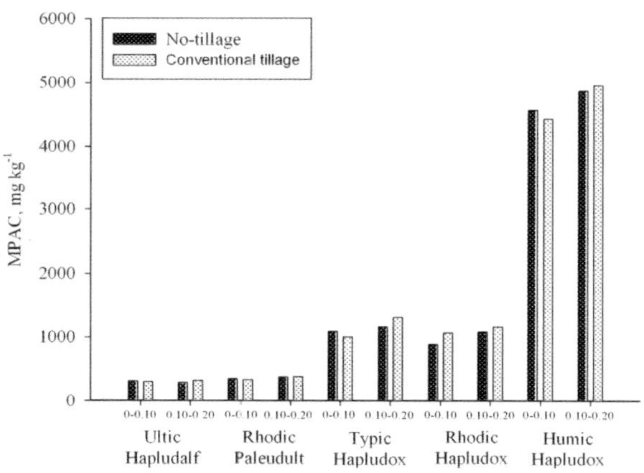

Figure 3. Maximum phosphorus adsorption capacity (MPAC) in two layers of each soil for both soil management systems (values of MPAC did not differ significantly – Tukey's test at the 5% level – between soil depth neither between tillage system, in all soils).

Regarding the relationships between the MPAC and the influential soil parameters, the lower MPACs of the Ultic Hapludalf and Rhodic Hapludox resulted from the kaolinitic mineralogy ($Gb/(Gb+Kt) \leq 0.10$) combined with the low concentrations of iron oxides ($Fe_d < 15$ g kg^{-1}) and the consequently reduced SSA in these soils (SSA < 20 m^2 g^{-1}). The Typic and Rhodic Hapludox showed intermediate MPAC values compared with the other soils, but these soils were characterized by distinct combinations of mineralogical attributes. The greater concentration of pedogenic Fe oxides ($115 < Fe_d < 120$ g kg^{-1}) in the Rhodic Hapludox, due to the predominance of hematite and maghemite minerals, was compensated for in the Typic Hapludox by the concentration of goethite and the greater proportion of gibbsite [$Gb/(Gb+Kt) = 0.69$]. The high MPAC of the Humic Hapludox resulted from its gibbsitic mineralogy [$Gb/(Gb+Kt) = 0.78$] combined with a high concentration of Fe oxides ($Fe_d > 100$ g kg^{-1}), mainly due to the presence of goethite minerals, which resulted in the largest SSA of all the soils ($60 <$ SSA < 65 m^2 g^{-1}).

Table 5. The 'r_s' and 'p' values of the correlation analyses of variables that affect the maximum phosphorus adsorption capacity.

	MPAC	SSA	Organic C
MPAC	1		
SSA	0.88[1] <0.0001[2]	1	
Organic C	0.85 <0.0001	0.84 <0.0001	1
Fed	0.73 <0.0001	0.86 <0.0001	0.597 <0.0001
Goethite	0.57 <0.0001	0.38 0.0027	0.62 <0.0001
Hematite	0.55 <0.0001	0.76 <0.0001	0.40 0.0016
Maghemite	0.13 0.3102	0.34 0.0071	-0.06 0.6164
Ferrihidrite	0.81 <0.0001	0.74 <0.0001	0.84 <0.0001
Gb/(Gb+Kt)	0.65 <0.0001	0.49 <0.0001	0.70 <0.0001
Gt/(Gt+Hm)	0.36 0.0047	0.10 0.4563	0.52 <0.0001
Clay	0.78 <0.0001	0.68 <0.0001	0.85 <0.0001

[1]r_s: Spearman correlation coefficient; [2]p: level of significance. MPAC: maximum phosphorus adsorption capacity; SSA: specific surface area; Fed: concentration of Fe relative to the total pedogenic Fe oxides; Gb, gibbsite; Kt, kaolinite; Gt, goethite; Hm, hematite.

Effect of soil management systems on soil phosphorus adsorption capacity

The MPACs of all the soils were similar at the depths 0.0–00.10 m and 0.10–0.20 m (Figure 3), even with the greater concentration of organic C in the 0.00–0.10 m layers of the no-tilled Rhodic Paleudult and Humic Hapludox (Figure 1). An effect of the different soil management systems on MPAC was also not observed in two Oxisols and one Alfisol in southern Brazil in the study of Rheinheimer et al. (2003). These results contradict other studies indicating that the interactions between mineral surfaces and organic compounds reduce the P adsorption capacity of a soil (ALMEIDA et al., 2003; SCHWERTMANN et al., 1986), although these authors did not specifically evaluate the effect of management systems on MPAC; instead, they determined the MPAC of different soil classes.

The results of the present study may have been influenced by the depth of the surface layer assessed (0.00–0.10 m), thus supporting the explanation provided by Pavinato et al. (2010), who observed a change in the characteristics of P adsorption in the zone near the surface. The accumulation of organic C in the NT soil, which is normally highest in the first centimeters from the soil surface (0.00–0.025 m and 0.025–0.05 m), may have been diluted by sampling the 0.00–0.10 m layer. Additionally, the soil grinding procedure to obtain the ADFS destroys the aggregates formed under NT, thereby exposing the P adsorption sites that would be less exposed under the natural conditions of the management system.

Conclusion

The diversity of the mineralogical composition of the soils resulted in significant differences in the maximum phosphorus adsorption capacities among the soils. The P adsorption capacities of soils followed the sequence: Ultic Hapludalf ≈ Rhodic Paleudult < Typic Hapludox ≈ Rhodic Hapludox < Humic Haplodox, and the increase was positively related to the Fe in pedogenic Fe oxides, the concentrations of goethite and ferrihydrite and the gibbsite to kaolinite ratio, which were instrumental in increasing the specific surface area of soils. The maximum phosphorus adsorption capacity of each soil was not affected by long-term no-tillage system.

Acknowledgements

The authors thank CNPq for providing fellowships to A.V. Inda and C. Bayer, and financial support under Universal Edital (process 470718/2009-9; 481694/2011-0). This work was also partly financed by the Spain's Ministry of Science and Innovation (Project CGL 2010-15067) and the European Regional Development Fund.

References

AFIF, E.; BARRÓN, V.; TORRENT, J. Organic matter delays but does not prevent phosphate sorption by Cerrado soils from Brazil. **Soil Science**, v. 159, n. 3, p. 207-211, 1995.

ALMEIDA, J. A.; TORRENT, J.; BARRÓN, V. Soil color, phosphorus pools and phosphate adsorption in latosols developed from basalt in the south of Brazil. **Revista Brasileira de Ciência do Solo**, v. 27, n. 6, p. 985-1002, 2003.

BARBIERI, D. M.; MARQUES JÚNIOR, J.; ALLEONI, L. R. F.; GARBUIO, F. J.; CAMARGO, O. A. Hillslope curvature, clay mineralogy, and phosphorus adsorption in an Alfisol cultivated with sugarcane. **Scientia Agricola**, v. 66, n. 6, p. 819-826, 2009.

BIGHAM, J. M.; FITZPATRICK, R. W.; SCHULZE, D. Iron oxides. In: DIXON, J. B.; SCHULZE, D. G. (Ed.). **Soil mineralogy with environmental applications**. Madison: SSSA, 2002. p. 323-366.

BROGGI, F.; OLIVERIRA, A. C.; FREIRE, F. J.; FREIRE, M. B. G. S.; NASCIMENTO, C. W. A. Adsorption and chemical extraction of phosphorus as a function of soil incubation time. **Revista Brasileira de Engenharia Agrícola e Ambiental**, v. 14, n. 1, p. 32-38, 2010.

BROWN, G.; BRINDLEY, G. W. X-ray diffraction procedures for clay mineral identification. In: BRINDLEY, G. W.; BROWN, G. (Ed.). **Crystal structures of clays minerals and their X-ray identification**. London: Mineralogical Society, 1980. p. 305-360.

CELI, L.; DE LUCA, G.; BARBERIS, E. Effects of interaction of organic and inorganic P with ferrihydrite and kaolinite-iron oxide systems on iron release. **Soil Science**, v. 168, n. 7, p. 479-488, 2003.

CESSA, R. M. A.; CELI, L.; VITORINO, A. C. T.; NOVELINO, J. O.; BARBERIS, E. Specific surface área and porosity of the clay fraction and phosphorus adsorption in two rhodic ferralsols. **Revista Brasileira de Ciência do Solo**, v. 33, n. 5, p. 1153-1162, 2009.

CURI, N.; FRANZMEIER, D. P. Toposequence of Oxisols from the central plateau of Brazil. **Soil Science Society of America Journal**, v. 48, n. 2, p. 341-346, 1984.

EMBRAPA-Empresa Brasileira de Pesquisa Agropecuária. Centro Nacional de Pesquisa de Solos. **Manual de métodos de análises de solos**. 2. ed. Rio de Janeiro: Embrapa, CNPS, 1997.

FONTES, M. P. F.; WEED, S. B. Phosphate adsorption by clays from Brazilian Oxisols: relationships with specific surface area and mineralogy. **Geoderma**, v. 72, n. 1-2, p. 37-51, 1996.

JOHNSON, S.; LOEPPERT, R. H. Role of organic acids in phosphate mobilization from iron oxide. **Soil Science Society of America Journal**, v. 70, n. 1, p. 222-234, 2006.

KÄMPF, N.; SCHWERTMANN, U. The 5M - NaOH concentration treatment for iron oxides in soils. **Clays and Clay Minerals**, v. 30, n. 6, p. 401-408, 1982.

LAIR, G. J.; ZEHETNER, F.; KHAN, Z. H.; GERZABEK, M. H. Phosphorus sorption-desorption in alluvial soils a young wethering sequence at the Danube River. **Geoderma**, v. 149, n. 1-2, p. 39-44, 2009.

LIU, F.; HE, J.; COLOMBO, C. Competitive adsorption of sulfate and oxalate on goethite in the absence or presence of phosphate. **Soil Science**, v. 164, n. 3, p. 180-189, 1999.

MEHRA, O. P.; JACKSON, M. L. Iron oxides removal from soil and clays by a dithionite-citrate system buffered with sodium bicarbonate. **Clays and Clay Minerals**, v. 7, n. 1, p. 317-327, 1960.

MIKUTTA, C.; LANG, F.; KAUPENJOHANN, M. Kinetics of phosphate sorption to polygalacturonate-coated goethite. **Soil Science Society of America Journal**, v. 70, n. 2, p. 541-549, 2006.

MOTTA, P. E. F.; KÄMPF, N. Iron oxide properties as support to soil morphological features for prediction of moisture regimes in Oxisols of Central Brazil. **Zeitschrift für Pflanzenernährung und Bodenkunde**, v. 155, n. 5, p. 385-390, 1992.

MURPHY, J.; RILEY, J. P. A modified single solution method for the determination of phosphate in natural waters. **Analytica Chimica Acta**, v. 27, n. 1, p. 31-36, 1962.

NITZSCHE, R. P.; PERCIVAL, J. B.; TORRANCE, J. K.; STIRLING, J. A. R.; BOWEN, J. T. X-ray diffraction and infrared characterization of Oxisols from central and southeastern Brazil. **Clay Minerals**, v. 43, n. 4, p. 549-560, 2008.

NORTH, P. F. Towards an absolute measurement of soil structural stability using ultrasound. **European Journal of Soil Science**, v. 27, n. 4, p. 451-459, 1976.

NOVAIS, R. F.; SMYTH, T. J. **Fósforo em solo e planta em condições tropicais**. Viçosa: UFV, 1999.

OLSEN, S. R.; WATANABE, F. S. A method to determine a phosphorus adsorption maximum of soil as measured by the Langmuir isotherm. **Soil Science Society of America Proceedings**, v. 21, n. 2, p. 144-149, 1957.

PAVINATO, P. S.; DAO, T. H.; ROSOLEM, C. A. Tillage and phosphorus management effects on enzyme-labile bioactive phosphorus availability in Cerrado Oxisols. **Geoderma**, v. 156, n. 3-4, p. 207-215, 2010.

PAVINATO, P. S.; MERLIN, A.; ROSOLEM, C. A. Phosphorus fraction in Brazilian Cerrado soils as affected by tillage. **Soil and Tillage Reseach**, v. 105, n. 1, p. 149-155, 2009.

QUIRK, J. P. Significance of surface áreas calculated from water vapor sorption isotherms by use of the B.E.T equations. **Soil Science**, v. 80, n. 6, p. 423-430, 1955.

RANNO, S. K.; SILVA, L. S.; GATIBONI, L. C.; RHODEN, A. C. Phosphorus adsorption capacity in lowland soils of Rio Grande do Sul state. **Revista Brasileira de Ciência do Solo**, v. 31, n. 1, p. 21-28, 2007.

RHEINHEIMER, D. S.; ANGHINONI, I.; CONTE, E. Phosphorus sorption in soil in relation to the initial content and soil management. **Revista Brasileira de Ciência do Solo**, v. 27, n. 1, p. 41-49, 2003.

SCHAEFER, C. E. G. R.; FABRIS, J. D.; KER, J. C. Minerals in the clay fraction of Brazilian Latosols (Oxisols): a review. **Clay Minerals**, v. 43, n. 1, p. 137-154, 2008.

SCHULZE, D. G. The influence of aluminum on iron oxides: VIII. Unit-cell dimensions of Al-substituted goethites and estimation of Al from them. **Clays and Clay Minerals**, v. 32, n. 1, p. 36-44, 1984.

SCHWERTMANN, U. Differenzierung der eisenoxide des bodens durch extraktion mit ammoniumoxalat-lösung. **Zeitschrift für Pflanzenernährung, Düngung und Bodenkunde**, v. 105, n. 3, p. 194-202, 1964.

SCHWERTMANN, U.; FECHTER, H. The influence of aluminium on iron oxides: XI. Aluminium-substituted maghemite in soils and its formation. **Soil Science Society of America Journal**, v. 48, n. 6, p. 1462-1463, 1984.

SCHWERTMANN, U.; FITZPATRICK, R. W.; TAYLOR, R. M.; LEWIS, D. G. The influence of aluminum on iron oxides. Part II. Preparation and properties of Al-substituted hematites. **Clays and Clay Minerals**, v. 27, n. 2, p. 105-112, 1979.

SCHWERTMANN, U.; KODAMA, H.; FISCHER, W. R. Mutual interactions between organic and iron oxides. In: HUANG, P. M.; SCHNITZER, M. (Ed.). **Interactions of soil minerals with natural organics and microbes**. Madison: SSSA, 1986. p. 223-250.

SCHWERTMANN, U.; SCHULZE, D. G.; MURAD, E. Identification of ferrihydrite in soils by dissolution kinetics, differential x-ray diffraction and Mössbauer spectroscopy. **Soil Science Society of America Journal**, v. 46, n. 4, p. 869-875, 1982.

SILVA NETO, L. F.; INDA, A. V.; BAYER, C.; DICK, D. P.; TONIN, A. T. Iron oxides in tropical and subtropical Brazilian oxisols under long-term no-tillage. **Revista Brasileira de Ciência do Solo**, v. 32, n. 5, p. 1873-1881, 2008.

SOIL SURVEY STAFF. **Keys to Soil Taxonomy**. 11th ed. Washington, D.C.: United States Department of Agriculture. Natural Resources Conservation Service, 2010.

TORRENT, J.; BARRÓN, V.; SCHWERTMANN, U. Phosphate adsoption and desorption by goethites differing in crystal morphology. **Soil Science Society of America Journal**, v. 54, n. 4, p. 1007-1011, 1990.

TORRENT, J.; SCHWERTMANN, U.; BARRÓN, V. Fast and slow phosphate sorption by goethite-rich natural materials. **Clays and Clay Minerals**, v. 40, n. 1, p. 14-21, 1992.

TORRENT, J.; SCHWERTMANN, U.; BARRÓN, V. Phosphate sorption by natural hematites. **European Journal of Soil Science**, v. 45, n. 1, p. 45-51, 1994.

7

Variability in soil physical properties in landslide-prone areas

Letícia d'Agosto Miguel Fonseca[1*], João Luiz Lani[2], Elpídio Inácio Fernandes Filho[2], Gerson Rodrigues dos Santos[3], Williams Pinto Marques Ferreira[4] and Adriana Maria Rocha Trancoso Santos[3]

*[1]Instituto Nacional de Pesquisas Espaciais, Av. dos Astronautas, 12227-010, São José dos Campos, São Paulo, Brazil. [2]Departamento de Solos, Centro de Ciências Agrárias, Universidade Federal de Viçosa, Viçosa, Minas Gerais, Brazil. [3]Departamento de Estatística, Universidade Federal de Viçosa, Viçosa, Minas Gerais, Brazil. [4]Empresa Brasileira de Pesquisa Agropecuária, Embrapa Café, Viçosa, Minas Gerais, Brazil. *Author for correspondence. E-mail: Leticiafonseca.geo@gmail.com*

ABSTRACT. The present study aimed to evaluate the relationship between the physical properties of the soil and relief in areas that are susceptible to landslides in the sub-basin of the Córrego do Yung, a creek in the urban district Três Moinhos of Juiz de Fora, Minas Gerais. Depth measurements of the *solum* were made at the soil sampling points, and the subsurface material was collected and analyzed for the two factors of particle size and type of horizon. In the laboratory, we determined the hydraulic conductivity, bulk and particle density, and total porosity, macroporosity and microporosity. The values of the physical properties throughout the sub-basin were predicted and mapped. Thematic maps were generated and showed a relation to each other: the clay was related to a higher hydraulic conductivity, higher porosity and lower soil density. In the steeper areas, the greater silt contents were mapped in the deeper layers and indicated the thin thickness of horizon B in these areas. A close relation of the curvature with the soil thickness was detected and was greater in the flat areas with less slope. The geostatistical analysis showed that the range of values for the properties of macroporosity, microporosity and silt was the lowest in the deeper layers, indicating low structural continuity in potentially erodible areas.

Keywords: physical soil, geostatistics, landslide.

Introduction

Landslides have always been a cause of great concern, particularly during the rainy season in urban areas. Several landslides occurred in recent years in different regions of Brazil, such as in the great disasters of the landslides in Santa Catarina in 2008 and in the highland region of Rio de Janeiro in 2011.

To define the conditions of susceptibility to landslides, the hydrological processes associated with soil and relief, which are fundamental in environmental studies and risk management areas, must be understood.

Thus, information on the soil properties are important, as is information about the relief forms that affect the water infiltration and storage capacity in the soil, which influences the possibility of the occurrence of landslides (Petschko, Brening, Bell, Goetz, & Glade, 2014). These combined factors of the spatial dynamics can interfere with the kinetic rain energy (Aquino, Silva, Freitas, Curi, & Avanzi, 2012).

Studies on the issue of landslides classify soil depth and hydraulic conductivity as key information in slope stability analyses, though these factors are the most difficult to measure. However, other essential data can also be obtained through remote sensing, e.g., elevation model, landslide inventory and land use maps (Western, Castellanos, & Kuriakose, 2008).

Experiments relating erosion and soil texture showed that soils with high silt amounts are more susceptible to erosion than are clayey and sandy soils under the same conditions due to greater runoff. Additionally, the slope and the presence of vegetation influence erosion intensity (Sadeghi, Moosavi, Karami, & Behnia, 2012).

In soils classified as Cambisols, the degree of erodibility is higher due to the steepness of the terrain, the silt content and the thickness of the B horizon (Vestena, & Kobiyama, 2014). In the area of "Mares de Morros," the C horizon tends to be very deep, and the sizes of the solum and the soil are disproportionate, indicating rejuvenation of the geomorphic process in this domain (Rezende, & Resende, 1996). According to these authors, in the steeper portions, which usually have an exposed C horizon, the recovery of the vegetation cover on the soils is more difficult due to the lack of nutrients and seasonal drought, thus intensifying the erosive processes.

The relationships between soil properties, such as soil texture, porosity and bulk density with hydraulic conductivity (K_0), and soil water retention curve (SWRC) in areas that are affected by landslide processes indicate that these properties are satisfactory for understanding the interaction between the soil structure and the hydraulic behavior (Bogner et al., 2014). One example is saturated hydraulic conductivity, which depends on water leaching due to the macroporosity, which, in turn, is related to soil texture, particle arrangement (structure) and bulk density. The free passage of water is crucial to reduce runoff (Gomes et al., 2011).

The spatial continuity of the soil physical properties was confirmed (Kamimura, Santos, Oliveira, Dias, & Guimarães, 2013); i.e., spatial models represent ideal tools for data analysis by geostatistical spatial modeling (Silva, Mello, Curi, & Oliveira, 2008). These authors further emphasize rainfall erosion, which is characterized and modeled by properties with continuous spatial behavior, including soil-related data, vegetation cover, rainfall erosivity, and topography. Soil erosion can be mapped by spatial models, provided that the studies on the spatial dependence of data and the modeling of the experimental semivariogram are consistent. The soil slope, relief, geology, depth and resistance, land use and vegetation cover, rainfall, and mapped points of occurrence of landslides are the main GIS factors used in these studies (Klimes & Escobar, 2010; Costanzo, Rotigliano, Irigaray, Jiménez-Perávlvarez, & Chaón, 2012; Di Martire, Rosa, Pesca, Santangelo, & Calcaterra, 2012; Zizioli, Meisina, Valentino, & Montrasio, 2013).

Thus, soil research focused on these landslide and erosion processes is of fundamental interest to professionals from various fields of knowledge related to this particular field of study. In the city of Juiz de Fora, urban sprawl has increased significantly in recent years. Every year, landslides in various parts of the city are recorded by the civil defense, especially in the rainy season. One of the regions with the highest number of occurrences is the urban district Três Moinhos.

Thus, this study defined indicators in a landslide-prone area in the metropolitan area of Juiz de Fora (district Três Moinhos) to support the authorities in charge of managing this risk area.

Material and methods

Study area

The study was conducted in the sub-basin of the Yung creek, in the urban district Três Moinhos, Juiz de Fora, in the mesoregion of Zona da Mata, Minas Gerais. The total area of the sub-basin is 35 ha, and it is in mountainous terrain with ravine slips and gullies. The altimetry variation of the area is between 0 and 900 m. According to Fonseca, Lani, Fernandes, Marques, and Ferreira (2016), the soil of the sub-basin was classified as dystrophic Tb Haplic Inceptisol.

The district Três Moinhos is located in a development region of Linhares, in the Yung basin, in the eastern region of the city. It is characterized by residences that have no basic infrastructure and are irregularly distributed. The terrain is hilly, and many points are vulnerable to landslides. (Prefeitura de Juiz de Fora [PJF], 2004). This is the planning region with the largest number of points of risk areas mapped by the civil defense and occurrences of landslides, which comprised 256 events that occurred at 27 locations over the past eight years; the landslides were concentrated in the Western neighborhoods of Três Moinhos and Bom Jardim, which are inserted in the Yung basin (PJF, 2004).

The presence of deep valleys associated with hillsides with steep slopes, known as "Mares de Morros" (Ab'Saber, 1996), exposure to high rainfall and the characteristics of mountainous tropical climate are the main factors that have led to the intense surface dynamic of the region, with soil erosion and sedimentation, flooding and landslide areas (PJF, 2004).

Through geotechnical characterization tests and erodibility resistance, it can be concluded that in the sub-basin of the Yung creek, the areas with the greatest erosion potential are those that expose the C horizon. However, because of the rugged relief, constant erosion processes on horizon B occur by washouts, though this horizon presents greater resistance (Fonseca et al., 2016).

Sample collection and preparation

The sampling points were first localized in the digital environment in ArcGIS 10.1, hexagonal mesh, split into distant points of 100 m in a horizontal line, resulting in 43 points across the study area (Figure 1). The hexagonal mesh was used as the most representative, according to the study of Yfantis, Flatman, and Behar (1987). The authors proved that by analyzing the same variable, this sampling grid performed better in terms of the maximum mean square error than did the triangular and square mesh.

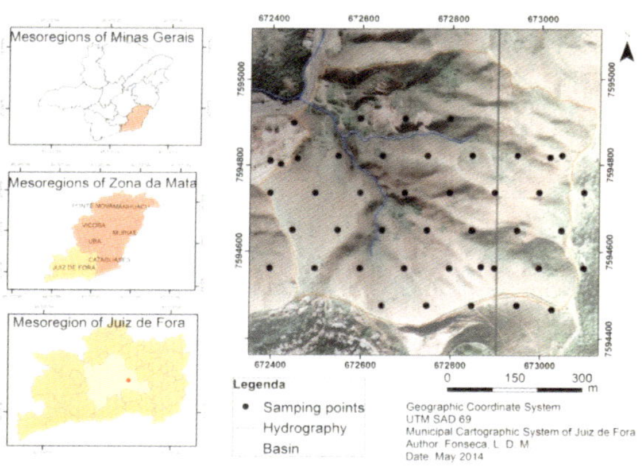

Figure 1. Distribution map of sampling points in the Yung sub-basin in the urban district Três Moinhos, Juiz de Fora, Minas Gerais State

The points were inserted with their respective coordinates in the GPS navigation model "Garmin map 62 Trilha," using the TrackMaker software. The search function of GPS points (function go to) was essential for the localization of the sampling points in the field. Once a point was found, disturbed samples were collected from the surface (0 - 0.10 m layer) and soil samples with an undisturbed structure using a volumetric ring (height 0.045 m, diameter 0.0405 m).

For measurements of the soil thickness and sampling the subsurface material, an auger with a length of 3 m was used. In the deeper layers, two factors were investigated: the particle size and type of horizon.

The collected soil material was packed in plastic bags, crumbled, air-dried and sieved (< 2 mm) (ADFS). The soils collected in rings (auger) were wrapped in plastic film to preserve the structure during transport and physical analysis in the laboratory. The hydraulic conductivity, bulk and particle density, total porosity, macroporosity and microporosity were determined in these samples.

Laboratory analyses

The pipette method was used for particle size analysis. A 0.1 mol L^{-1} NaOH solution was used as the chemical dispersant under mechanical stirring (Empresa Brasileira de Pesquisa Agropecuária [EMBRAPA], 1997; Ruiz, 2006).

The bulk density (BD) was determined by the relationship between the dry soil mass and the internal ring volume. The particle density (PD) was determined by the volumetric flask method. Total porosity (TP) was defined as the ratio between the bulk density and particle density using the expression TP= [1-(Ds / PD)] (EMBRAPA, 1997). The hydraulic conductivity was performed using the "Mariotte bottle" method. The soil macroporosity (Macro) was computed as the difference between TP and microporosity (Micro), which is the water content retained at a matric potential of 6 kPa (EMBRAPA, 1997).

Terrain analysis

The terrain variables of the digital elevation model (DEM) were used, which is based on a plan-altimetric topographic survey, with contour lines with 1-m intervals, provided by the urban administration of Juiz de Fora and developed from an aero-photogrammetric survey of 2007.

Geostatistical analysis

The adjustment of the semivariogram was the main stage modeled by geostatistics (Yamamoto & Landim, 2013). The distances with better visual setting were tested to further quantify the fitting of the theoretical model. The exponential, Gaussian and spherical models were adjusted using the least squares method for almost all of the physical properties, except the data obtained for coarse sand at the surface, which was fitted by the maximum likelihood method.

After choosing the semivariogram models, interpolations were performed by the Simple Kriging (Ks) method to predict and map the values of the physical properties across the watershed. Santos, Oliveira, Louzada, and Santos (2011) found that Ks is more accurate than the universal and ordinary kriging when comparing the mean kriging variance between the observations and the predicted values obtained by self-validation.

Kriging was adopted for tending to annihilate the mean residual error and reduce the variance of errors. For this purpose, a self-validation procedure called jack-knifing is used. According to Vieira (2000), this procedure involves the measurement of the points observed in the field, "hiding the variable," so that it is estimated by kriging as if it were a lost point, and assessing the performance to estimate the known values. The observed residual errors are generally marginal, ensuring the reliability of the values estimated by this interpolator.

Pedotransfer function

Because the data of hydraulic conductivity (K_0) were not spatially adjusted, the relationship between the soil physical characteristics and K_0 was tested by multiple linear regressions with program R. When the variables are soil properties, these regressions are called pedotransfer functions.

Pedotransfer functions can be defined as regressions used to obtain estimates of soil properties from other more easily measurable characteristics (Medeiros, Cooper, Rosa, Grimaldi, & Coquet, 2014). The data for the clay, bulk density (BD) and macroporosity (Macro) were significant in the regression results in equation (1):

$$K_0 = -6.74 * BD + 51.74 * Macro + 0.01 * Clay \quad (1)$$

The equation showed the best values of the multiple linear regression fitting, with $r^2 = 0.6$ and p-value = $2.9\ 10^{-8}$. Therefore, K_0 was mapped in the area of the sub-basin using the equation established, performing operations between images consisting of a cell matrix (rasters).

Results

Properties of the Yung subwatershed

Where the flat curvature (perpendicular to the slope) directly influences the flow velocity and, consequently, soil erosion and sediment transport, the classes were divided into concave, flat and convex, as defined by an analysis of the relief.

The plan curvature showed that of the 35ha area of the basin, 40% is concave, 22% is flat and 34% is convex slopes. The curvature was also closely related to soil thickness. In the flat areas, the mean soil depth was higher than in the other relief forms.

The data of the *solum* measured *in situ* at the sampled points were interpolated to create the thematic map. The legend of the depth map (Figure 2) indicates that in the mapped areas, there was no change of horizon below 3 m; thus, the surface of the C horizon is below a depth of 3 m.

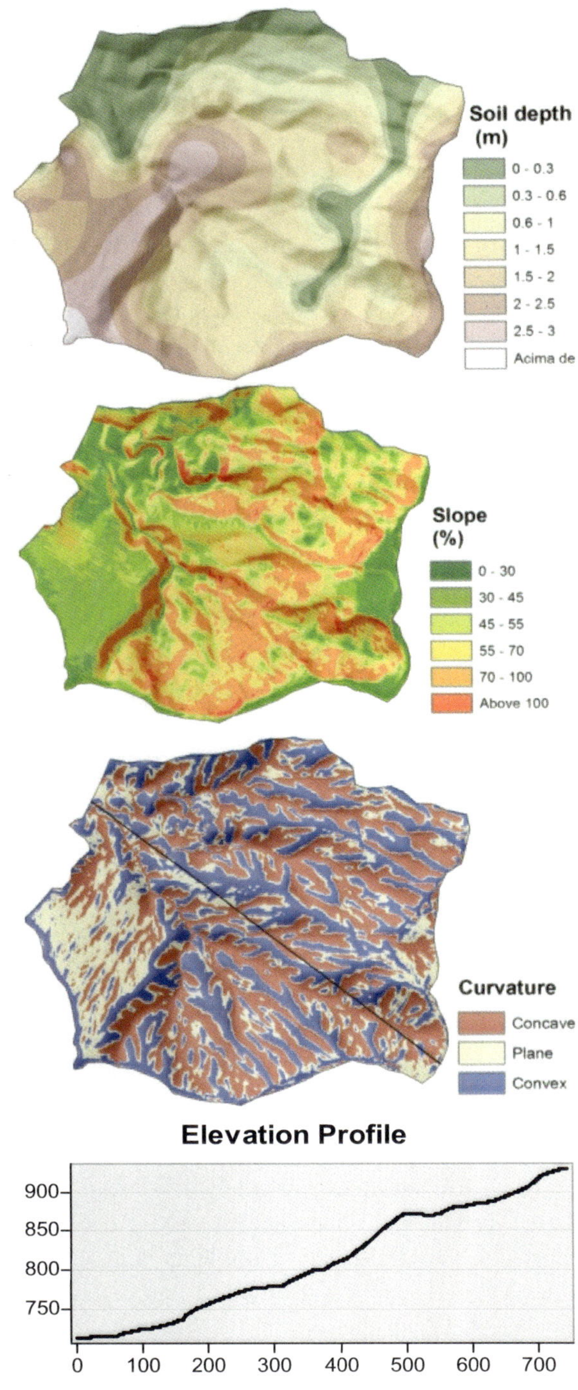

Figure 2. Terrain properties and profile graph of the Yung subwatershed, Juiz de Fora, Minas Gerais State.

It was found that soil depth was directly related to slope; the lowest mean slopes were observed in areas with deeper soils (> 2 m), and this class occupies the only 4h of the 35h basin. The majority of the areas (70% of the basin) have a soil thickness of 1.5 m, and these are the areas with the highest mean slope.

Spatial distribution of soil physical properties

According to Vieira (2000), statistics and geostatistics are complementary, and Table 1 shows the results of the studied data.

The data of the soil thickness and hydraulic conductivity were not spatially dependent with a pure nugget effect. According to Yamamoto and Landim (2013), the nugget effect can be a result of the spatial phenomenon variability under study, such as the sampling scale, so the data of spatial distribution and independence have random characteristics. For the other analyzed properties, the models were fitted to the data, thus meeting the assumptions of geostatistical modeling, such as intrinsic stationarity, and second-order stationarity.

The results of the semivariogram parameters (Table 2) show high variance values for some properties, namely, clay, silt and coarse sand. The high values are associated with the sensitivity of the variable to the collected distances, indicating that these properties must require denser sampling for optimal fitting. However, all properties were adjusted in a variogram with a sill, indicating a correlation between the values.

The ranges of values (i.e., the extent of spatial dependence) for the properties of macroporosity, microporosity and silt in the deeper layers were the lowest, suggesting that these variables are spatially correlated at short distances. This provides evidence of the low structural continuity of the surface and thickness of the C horizon in the potentially erodible areas.

The highest range values were observed for the surface clay and soil thickness. Therefore, a point at the surface influences the clay content to a radius of up to 540 m in the surrounding and soil thickness to a depth of 289 m.

Estimates of the mean and variance of the residues of the self-validation should be close to 0 and 1, respectively (Table 3), because according to Vieira (2000), the mean of the residues is the sum of the residues between of the kriging values and the observed points; i.e., theoretically, these measures would be 0 and 1. The results of the self-validation were satisfactory for all interpolated properties.

The semivariogram shows a model of the spatial correlation structure until the range, from which the semivariance becomes constant, restricts the area of spatial influence. The spherical model was adjusted for soil density, coarse sand at the surface, clay, fine sand and coarse sand in the deeper layers.

Table 1. Statistical data of the soil properties.

Property	Minimum	1° Quartile	Median	Mean	3° Quartile	Maximum	CV	SD
			Surface					
Clay	90.00	330.00	420.00	450.00	590.00	820.00	39.98	179.89
Silt	10.00	60.00	80.00	114.60	120.00	670.00	106.22	121.76
Fine sand	50.00	105.00	130.00	122.60	140.00	180.00	23.43	28.71
Coarse sand	70.00	215.00	330.00	317.00	415.00	610.00	45.38	143.83
Bulk density	0.99	1.23	1.28	1.31	1.41	1.60	10.71	0.14
Macroporosity	0.06	0.12	0.15	0.16	0.20	0.22	30.79	0.05
Microporosity	0.18	0.26	0.34	0.33	0.38	0.49	23.01	0.08
			Depth					
Clay	160.00	337.50	445.00	486.90	640.00	910.00	40.61	197.74
Silt	0.00	20.00	80.00	101.50	150.00	400.00	97.59	99.09
Fine sand	10.00	90.00	100.00	105.30	120.00	170.00	31.17	32.82
Coarse sand	50.00	160.00	265.00	297.80	412.50	670.00	56.26	167.54

CV = variance coefficient; SD = standard deviation.

Table 2. Main data from the fitted semivariograms.

	Descriptive data				Fitted semivariogram				
Properties	N	Mean	Var	Max. distance	Model	Contrib.	Range	Pepita	Methods
				Surface					
Clay	43	377.00	32,360.00	487	Exponential	27,540.50	541.70	6,281.00	WLS
Fine sand	43	122.60	824.30	367	Exponential	746.20	254.80	90.00	OLS
Silt	43	114.60	14,825.50	650	Gaussian	13,218.70	237.60	1,000.00	WLS
Coarse sand	43	317.00	20,688.30	612	Spherical	14,129.00	212.40	6,099.00	ML
Macroporosity	41	0.20	0.00	755	Exponential	0.00	117.60	0.00	WLS
Microporosity	41	0.30	0.00	570	Gaussian	0.00	140.40	0.00	WLS
Bulk density	41	1.30	0.00	562	Spherical	0.00	558.20	0.00	OLS
				Depth					
Clay	36	486.90	39,101.80	632	Spherical	35,211.40	289.60	4,882.00	OLS
Silt	39	101.50	9,818.60	595	Gaussian	8,776.60	155.40	1,373.00	OLS
Fine sand	36	105.30	1,077.10	617	Spherical	990.60	247.60	119.00	OLS
Coarse sand	36	297.80	28,069.20	642	Spherical	25,412.40	242.70	3,250.00	OLS

n = number of samples; Var = variance of the data; Max distance = maximum distance adopted in the semivariogram.; Contrib. = Contribution; WLS = method of weighted least squares; OLS = ordinary least squares method; ML = maximum likelihood.

The fittings of the Gaussian and exponential models for the other properties were tested. The surface silt, thickness and microporosity were better fitted to the Gaussian model, and the other properties were better fitted to the exponential model (Figure 3), as shown by Kamimura et al. (2013).

After choosing the semivariograms, the data were interpolated by universal kriging, and the resulting grids divided into 1,000 horizontal and 1,000 vertical lines, corresponding to a pixel of 0.90 m; this operation was carried out for all properties. The images in ASC format were included in ArcGIS for conversion into GRID and layout preparation.

Table 3. Main fitting quality indicators obtained in the validation of the adjustments for each property.

Property	Regression data		Self-validation	
	Intercept	Predicted	Mean error	Error standard deviation
		Surface		
Clay	1.oo	-0.48	-0.01	1.08
Fine sand	11.18	0.91	0.06	1.06
Silt	70.81	0.40	0.04	2.21
Coarse sand	-24.90	1.08	0.00	0.99
Macroporosity	0.25	-0.60	0.00	1.01
Microporosity	0.08	0.77	-0.01	1.07
Bulk density	0.42	0.68	0.01	1.39
		Depth		
Clay	42.43	0.91	0.00	1.11
Silt	142.43	-0.40	-0.27	1.35
Fine sand	39.00	0.63	0.01	1.20
Coarse sand	94.00	0.67	-0.01	1.20

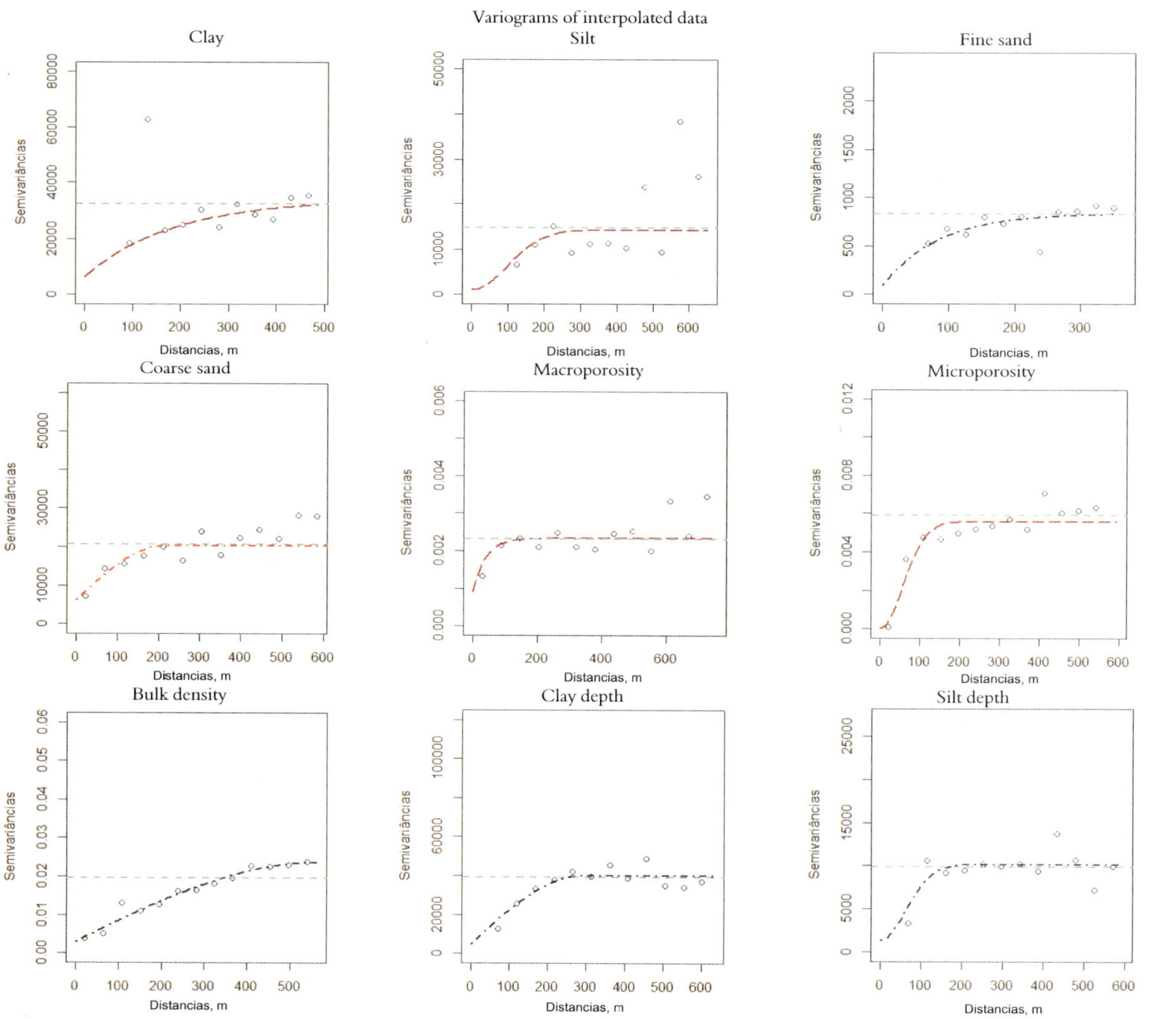

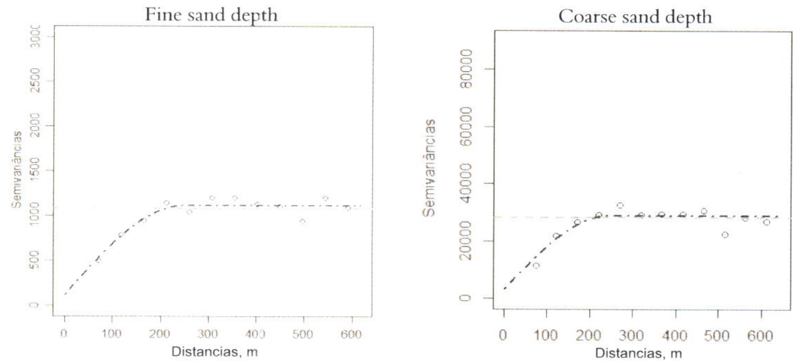

Figure 3. Semivariogram for the studied physical properties.

Discussion

The widest ranges of clay content (500 - 820 g kg^{-1}) were observed in the areas with flatter tops in the Eastern part of the basin, as highlighted on the map in red and orange, where the silt and sand contents were lowest (Figure 4). The vegetation was less degraded, and the soil structure of horizon A was granular, with a moderate degree of development, which is a structure favored by the presence of organic matter.

The highest silt values were found in the northwest of the basin, corresponding to the areas that were identified in the field as degraded and eroded, exposing C horizon. Fine sand, although the numerical contents were not high, was the property whose pronounced spatial distribution had the widest range (120 - 180 g kg^{-1}) in most parts of the basin.

The areas with the highest silt contents in the deeper layers, considering the ranges 100 - 397 g kg^{-1} with beige, green and dark green soil material (Figure 5), were mapped in the areas with steeper terrain and shallower *solum*. In the same areas, high values of fine sand (100 - 170 g kg^{-1}) and coarse sand (250 - 670 g kg^{-1}) were detected.

High clay contents (550 - 900 g kg^{-1}) were observed in the deepest soil layers, where no C horizon was found by drilling to a depth of 3 m (Figure 6)

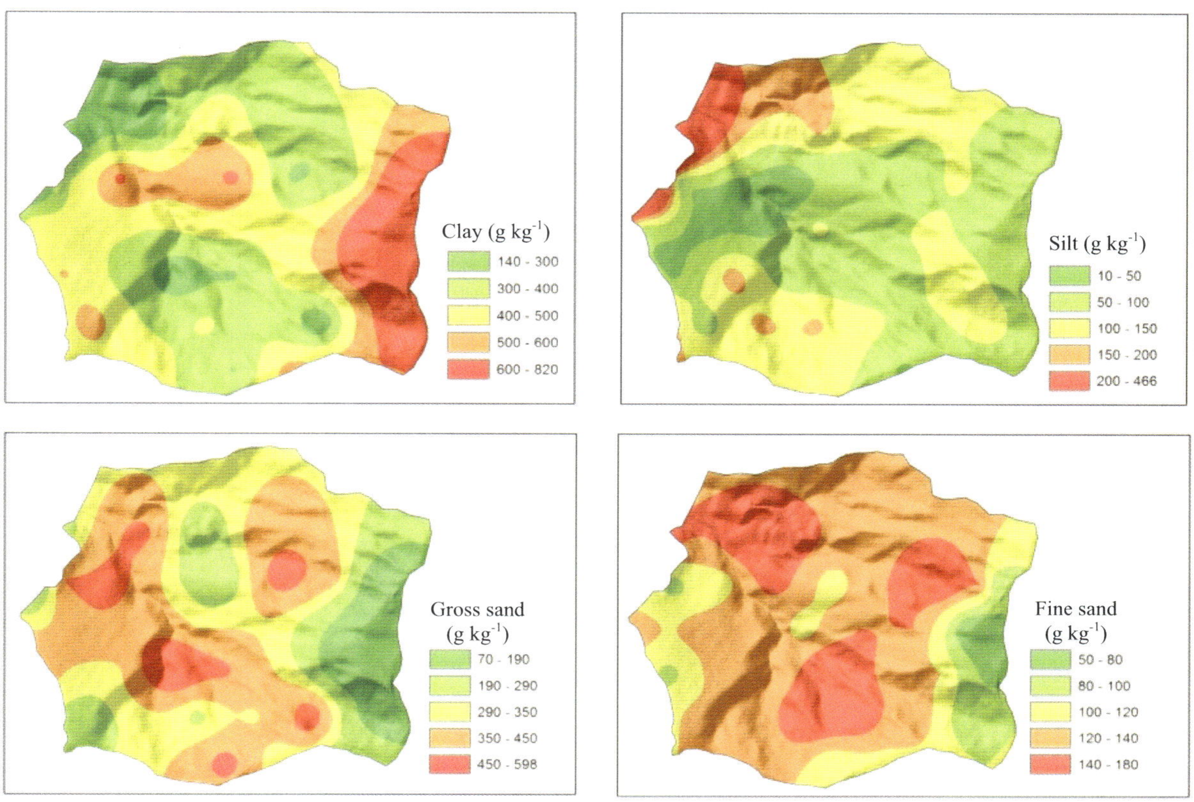

Figure 4. Spatialization of soil surface texture in the Yung sub-basin, Juiz de Fora, Minas Gerais State.

Microporosity was low at the observed points, except at a few points at the top, where the topography is smooth and the vegetation is only slightly degraded. It can be concluded that due to the steep slopes of this basin, the erosion process of the *solum* is constant, causing the degradation of the soil structure in some areas (Figure 6).

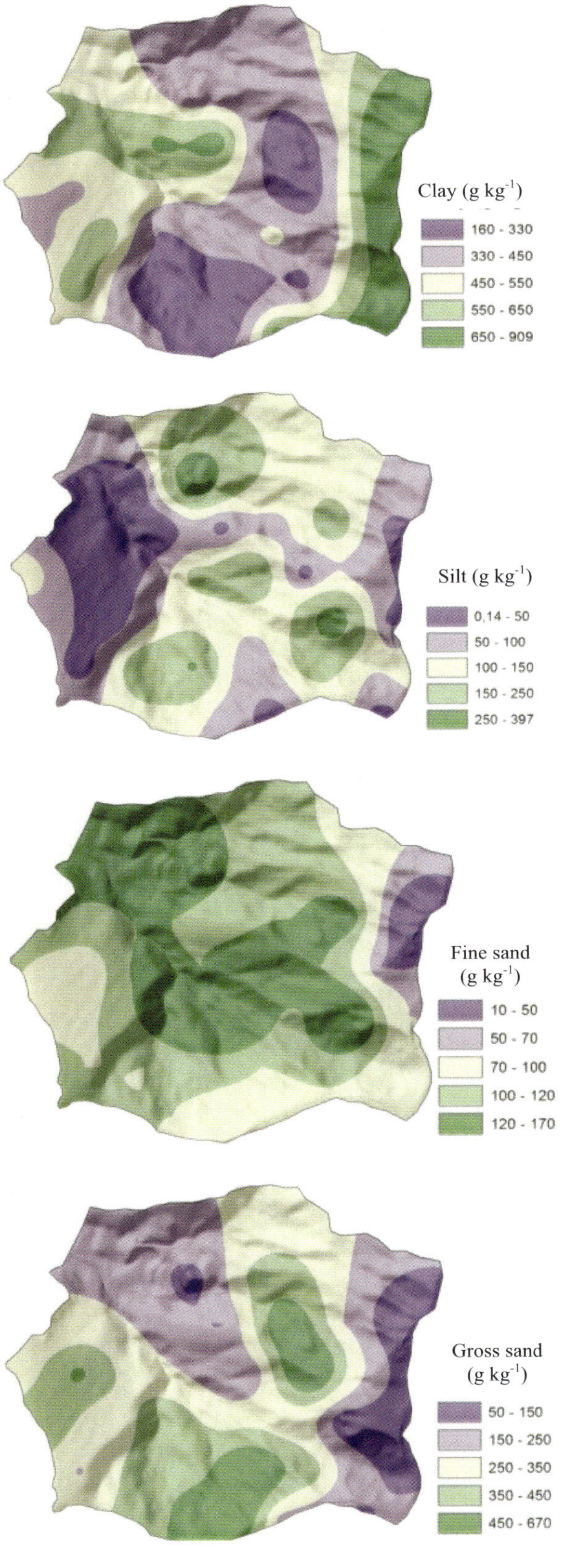

Figure 5. Spatialization of the soil texture in deeper layers in the Yung sub-basin, Juiz de Fora, Minas Gerais State.

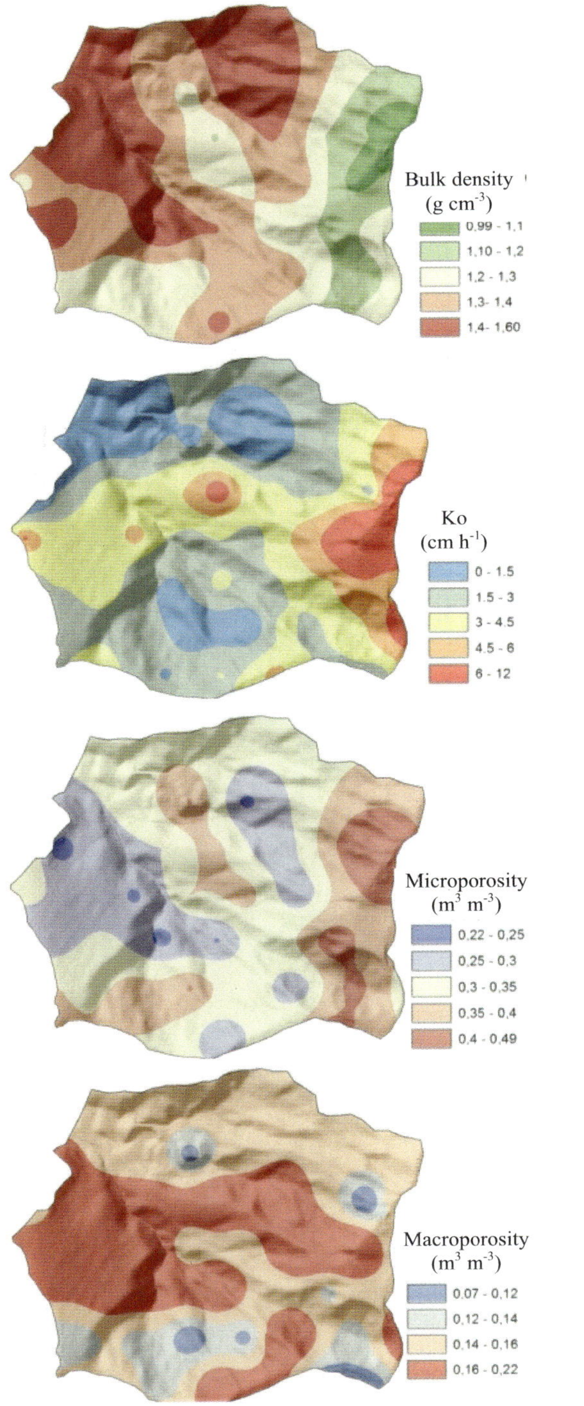

Figure 6. Spatial-physical properties in the sub-basin of the Yung creek, Juiz de Fora, Minas Gerais State.

The map that indicated higher hydraulic conductivity represented the less-degraded vegetation-covered areas in the basin. These areas are characterized by higher infiltration and lower runoff rates.

The correlations between the hydraulic conductivity and the soil properties were also investigated. Despite the low values of the correlations in Table 4, the best-fitted physical properties were the bulk density and fine sand, which had negative correlations, and clay, total porosity and macroporosity, which had positive correlations, as reported by Sampaio, Menezes, Furtini, and Ribeiro (2006).

Table 4. Analysis of the correlation between hydraulic conductivity and soil properties in the *solum*.

Properties	K_0 x		K_0 x Sand		K_0 x				
	Silt	Clay	Coarse	Fine	BD	Pt	Micro	Macro	PD
Correlation	-0.28	0.40	-0.18	-0.37	-0.41	0.50	0.12	0.37	0.24

BD = soil density; Pt = total porosity; Micro = microporosity; Macro = macroporosity; PD = particle density.

The correlation analysis between the soil depth and texture (Table 5) showed that the best-correlated textures were clay, silt and coarse sand. The shallower the depth of the soil is, the higher the silt is and the smaller the clay proportion is because C horizon will be closer to the surface. This finding reinforces the character of spatial dependency between these properties.

Table 5. Analysis of correlation between soil thickness and texture of soil from the surface of C horizon.

Property	Depth x			
	Sand Coarse	Sand Fine	Silt	Clay
Correlation	0.45	0.29	0.44	-0.64

Conclusion

Thematic maps were related to each other: clay texture was related to higher hydraulic conductivity, higher porosity and lower soil density. The areas mapped with higher silt content were located in degraded areas of the sub-basin.

In the steepest areas, higher silt contents were mapped, corresponding to the surface of C horizon and indicating the thin thickness of B horizon.

The range of the values of macroporosity, microporosity and silt in the deeper layers were the lowest, thus confirming previous information about the low structural continuity of the soil surface and the depth of horizon B in potentially erodible areas.

Acknowledgements

The authors acknowledge FAPEMIG (Research Support Foundation of Minas Gerais) and CAPES (the Foundation for the Coordination and Improvement of Higher Level or Education Personnel) for providing financial support for this study and the Federal University of Viçosa (UFV).

References

Ab'saber, A. (1986). Domínios morfoclimáticos e solos do Brasil. In V. V. H. Alvarez, L. E. F. Fontes, & M. P. F. Fontes (Eds.), *Os solos nos grandes domínios morfoclimáticos do Brasil e o desenvolvimento sustentável* (p. 1-18). Viçosa, MG: Sociedade Brasileira de Ciência do Solo.

Aquino, R. F., Silva, L. M. N., Freitas, D. A. F., Curi, N., & Avanzi, J. C. (2012). Soil losses from typic Cambisols and Red Latosol as related to three erosive rainfall patterns. *Revista Brasileira de Ciência do Solo, 37*(1), 213-220.

Bogner, C., Bauer, F., Trancon Y Widemann, B., Viñan, P., Balcazar, L., & Huwe, B. (2014) Quantifying the morphology of flow patterns in landslide-affected and unaffected soils. *Journal of Hydrology, 511*(1), 460-473.

Costanzo, D., Rotigliano, E., Irigaray, C., Jiménez-Perávlvarez, J. D., & Chaón, J. (2012). Factors selection in landslide susceptibility modelling on large scale following the GIS matrix method: application to the river Beiro basin (Spain). *Natural Hazards Earth System Science, 12*(1), 327-340.

Di Martire, D., Rosa, M., Pesca, V., Santangelo, M. A., & Calcaterra, D. (2012). Landslide hazard and land management in high-density urban areas of Campania region, Italy. *Natural Hazards Earth System Science, 12*(1), 905-926.

Empresa Brasileira de Pesquisa Agropecuária [EMBRAPA]. (1997). *Centro Nacional de Pesquisa de Solos* (Manual de métodos de análise de solo, 212). Rio de Janeiro, RJ: Embrapa.

Fonseca, L. D. M., Lani, J. J., Fernandes Filho, E. I., Marques, E. A. G. M., & Ferreira, W. P. F. (2016). Avaliação da erodibilidade de um Cambissolo em Juiz de Fora, Minas Gerais. *Geotecnia, 136*(1), 143-156.

Gomes, M. A., Lani, J. J., Costa, L. M., Pontes, L. M., Figueiredo, N. A., & Bardales, N. G. (2012). Solos, manejo e aspectos hidrológicos na bacia hidrográfica do Araújos, Viçosa – MG. *Revista Árvore, 36*(1), 93-102.

Kamimura, K. M., Santos, G. R., Oliveira, M. S., Dias, M. S., & Guimarães, P. T. G. (2013). Variabilidade espacial de atributos físicos de um Latossolo Vermelho-Amarelo sob lavoura cafeeira. *Revista Brasileira de Ciência do Solo, 3*(4), 877-888.

Klimes, J., & Escobar, V. R. (2010). A landslide susceptibility assessment in urban areas based on existing data: an example from the Iguaná Valley, Medellín City, Colombia. *Natural Hazards Earth System Science*, *10*(1), 2067-2079.

Medeiros, J. C., Cooper, M., Rosa, J. D., Grimaldi, M., & Coquet, Y. (2014). Assessment of pedotransfer functions for estimating soil water retention curves for the Amazon region. *Revista Brasileira de Ciência do Solo*, *38*(3), 730-743.

Petschko, H., Brening, A., Bell, R., Goetz, J., & Glade, T. (2014). Assessing the quality of landslide susceptibility maps – case study Lower Austria. *Natural Hazards Earth System Science*, *14*(1), 95-118.

Prefeitura de Juiz de Fora [PJF]. (2004). *Plano Diretor de Desenvolvimento Urbano de Juiz de Fora* (394p). Juiz de Fora, MG: FUNALFA Edições.

Rezende, S. B., & Resende, M. Solos dos Mares de Morros: Ocupação e uso. (1986). In V. V. H. Alvarez, L. E. F. Fontes, & M. P. F. Fontes (Eds.), *Os solos nos grandes domínios morfoclimáticos do Brasil e o desenvolvimento sustentável* (261-289). Viçosa, MG: Sociedade Brasileira de Ciência do Solo.

Ruiz, H. A. (2006). *Métodos de análises físicas do solo*. Viçosa, MG: Universidade Federal de Viçosa. Dep artamento de Solos.

Sadeghi, S. H., Moosavi, V., Karami, A., & Behnia, N. (2012). Soil erosion assessment and prioritization of affecting factors at plot scale using the Taguchi method.*Journal of Hydrology, 448*(1), 174-180.

Sampaio, F. M. T., Menezes, S. M., Furtini, M. B., & Ribeiro, K. D. (2006). Correlação entre a permeabilidade e os atributos físicos do solo na sub-bacia do córrego Centenário da cidade de Lavras – MG. *Ciência e Agrotecnologia*, *30*(4), 798-803.

Santos, G. R., Oliveira, M. S., Louzada, J. M., & Santos, A. M. R. T. (2011). Krigagem Simples versus Krigagem Universal: qual o preditor maisxzsssssa preciso?. *Revista Energia na Agricultura, 26*(2), 49-55.

Silva, A. M. D., Mello, C. R., Curi, N., & Oliveira, P. M. (2008). Simulação da variabilidade espacial da erosão hídrica em uma sub-bacia hidrográfica de Latossolos no sul de Minas Gerais. *Revista Brasileira de Ciência do Solo, 32*(5), 2125-2132.

Vestena, L. R., & Kobiyama, M. (2014). Identificação e mapeamento dos processos erosivos na bacia hidrográfica do rio Caeté, no município de Alfredo Wagner/SC. *Revista Brasileira de Geomorfologia*, *15*(1), 23-33.

Vieira, S. R. (2000). Geoestatística em estudo de variabilidade espacial do solo. In R. F. Novais, V. V. H. Alvarez, & C. E. G. R. Schaefer (Eds.), *Tópicos em ciência do solo* (1, 1-55). Viçosa, MG: Sociedade Brasileira de Ciência do Solo.

Western, C. J. V., Castellanos, E., & Kuriakose, S. L. (2008). Spatial data for landslide susceptibility, hazard, and vulnerability assessment: An overview. *Engineering Geology*, *102*(4), 112-131.

Yfantis, E. A., Flatman, G. T., & Behar, J. V. (1987). Efficiency of Kriging estimation for Square, Triangular and Hexagonal Grids. *Matematical Geology*, *19*(3), 183-205.

Yamamoto, J. K. & Landim, P. M. B. (2013). *Geoestatística: conceitos e aplicações*. São Paulo, SP: Oficina de Textos.

Zizioli, D., Meisina, C., Valentino, R., & Montrasio, L. (2013). Comparison between different approaches to modeling shallow landslide susceptibility: a case history in Oltrepo Pavese, Northern Italy. *Natural Hazards Earth System Science*, *13*(1), 559-573.

8

Enteropathogenic bacterial contamination of a latosol following application of organic fertilizer

Pedro Alexandre Varella Escosteguy[1*], Bernardo da Silva Tissot[2], Rafael Mariath Cherubini[1], Cristiano Pauletto[1] and Simone Meredith Scheffer-Basso[1]

*[1]Faculdade de Agronomia e Medicina Veterinária, Universidade de Passo Fundo, BR-285, km 171, 99001-970, Passo Fundo, Rio Grande do Sul, Brazil. [2]Programa de Pós-graduação em Agronomia, Faculdade de Agronomia e Medicina Veterinária, Universidade de Passo Fundo, Passo Fundo, Rio Grande do Sul, Brazil. *Author for correspondence. E-mail: escosteguy@upf.br*

ABSTRACT. Poultry manure is used as fertilizer *in natura*, but little is known about whether it contaminates the soil with pathogenic organisms. The aim of this study was to assess the effects of organic, organomineral and mineral fertilizers on soil contamination by enteric pathogens, using poultry manure as the organic fertilizer. Manure was applied in field experiments at rates of 7.0 ton. ha^{-1} (maize crop, 2008/2009), 8.0 ton. ha^{-1} (wheat crop, 2009) and 14 ton. ha^{-1} (maize crop, 2010/2011). Organomineral fertilizer was applied at the same rates but was comprised of 50% manure and 50% mineral fertilizer. At 30 and 70 days after fertilization, the organic fertilizer and the upper 0-5 cm layer of the soil were tested for the presence of helminth eggs and larvae and enteropathogenic bacteria. Fecal and non-fecal coliforms (*Escherichia coli* and *Clostridium perfringes*) were found in the organic fertilizer, but neither *Salmonella* spp. nor enteroparasites were detected. The population of enteropathogenic bacteria in the soil was similar among the treatments for all crops at both evaluation times. The population of thermotolerant coliforms in the organic fertilizer was larger than the maximum level allowed in Brazil, but neither the organic or nor the organomineral fertilizer contaminated the soil.

Keywords: soil contamination, soil quality, *Triticum aestivum*, *Zea mays*.

Introduction

The application of agricultural residues as organic fertilizer has grown in Brazil and worldwide. As a result of the increased generation of these residues, their application as fertilizer has increasingly been the subject of environmental (DELGADO et al., 2012; GALLOWAY et al., 2003; HOULBROOKE et al., 2004) and agronomic (MAGUIRE et al., 2011) studies. The generation of poultry manure is increasing as the egg industry grows in response to domestic and international markets, enabling the application of raw (untreated) manure as fertilizer for agricultural soils (DEMIR et al., 2010; FIGUEROA et al., 2012). Studies by Demir et al. (2010) and Figueroa et al. (2012) demonstrate the beneficial effects of the application of organic fertilizers including of the application of poultry manure in high doses (tons per ha), which has a low concentration of nutrients compared to mineral fertilizers.

However, the application of high doses of organic fertilizers has the potential to contaminate soil and water with enteropathogenic microorganisms and enteroparasites (HAWKE; SUMMERS, 2006; MACIOROWSKIA et al., 2004; UNC; GOSS, 2004). Although contamination by these organisms is of concern given the risk associated with the transmission of infectious and parasitic diseases through the food chain (MACIOROWSKIA et al., 2007; SILVA et al., 2005), other studies show that enteric organisms either do not survive or have a high rate of decline in fertilized soil (GAGLIARDI; KARNS, 2002; GARCÍA-ORENES et al., 2007; TALLON et al., 2007; ZIEMER et al., 2010). The discrepancies in the literature about enteropathogenic bacterial contamination of soil are related to the many different factors that affect the survivability of enteric organisms in the soil, including the type of organic fertilizer; fertilized crop, management and dosage; the use of mitigation practices; and environmental factors such as soil type, pH, humidity and temperature (MACIOROWSKIA et al., 2004; FORSLUND et al., 2011). However, the survival of pathogenic organisms in latosols fertilized with poultry manure has not been studied in Brazilian soils and climate conditions, and there are only a few reports from other countries (McMURRY et al., 1998). This lack of data makes predicting the environmental impacts of fertilization with poultry manure difficult; however, these impacts are of great concern (MACIOROWSKIA et al., 2004; SANTOS et al., 2010).

The aim of this study was to assess the effects of organic and organomineral fertilization with raw poultry manure on soil contamination by enteropathogenic organisms.

Material and methods

The study was conducted at the University of Passo Fundo (Universidade de Passo Fundo) and the field experiments were located in the municipality of Coxilha, Rio Grande do Sul, Brazil (28° 10.475' S, 52° 20.840' W). The region has an average altitude of 676 m, an average annual rainfall of 1,786 mm and an average annual temperature of 17.5°C. The soil in the study area is a dystrophic Red Latosol and has been managed as a no-till farming system for 19 years.

Prior to the experiment, soil was sampled and analyzed according to Tedesco et al. (1995). Soil attributes were as follows: 29 and 172 mg dm^{-3} extractable P and K (Mehlich-I), respectively; 0.2, 6.1 and 2.5 $cmol_c$ dm^{-3} exchangeable Al, Ca and Mg, respectively; 26 mg dm^{-3} exchangeable Mn (potassium chloride, 1 mol L^{-1}); 5.3 pH in water (soil suspension:extractor 1:1) and 5.5 SMP index (soil suspension:extractor 1:1.5); 17 mg dm^{-3} available S (calcium phosphate, 500 mg P L^{-1}); 0.7 mg dm^{-3} available B (hot water); 0.9 and 2.4 mg dm^{-3} available Zn and Cu, respectively (hydrochloric acid, 0.1 mol L^{-1}); 31 g dm^{-3} organic matter (sulfochromic solution method and spectrophotometric quantification of reduced chromium); and 470 g dm^{-3} clay (densimeter).

The manure used as organic fertilizer was generated for five days on mats inside a poultry house and stored in the poultry house until being transported to the study area. Prior to fertilizer application, the fertilizer material was stored in stacks on the field. A 2.0 dm^{-3} sample of the fertilizer was collected from the stacks and analyzed for pathogens and physicochemical soil properties. Subsamples were taken from five locations in the stacks at a depth of 0.40 m using a shovel. The subsamples were pooled and packaged in sterile plastic bags inside Styrofoam boxes prior to being transported to the laboratory. The analyses of pathogenic organisms began on the day of sample collection.

To test the manure for the presence of viable helminth eggs, 4 g of wet manure sample were processed according to the Willis-Mollay Flotation Technique using a saturated NaCl solution with a density of 1,182 (m v^{-1}). The analysis of parasite larvae was performed using the Barmann method with an optical microscope with a 10x eyepiece and a 3.2x objective. The multiple-tube method was used for the analysis of total coliforms and thermotolerant coliforms, and the most probable number (MPN) was determined. The initial dilution was 1:10 (10^{-1}) in 10 g of wet sample diluted in 90 mL of 0.1% peptone water. Subsequent dilutions went up to 10^{-5}, using five tubes per dilution (EATON et al., 2005). For the *Salmonella* spp. assay, a 10 g aliquot of the wet sample was homogenized in 90 mL of 1% peptone water and incubated at 36±1°C for 18 to 24 hours. After pre-enrichment, the sample was inoculated in selective broths and plated onto selective and differential agars; genetically compatible colonies were confirmed through biochemical and serological tests (DOWNES; ITO, 2001).

The population of thermotolerant coliforms found in the manure was compared to the limits set by the Normative Instruction no. 27/2006 of the Department of Agriculture, Livestock and Food Supply (Ministério da Agricultura, Pecuária e Abastecimento - MAPA) (BRASIL, 2006). Assays to test for the presence of total coliforms and *Salmonella* spp. were performed to complement the information from the other contamination indicators. Total coliforms and *Salmonella* spp. have been used as indicators of fecal contamination (GARCÍA-ORENES et al., 2007; MACIOROWSKIA et al., 2004). *E. coli* and *C. perfringes* were also tested since for because they are commonly found in poultry manure (SCHOCKEN-ITURRINO et al., 2010).

For the physicochemical analyses, the samples of organic fertilizer were dried at 65°C in a convection oven for 16 h, ground and sieved (10 mesh) and then analyzed in duplicates according to Tedesco et al. (1995). Organic C was quantified by the Walkley-Black method with external heating. Total N, P, K, Ca and Mg were determined by acid digestion (H_2O_2 + H_2SO_4) and a digestion mixture. Mineral N content was extracted with 1 mol L^{-1} KCl. Sulfur (S) and micronutrient (Zn, Fe, Mn and Cu) content were determined using nitro-perchloric digestion, except for B content, which was extracted through dry combustion in a muffle furnace at 600°C for 1 hour (Table 1).

The field treatments included three types of fertilization: organic, organomineral and mineral fertilization (Table 2). Fertilizer dosage was based on the concentration of total N in the organic fertilizer, the organic matter content of the soil, the expected grain yield of the fertilized crop and the type of plant residue from the previous crop (CQFS, 2004).

Table 1. Total concentration of some physicochemical properties in poultry manure used as organic fertilizer for grain crops. The results are expressed in wet mass of fertilizer.

Property	Maize (2008/2009)	Wheat (2009)	Maize (2010/2011)
Organic C (g kg^{-1})	120.0	87.1	68.0
Total N (g kg^{-1})	19.75	16.0	10.0
Mineral N (g kg^{-1})	1.61	0.50	0.90
P_2O_5 (g kg^{-1})	11.5	8.0	7.0
K_2O (g kg^{-1})	7.3	4.0	7.0
Ca (g kg^{-1})	24.7	17.2	25.5
Mg (g kg^{-1})	1.2	1.1	1.2
S (mg kg^{-1})	0.8	0.3	0.6
Zn (mg kg^{-1})	72.42	42.10	73.21
Fe (mg kg^{-1})	305.66	428.80	314.56
B (mg kg^{-1})	32.28	17.49	30.53
Mn (mg kg^{-1})	61.4	58.33	67.20
pH in $CaCl_2$	8.1	8.7	8.3
Water (g kg^{-1})	670.0	741.0	739.5

The treatments were applied to two crops of maize (*Zea mays*) and one crop of wheat (*Triticum aestivum* L.) planted in succession. After the wheat crop, soybean (*Glycine max* (L.) Merrill) was planted, followed by radish (*Raphanus sativus* L.). The fertilization treatments were not applied to the soybean and radish crops. After the radish cultivation, the treatments were reapplied to a second crop of maize.

Table 2. Fertilizer type and content of nitrogen (N), phosphorus (P_2O_5) and potassium (K_2O) for fertilizers applied to grain crops grown in a red latosol.

Fertilizer/Nutrient	Mineral	Organomineral		Organic
	Maize cultivation (2008/2009)			
	Sowing			
Fertilizer	NPK 08-25-20	NPK 08-25-15	Manure	Manure
Fertilizer (kg ha^{-1})	250	125	3,500	7,000
N, P_2O_5 and K_2O (kg ha^{-1})	20, 62 and 50	10, 31 and 25	69, 32[1] and 26	138, 64[1] and 51
		Top-dressing		
N (kg ha^{-1} as urea)	120	60		0
		Sowing + top-dressing		
N, P_2O_5 and K_2O (kg ha^{-1})	140, 62 and 50	139, 63 and 51		138, 64 and 50
	Wheat cultivation (2009)			
	Sowing			
Fertilizer	NPK 15-25-15	NPK 15-25-15	Manure	Manure
Fertilizer (kg ha^{-1})	200	100	4,000	8,000
N, P_2O_5 and K_2O (kg ha^{-1})	30, 50 and 30	15, 25 and 15	64, 26[1] and 16	128, 51[1] and 32
		Top dressing		
N (kg ha^{-1} as urea)	90	41		0
		Sowing + top-dressing		
N, P_2O_5 and K_2O (kg ha^{-1})	120, 50 and 30	120, 51 and 31		128, 51[1] and 32
	Maize Cultivation (2008/09)			
	Sowing			
Fertilizer	DAP and KCl	DAP and KCl	Manure	Manure
Fertilizer (kg ha^{-1})	170 and 164	85 and 82	7,000	14,000
N, P_2O_5 and K_2O (kg ha^{-1})	31, 78 and 98	15.5; 39 and 49	64, 26[1] and 16	140, 78[1] and 98
		Top-dressing		
N (kg ha^{-1} as urea)	109	60.5		0
		Sowing + top-dressing		
N, P_2O_5 and K_2O (kg ha^{-1})	140, 78 and 98	140, 78 and 98		140, 78[1] and 98

[1]Amount of available P_2O_5 as estimated with the efficiency index of 0.8.

The experimental design consisted of randomized blocks with four replicates. The area of the plots was 240 m^2 (20 x 12 m) with 2.5-m alleys between the plots and a 1.25-m border strip at either end of each plot. The amount of organic fertilizer varied with crop type, corresponding on a wet basis, to 7.0 t ha^{-1} (maize 2008-2009), 8.0 t ha^{-1} (wheat) and 14.0 t ha^{-1} (maize 2010-2011). The quantities of mineral fertilizer used in the mineral and organomineral fertilizer treatments were designed to match the amount of N, P and K applied in the organic fertilizer treatments (Table 2). The manure was applied with a truck equipped with broadcast spreader with a 12-m wide swath. Fertilizer application was performed with a seeder fertilizer machine in the planting furrows (P and K on maize) or manually by top-dressing (Urea on both maize and wheat, and P and K on this last crop).

At 30 and 70 days after fertilizer application, soil samples were taken for the enteropathogenic bacteria assays. Because of the limited vertical mobility of these bacteria in clayey soils (TALLON et al., 2007) and because the fertilizer was applied by top-dressing without incorporation into the soil, a shallow soil layer (0-5 cm) was sampled. The indicators of fecal contamination for the soil samples were also used to test for contamination in the organic fertilizer, as there is no specific national legislation indicating which microorganisms should be monitored in soils contaminated with animal manure.

Results and discussion

Enteroparasites and enteropathogenic bacteria in organic fertilizers

No helminth larvae or eggs were observed in the organic fertilizers applied to any of the crops (Table 3). The results shown in Table 3 are consistent with the quality standards for organic fertilizers, which allow a maximum of one viable egg per 4 g of dry matter (BRASIL, 2006). These results differ from those of a study by Santos et al. (2010), who noted the occurrence of helminth eggs in poultry fertilizers. The inconsistency between the results of this study and the results reported by Santos et al. (2010) may be related to temporal variability in parasite colonization of poultry manure or to characteristics of the fertilizer sample used by these authors. In the Santos et al. (2010) study, the organic fertilizer was stored for 120 days in compost heaps, whereas in the experiment in this study, the manure was deposited on the soil a few days after it was produced. Furthermore, Santos et al. (2010) warn that the presence of helminth eggs in their manure samples should be interpreted with caution because an error could have resulted from a flaw in their sampling procedures. The samples most likely represented the outer layers of the compost heap, which remain cooler in temperature than the central layers and are therefore less damaging to enteropathogenic organisms.

Table 3. Presence or absence of helminth viable eggs and larvae in poultry manure prior to the application of the manure as an organic fertilizer for three grain crops.

Crop (agricultural year)	Total solids (%)	Eggs	Larvae
Maize (2008/2009)	33.0	Absent	Absent
Wheat (2009)	18.0	Absent	Absent
Maize (2010/2011)	26.1	Absent	Absent
Mean	25.7	Absent	Absent

The MAPA normative instruction no. 27/2006 restricts the population levels of enteropathogenic microorganisms and prohibits the occurrence of enteroparasites in organic fertilizers (BRASIL, 2006). Unlike helminth eggs and larvae, enteropathogenic bacteria were found in the samples of the organic fertilizers applied to the three grain crops (Table 4). Among the detected bacteria, there was a predominance of total coliforms and thermotolerant coliforms, with the latter group representing the majority of the coliforms. The presence of coliform bacteria is typical in animal manure, which is why thermotolerant coliforms are the preferred indicators of enteric organisms of homeothermic animals in organic fertilizer analyses. In general, thermotolerant coliforms comprise over 90% of total coliforms in animal manure, including poultry manure (SCHOCKEN-ITURRINO et al., 2010).

The most probable number (MPN) of thermotolerant coliforms in the organic fertilizers ranged from 3.3 x 10^5 (maize cultivation, 2008/2009) to > 9.0 x 10^7 (wheat cultivation, 2009), with an estimated mean of 3.3 x 10^5 per gram of dry matter (DM). These values are higher than the maximum values for organic fertilizers allowed by the MAPA in Brazil (BRASIL, 2006) and were found in all of the samples (Table 4). Although these MPN values were expected, the values are considered to be high because they exceed 30% of the maximum allowed value (Table 4). Moreover, these values are substantially higher than the reference value of quality for Class 1 watercourses in Brazil, which is less than 200 MPN for a 100 mL water sample.

Table 4. Enteropathogenic bacteria in poultry manure[1] used as organic fertilizer for grain crops.

Crop (year)	Total coliforms	Thermotolerant coliforms	*Escherichia coli*	*Salmonella* spp.	*Clostridium perfringes*
	....................MPN[3] g^{-1}		....	In 10 g	CFU[4] g^{-1}
Maize (2008/2009)	5.2×10^6	3.3×10^5	6.1×10^2	Absence	6.4×10^3
Wheat (2009)	$> 9.0 \times 10^7$	$> 9.0 \times 10^7$	7.5×10^3	Absence	1.7×10^6
Maize (2010/2011)	1.1×10^7	6.7×10^6	8.4×10^3	Absence	8.8×10^4
Mean	$> 3.5 \times 10^7$	$> 3.2 \times 10^7$	5.5×10^3	Absence	5.9×10^5
MVA[2]	-	1.0×10^3	-	Absence	-

[1]Results expressed as dry mass of organic fertilizer. Water content (kg t^{-1}): 670 (maize, 2008/2009), 741 (wheat, 2009) and 739.5 (maize, 2010/2011). [2]Maximum value allowed in organic fertilizer according to Normative Instruction no. 27/2006 (BRASIL, 2006). [3]Most probable number. [4]Colony-forming unit.

E. coli and *C. perfringes* were also found in all of the analyzed samples (Table 4). The populations of these two species were smaller that the populations of thermotolerant coliforms, but had relatively high mean values (MPN $g^{-1} = 5.5 \times 10^3$ and 5.9×10^5 for *E.coli* and *C. perfringes*, respectively). *E. coli* is one of the main specie in the coliform group and is typically used in the control of drinking water quality. This bacterium is found in the intestinal tract of birds and rarely propagates in other environments. In contrast, *Salmonella* spp. was the only species absent from all of the analyzed samples, which is acceptable under the MAPA standard (Table 4).

Enteropathogenic bacteria in soil

The MPNs of thermotolerant coliforms in the organic fertilizers were higher than the maximum value allowed by the MAPA (Table 4). Thus, these fertilizers should be treated before application to agricultural soils to avoid risk of environmental contamination (TIQUIA; TAM, 2002; SANTOS et al., 2010). However, poultry manure has been applied *in natura* to grain crops and is necessary to determine the survivability of enteropathogenic bacteria that are transferred to soils. The soil analyses in this study indicated that thermotolerant coliform populations were significantly smaller in the fertilized soils compared to the organic fertilizers (Table 4) for all three crops at both evaluation times (Tables 5 and 7).

Although thermotolerant coliforms and *E. coli* were found in the organic fertilizers, these bacteria were not detected (MPN $< 1.8 \times 10^2$ g^{-1}) in the soil samples collected from any of the three crops at either evaluation time (Tables 5 and 7). The bacteria did not survive in the fertilized soil, even in the wheat plots, which received an application of organic fertilizer with a higher density of thermotolerant coliforms (MPN $> 9.0 \times 10^7$ g^{-1} dry matter [DM]) than the organic fertilizer applied to maize (MPN $\leq 6.7 \times 10^6$ g^{-1} DM) (Table 4).

In addition to high levels of thermotolerant coliforms, the organic fertilizer used on the wheat crop contained a higher amount of colony-forming units of *C. perfringes* than the fertilizer applied to maize (Table 4). However, at 30 days after fertilization (DAF), the difference in population density between the soils fertilized with organic fertilizer or organomineral fertilizer (CFU $< 2{,}0$ g^{-1}) and the soils treated with mineral fertilizer was negligible and was determined as not significant by a t test. The same result was observed for total coliforms at 70 DAF (Table 6). *Salmonella* was not found in the soil samples from any of the treatments, which was expected given the absence of *Salmonella* in the organic fertilizer (unlike the other bacteria in this study, which were present in the organic fertilizer) (Table 4).

Among the bacteria that were tested for, only total coliforms and C. perfringes were detected in the soil samples. The MPNs of these organisms were low and generally did not differ among fertilization treatments for the three crops and the two evaluation times. The population of total coliforms decreased during the period from 30 DAF to 70 DAF, and C. perfringes was not detected in the samples at 70 DAF (Tables 5, 6 and 7).

Table 5. Enteropathogenic bacteria in a latosol 30 and 70 days after the application of different types of fertilizer to maize (agricultural year 2008/2009).

Type of fertilization	Total coliforms	Thermotolerant coliforms	*Escherichia coli*	*Salmonella* spp.	*Clostridium perfringes*
	....................MPN[1] g^{-1}		....	In 10 g	CFU[2] g^{-1}
		30 days after fertilization			
Organic	$11.5 \times 10^{4\ ns}$	$< 1.8 \times 10^2$	$< 1.8 \times 10^2$	Absence	$1.8 \times 10^{2\ ns}$
Organomineral	6.4×10^4	$< 1.8 \times 10^2$	$< 1.8 \times 10^2$	Absence	1.9×10^2
Mineral	3.2×10^4	$< 1.8 \times 10^2$	$< 1.8 \times 10^2$	Absence	1.9×10^2
		70 days after fertilization			
Organic	$0.5 \times 10^{4\ ns}$	$< 1.8 \times 10^2$	$< 1.8 \times 10^2$	Absence	$<1.0 \times 10^1$
Organomineral	1.4×10^4	$< 1.8 \times 10^2$	$< 1.8 \times 10^2$	Absence	$<1.0 \times 10^1$
Mineral	3.0×10^4	$< 1.8 \times 10^2$	$< 1.8 \times 10^2$	Absence	$<1.0 \times 10^1$

[ns]Non-significant difference between means in the column when comparing two treatments by a t test ($p < 0.05$). [1]Most probable number. [2]Colony-forming unit.

Table 6. Enteropathogenic bacteria in a latosol 30 and 70 days after the application of different types of fertilizer to wheat (agricultural year 2009).

Type of fertilization	Total coliforms	Thermotolerant coliforms	*Escherichia coli*	*Salmonella* spp.	*Clostridium perfringes*
	....................MPN[1] g^{-1}		....	In 10 g	CFU[2] g^{-1}
		30 days after fertilization			
Organic	$4.1 \times 10^{4\ ns}$	$< 1.8 \times 10^2$	$< 1.8 \times 10^2$	Absence	2.0×10^2 a
Organomineral	6.7×10^4	$< 1.8 \times 10^2$	$< 1.8 \times 10^2$	Absence	2.2×10^2 a
Mineral	7.1×10^4	$< 1.8 \times 10^2$	$< 1.8 \times 10^2$	Absence	0.2×10^2 b
		70 days after fertilization			
Organic	1.2×10^4 a	$< 1.8 \times 10^2$	$< 1.8 \times 10^2$	Absence	$<1.0 \times 10^1$
Organomineral	0.3×10^4 b	$< 1.8 \times 10^2$	$< 1.8 \times 10^2$	Absence	$<1.0 \times 10^1$
Mineral	0.3×10^4 b	$< 1.8 \times 10^2$	$< 1.8 \times 10^2$	Absence	$<1.0 \times 10^1$

[ns, a, b]Non-significant (ns) or significant (a, b) difference between means in the column when comparing two treatments by a t test ($p < 0.05$). [1]Most probable number. [2]Colony-forming unit.

Table 7. Enteropathogenic bacteria in a latosol 30 and 70 days after the application of different types of fertilizer to maize (agricultural year 2010/2011).

Type of fertilization	Total coliforms	Thermotolerant coliforms	*Escherichia coli*	*Salmonella* spp.	*Clostridium perfringes*
	....................MPN[1] g^{-1}		....	In 10 g	...CFU[2] g^{-1}...
		30 days after fertilization			
Organic	$5.4 \times 10^{4\ ns}$	$< 1.8 \times 10^2$	$< 1.8 \times 10^2$	Absence	$< 1.0 \times 10^1$
Organomineral	1.6×10^4	$< 1.8 \times 10^2$	$< 1.8 \times 10^2$	Absence	$< 1.0 \times 10^1$
Mineral	3.3×10^3	$< 1.8 \times 10^2$	$< 1.8 \times 10^2$	Absence	$< 1.0 \times 10^1$
		70 days after fertilization			
Organic	$1.6 \times 10^{4\ ns}$	$< 1.8 \times 10^2$	$< 1.8 \times 10^2$	Absence	$< 1.0 \times 10^1$
Organomineral	1.0×10^4	$< 1.8 \times 10^2$	$< 1.8 \times 10^2$	Absence	$< 1.0 \times 10^1$
Mineral	0.3×10^4	$< 1.8 \times 10^2$	$< 1.8 \times 10^2$	Absence	$< 1.0 \times 10^1$

[ns]Non-significant difference between means in the column when comparing two treatments by a t test ($p < 0.05$). [1]Most probable number. [2]Colony-forming unit.

The absence of thermotolerant coliforms, E. coli and Salmonella demonstrate that the organic and organomineral fertilizers did not contaminate the soil. These results indicate that the enteropathogenic bacteria in the organic fertilizer either did not survive in the fertilized soil (such as in the case of the thermotolerant coliforms and *E. coli*) or had a high rate of decline (such as in the case of the total coliforms and *C. perfringes*), regardless of the simultaneous use of mineral fertilizer.

Many studies suggest that the enteropathogenic bacteria in organic fertilizers cannot compete with native microbiota and have low survival rate in soils fertilized with organic residues (GARCÍA-ORENES et al., 2007; MACIOROWSKIA et al., 2004; TALLON et al., 2007). This may be due to the adaptation of the enteropathogenic bacteria to an environment that is very different from the soil, as animal intestines that are characterized by low redox potential, high moisture content and high nutrient concentrations relative to soil (MACIOROWSKIA et al., 2004). Furthermore, the soil attenuation capacity was sufficient to prevent the survival or growth of the enteropathogenic organisms added to the soil in the organic fertilizer. Attenuation processes in soil include filtration, in which bacteria are retained in micropores, and adsorption, in which microorganisms are held to the surface of soil colloids through electrostatic charge and surface tension mechanisms (TALLON et al., 2007). These processes are maximized in soils with clayey texture (AISLABIE et al., 2011; GAGLIARDI; KARNS, 2000; UNC; GOSS, 2004). Gagliardi and Karns (2000) found a decrease of > 90% in E. coli and Cryptosporidium parvum Oocysts populations in clayey soil, which they attributed to the retention of these organisms on soil solids. The decline of enteropathogenic bacteria in soils treated with organic fertilizers is also favored by top-dressing (AISLABIE et al., 2011) because this form of application exposes pathogens to UV radiation and desiccation, decreasing the population density of fecal bacteria (MACIOROWSKIA et al., 2004). It is likely that these attenuation processes occurred in the latosol soil in this experiment, given that the clay content was 470 g kg-1 and the organic fertilizer was applied by top-dressing without incorporation into the soil.

Conclusion

Organic and organomineral fertilizers made with poultry manure did not contaminate the soil with enteropathogenic bacteria after 30 days following application.

References

AISLABIE, J. A. B.; MCLEOD, M. A.; RYBURN, J. A.; MCGILL A. A.; THORNBURROW, D. Soil type influences the leaching of microbial indicators under natural rainfall following application of dairy shed effluent. **Soil Research**, v. 49, n. 3, p. 270-279, 2011.

BRASIL. Ministério da Agricultura, Pecuária e Abastecimento. Instrução Normativa n. 27, de 5 de junho de 2006. **Diário Oficial da República Federativa do Brasil**, Brasília, 9 de junho de 2006. Seção 1, p. 15.

CQFS-Comissão de Quimica e Fertilidade do Solo. **Manual de adubação e de calagem para os Estados do Rio Grande do Sul e de Santa Catarina**. 10th. ed. Porto Alegre: SBCS: Núcleo Regional Sul/UFRGS, 2004.

DELGADO, M.; RODRÍGUEZ, C.; MARTÍN, J. V.; IMPERIAL, M. R.; ALONSO, F. Environmental assay on the effect of poultry manure application on soil organisms in agroecosystems. **Science of the Total Environment**, v. 416, n. 3, p. 532-535, 2012.

DEMIR, K.; SAHIN, O.; KADIOGLU, Y. K.; PILBEAM, D. J.; GUNES, A. Essential and non-essential element composition of tomato plants fertilized with poultry manure. **Scientia Horticulturae**, v. 127, n. 1, p. 16-22, 2010.

DOWNES, F. P.; ITO, K. **Compendium of methods for the microbiological examination of foods**. 4th. ed. Washington, D.C.: American Public Health Association, 2001.

EATON, A. D.; CLESCERI, L. S.; RICE, E. W.; GREENBERG, A. E. **Standard methods for the examination of water and wastewater**. 21th. ed. Maryland: American Public Health Association, 2005.

FIGUEROA, E. A.; ESCOSTEGUY, P. A. V.; WIETHÖLTER, S. Dose de esterco de ave poedeira e suprimento de nitrogênio à cultura do trigo [Dose of poultry manure and nitrogen supply to wheat crops]. **Revista Brasileira de Engenharia Agrícola e Ambiental**, v. 16, n. 7, p. 714-720, 2012.

FORSLUND, A.; MARKUSSEN, B.; TOENNER-KLANK, L.; BECH, T. B.; JACOBSEN, O. S.; DALSGAARD, A. Leaching of *Cryptosporidium parvum Oocysts*, *Escherichia coli*, and a *Salmonella enterica Serovar Typhimurium* bacteriophage through intact soil cores following surface application and injection of slurry. **Applied and Environmental Microbiology**, v. 77, n. 22, p. 8129-8138, 2011.

GAGLIARDI, J. V.; KARNS, J. S. Leaching of *Escherichia coli* O157:H7 in diverse soils under various agricultural management practices. **Applied and Environmental Microbiology**, v. 66, n. 3, p. 877-883, 2000.

GAGLIARDI, J. V.; KARNS, J. S. Persistence of *Escherichia coli* O157:H7 in soil and on plant roots. **Environmental Microbiology**, v. 4, n. 2, p. 89-96, 2002.

GALLOWAY, J. N.; ABER, J. D.; ERISMAN, J. W.; SEITZINGER, S. P.; HOWARTH, R. H.; COWLING, E. B.; COSBY, B. J. The nitrogen cascade. **BioScience**, v. 53, n. 4, p. 341-356, 2003.

GARCÍA-ORENES, F.; ROLDÁN, A.; GUERRERO, C.; MATAIX-SOLERA, J.; NAVARRO-PEDREÑO, J.; GÓMEZ, I.; MATAIX-BENEYTO, J. Effect of irrigation on the survival of total coliforms in three semiarid soils after amendment with sewage sludge. **Waste Management**, v. 27, n. 12, p. 1815-1819, 2007.

HAWKE, R. M.; SUMMERS, S. A. Effects of land application of farm dairy effluent on soil properties: A literature review. **New Zealand Journal of Agricultural Research**, v. 49, n. 3, p. 307-320, 2006.

HOULBROOKE, D. J.; HORNE, D. J.; HEDLEY, M. J.; HANLY, J. A.; SNOW.; V. O. A review of literature on the land treatment of farm-dairy effluent in New Zealand and its impact on water quality. **New Zealand Journal of Agricultural Research**, v. 47, n. 4, p. 499-511, 2004.

MACIOROWSKIA, K. G.; JONES, F.T.; PILLAI, S. D.; RICKE, S. C. Incidence, sources, and control of food-borne *Salmonella* spp. in poultry feed. **World's Poultry Science Journal**, v. 60, n. 4, p. 446-457, 2004.

MACIOROWSKIA, K. G.; HERRERA, P.; PILLAI, S. D.; RICKE, S. C. Effects on poultry and livestock of feed contamination with bacteria and fungi. **Animal Feed Science and Technology**, v. 133, n. 1-2, p. 109-136, 2007.

MAGUIRE, R. O.; KLEINMAN, J. N. D.; BEEGLE, D. B. Novel manure management technologies in no-till and forage systems: Introduction to the special series. **Journal of Environmental Quality**, v. 40, n. 2, p. 287-291, 2011.

McMURRY, S. W.; COYNE, M. S. PERFECT, E. Fecal coliform transport through intact soil blocks amended with poultry manure. **Journal of Environmental Quality**, v. 27, n. 1, p. 86-92, 1998.

SANTOS, F. G.; ESCOSTEGUY, P. A. V.; RODRIGUES, L. B. Qualidade de esterco de ave poedeira submetido a dois tipos de tratamentos de compostagem **Revista Brasileira de Engenharia Agrícola e Ambiental**, v. 14, n. 10, p. 1101-1108, 2010.

SCHOCKEN-ITURRINO, R. P.; VITTORI, J.; BERALDO-MASSOLI, M. C.; DELPHINO, I.; PRISCILA, T. P. C.; DAMASCENO, R. *Clostridium perfringes* em rações e águas fornecidos a frangos de corte em granjas avícolas do interior paulista. **Ciência Rural**, v. 40, n. 1, p. 197-199, 2010.

SILVA, C. G. M.; ANDRADE, S. A. C.; STAMFORD, T. L. M. Ocorrência de *Cryptosporidium* spp. e outros parasitas em hortaliças consumidas *in natura*, no Recife. **Ciência e Saúde Coletiva**, v. 10, supl., p. 63-69, 2005.

TALLON, L. K.; SI, B. C.; KORBER, D.; GUO, X. Soil wetting state and preferential transport of *Escherichia coli* in clay soils. **Canadian Journal of Soil Science**, v. 87, n. 1, p. 61-72, 2007.

TEDESCO, M. J.; GIANELLO, C.; BISSANI, C. A.; BOHNEN, H. E; VOLKWEISS, S. J. **Análises de solo, plantas e outros materiais**. 2nd ed. Porto Alegre: Ufrgs, 1995.

TIQUIA, S. M.; TAM, N. F. Y. Characterization and composting of poultry litter in forced aeration piles. **Process Biochemistry**, v. 37, n. 8, p. 869-880, 2002.

UNC, A.; GOSS, M. J. Transport of bacteria from manure and protection of water resources. **Applied Soil Ecology**, v. 25, n. 1, p. 1-18, 2004.

ZIEMER, C. J.; BONNER, M. J.; COLE, D.; VINJÉ, J.; CONSTANTINI, V.; GOYAL, S.; GRAMER, M.; MACKIE, R.; MENG, X. J.; MYERS, G.; SAIF, L. J. Fate and transport of zoonotic, bacterial, viral, and parasitic pathogens during swine manure treatment, storage, and land application. **Journal of Animal Science**, v. 88, suppl., p. 84-94, 2010.

9

Weed control methods effect on the hydraulic attributes of a Latosol

Luiz Fernando Pires[1*], Cezar Francisco Araujo-Junior[2], Nivea Maria Piccolomini Dias[1], Moacir de Souza Dias Junior[3] and Elifas Nunes de Alcântara[4]

*[1]Laboratório de Física Aplicada a Solos e Ciências Ambientais, Universidade Estadual de Ponta Grossa, Ponta Grossa, Paraná, Brazil. [2]Instituto Agronômico do Paraná, Pólo de Londrina, Londrina, Paraná, Brazil. [3]Departamento de Ciência do Solo, Universidade Federal de Lavras, Lavras, Minas Gerais, Brazil. [4]Empresa de Pesquisa Agropecuária de Minas Gerais, Lavras, Minas Gerais, Brazil. *Author for correspondence. E-mail: lfpires@uepg.br; luizfpires@gmail.com*

ABSTRACT. Weed management plays a key role in the minimization of soil physical degradation processes such as compaction and hydric erosion. Different weed control managements can induce distinct changes in soil structure. One important soil physical attribute used for the analysis of modifications in soil structure is the soil water retention curve (SWRC). The objective of this work was to verify the use of physico-hydrical attributes to understand the effect of weed control managements on soil structure. Two soil layers (0–0.05 and 0.10–0.15 m) and six weed control managements divided into two groups were analyzed: I. no soil disturbance and ground cover (no weed control, post-emergence herbicide, mechanical mower); and II. soil disturbance and no ground cover (hand-hoe weeding, rotary tiller, pre-emergence herbicide). An area of native forest was used as reference. The results showed that the volumetric water capacity derived from the SWRC can be an interesting tool to evaluate the impact of weed control managements on soil structure. Evaluations of the air-filled porosity variation for different pressure heads also presented interesting findings. Distinct results of the weed control managements were found for the different depths analyzed in relation to the forest.

Keywords: mechanical weed control, interrow area, soil water retention, pore size distribution, air-filled porosity.

Introduction

Currently, the most important research on weed control management in cash crops emphasizes the agronomic effects of the competition of these weeds for water and nutrients and their relationship with decreases in crop yield. However, weed management plays a key role in minimizing soil physical degradation processes such as compaction and hydric erosion in tropical and subtropical regions (Araujo-Junior, Dias Junior, Guimarães, & Alcântara, 2011a; Araujo-Junior, Dias Junior, Guimarães, & Alcântara, 2011b). Additionally, weed control methods in coffee crop might improve soil organic matter quality and carbon stocks (Alcântara

& Ferreira, 2000; Watanabe et al., 2007; Martins, Araujo-Junior, Miyazawa, Vieira, & Milori, 2015).

Therefore, it is important to study the effects of different methods of weed control on soil structure, in both the row and interrow areas (Trintinalio, Tormena, Oliveira Júnior, Machado, & Constantin, 2005). Integrated management of weeds in coffee plantations is essential to achieve sustainable agricultural production systems. The objectives of integrated management are to reduce the losses caused by invasive plants, to control costs, energy, and other operations, to reduce cultivation, hydric erosion and accelerated wind, ensuring proper food production, to prevent damage by toxic plants and to maintain environmental quality by maximizing profit for the farmer (Victoria Filho, 2000).

The soil water retention curve (SWRC) represents a soil physical attribute that describes the amount of water retained in soil under a series of equilibria between the water in a soil sample and the water at chosen potentials (Klute, 1986). The traditional method of SWRC evaluation involves the establishment of a series of equilibria through a pressure plate apparatus or suction table. After pressure or suction application, at each equilibrium, the soil water content is determined through gravimetry (Libardi, 2005).

The SWRC shape and range are strongly affected by soil texture (Hillel, 1998). Soil structure also strongly affects the SWRC, mainly at high pressure heads (Klute, 1986; Stange & Horn, 2005). Changes in the soil porous system due to management practices or natural processes can lead to modifications in the amount and size of interaggregate pores, which greatly affect the shape and consequently, the slope of the SWRC (Silva, Alves, Sousa, & Fernandes, 2006; Cássaro, Borkowski, Pires, Rosa, & Saab, 2011). The soil organic matter content also influences the water retained at different potentials (Lal & Shukla, 2004).

Compaction is a common process that can occur with respect to the soil structure due to the influence of management systems. The soil volume changes induced by compaction tend to destroy the interaggregate pores and increase the number of pores of intermediate sizes (Hillel, 1998). As a consequence, the SWRC shape will be modified at the intermediate and low suction range.

In addition to the SWRC, the volumetric water capacity (C_θ), derived based on its shape after mathematical adjustment through empirical- or physical-based models, can be an interesting tool to analyze changes in the soil structure related to management systems (Ogunwole, Pires, & Shehu, 2015; Hillel, 1998). Soils with a distinct soil texture or structure will present different C_θ values. Therefore, the volumetric water capacity can be used to assess possible modifications in the soil porous space due to different management systems.

In addition to the fact that the soil porous space affects plant growth and root system development, its characterization can provide valuable information on the effects of weed control methods on the soil structure. Therefore, the aim of this study was to verify the use of physico-hydrical attributes to understand the impact of weed control managements between coffee rows (interrow area) on the soil structure.

Material and methods

Site description, characterization and soil sampling

The soil water retention data used in this study were obtained as part of a weed control methods study in the South of Minas Gerais, Brazil, involving mechanical, chemical and cultural methods (Alcântara & Ferreira, 2000; Araujo-Junior et al., 2011a; 2011b). A long-term field experiment (30 years) was carried out in a coffee plantation in São Sebastião do Paraíso County, State of Minas Gerais, Brazil (20°55'00" S, 47°07'10" W, ≈ 885 m a.s.l.). The average annual temperature of the study area is 20.8°C (27.6°C maximum, 14.1°C minimum), and the average annual rainfall is 1,470 mm (Alcântara & Ferreira, 2000). According to Köppen's classification, the climate of the region is the Cwa type, i.e., tropical highland, mesothermal with a dry winter.

The soil at the site is derived from basalt and is classified as a Dystroferric Red Latosol according to the Brazilian Soil Classification System (Santos et al., 2013), similar to a Ferralsol (FAO, 1998) and a Typic Haplorthox according to the USDA Soil Taxonomy (Soil Survey Staff, 1998).

At the beginning of the experiment in 1977, the field was planted with coffee trees (cultivar Catuaí Vermelho LCH 2077-2-5-99) with a 4 m spacing between coffee rows and a 1 m spacing between coffee plants. In December 2005, due to the decline of the coffee plant yield, the coffee shrubs were replaced with a new cultivar (Paraíso), without disturbing the inter row areas, with the same spacing between coffee rows and 0.70 m spacing between plants.

Analysis of a soil sample collected close to the experimental area under natural forest showed that the Dystroferric Red Latosol contains 570 g kg^{-1} clay, 230 g kg^{-1} silt and 200 g kg^{-1} sand (0 to 0.30 m depth), with a homogeneous structure throughout the profile. The soil has an average soil bulk density of 988±124 kg m^{-3} (n=9), an average total porosity of 0.676±0.056 m^3 m^{-3} (n=9), an average macroporosity of 0.352±0.086 m^3 m^{-3} (n=9) and exhibits a granular structure similar to coffee powder. Additional details pertaining to the site, soil and trial are provided by Alcântara and Ferreira (2000) and Araujo-Junior et al. (2011a; 2011b).

Experimental design, weed control and cover crops

In 1977, the weed control methods were laid out in a randomized complete block design with three replicates; each plot was 36 m long and 12 m wide. The experimental design was a split-plot with each weed control method used in three interrows as the main-plot factor, and the sampling layers (0–0.05 and 0.10–0.15 m) were used as the split-plot.

In the areas under the coffee canopy, the weeds were managed either by manual hoeing or with the application of herbicides. The type of weed management system, which was adopted satisfactorily in the coffee plantations over the 30-year period, influenced the number of operations needed, as well as the density and diversity of weeds found in the area at the time of the sampling (Araujo-Junior et al., 2011b).

The weed control methods were divided into two groups: I. no soil disturbance and ground cover + forest and II. soil disturbance and no ground cover + forest. The first group consisted of the following weed control: no weeding between coffee rows and under the coffee canopy – weed check (CHECK), post-emergence herbicide (POSH) and mechanical mower (MOW); the second group consisted of the following: hand-hoe weeding (MHW), rotary tiller (RTILL) and pre-emergence herbicide (PREH).

Soil sampling

In December 2007, thirty years after the installation of the experiment, soil samples were collected in the inter row area between coffee rows, 2 m from the stems of the coffee trees at depths of 0–0.05 and 0.10–0.15 m, totaling 180 soil samples (15 samples × 2 depths × 6 management systems). An additional fifteen undisturbed soil samples from each layer were collected from a Dystroferric Red Latosol under native forest (FRT) adjacent to the coffee cultivation. The undisturbed soil samples were collected using a cylindrical Uhland sampler and aluminum cylinders (volumetric rings), with a height of 2.54×10^{-2} m and a diameter of 6.35×10^{-2} m.

Soil water retention curve evaluation

After sample collection, the top and bottom of the samples were trimmed flat. The wetting procedure used to saturate the samples consisted of soaking them in a tray with the water level just below 1/3 of the top of the cylinders. This procedure was carried out over a period of 48 hours to allow saturation of the soil and to avoid the presence of entrapped air bubbles. After the wetting procedure, the soil samples were placed in contact with the porous media inside a suction table and were then placed in a pressure chamber.

The soil samples were subjected to the following pressure heads (h) in the suction table: -2, -4, and -6 kPa (Romano, Hopmans, & Dane, 2002), and they were subjected to -10, -33, -100, -500 and -1,500 kPa in the pressure chambers (Dane & Hopmans, 2002). The choice of h was based on the concepts of pore size to separate macropores and micropores, the water capacity and the permanent wilting point (Lal & Shukla, 2004).

After thermodynamic equilibrium was reached, the moist soil mass was evaluated using a precision balance (0.01×10^{-3} kg). The dry soil mass was obtained at the end of the SWRC evaluation by placing the samples in an oven at 105°C for 48 hours. The volumetric water content was determined by multiplying the gravimetric water content and the soil bulk density (Libardi, 2005). In this study, the water density was considered to equal 1,000 kg m^{-3}.

Data analysis

The SWRC experimental data were adjusted using the mathematical model proposed by van Genuchten (1980) in the SWRC Fit computer program (Seki, 2007):

$$S_e = \frac{1}{\left[1+(-\alpha h)^n\right]^m}, \qquad (1)$$

where S_e $(=\theta-\theta_r / \theta_s-\theta_r)$ represents the effective soil water saturation and α, n and m $(=1-1/n)$ are empirical parameters that govern the SWRC shape. The SWRC adjustments were obtained based on average values of θ (n=3).

To verify the quality of the SWRC mathematical adjustment, the root mean square error (RMSE) and the coefficient of determination (r^2) were analyzed. The highest RMSE and lowest r^2 evaluated for the different layers were obtained for the samples under the native forest, i.e., respective values of 0.034 and

0.994 for 0–0.05 m and 0.034 and 0.992 for 0.10–0.15 m. These results showed that the estimated parameters for the measured SWRCs were an excellent fit.

After the SWRC mathematical adjustments, the volumetric water capacity (C_θ) was obtained using the following equation (Radcliffe & Simùnek, 2010):

$$C_\theta = \frac{\alpha^n(\theta_s - \theta_r)mn(-h)^{n-1}}{\left[1+(-\alpha h)^n\right]^{m+1}}, \qquad (2)$$

where θ_r and θ_s denote the soil residual and saturated volumetric water contents, respectively. The equivalent cylindrical soil pore radii (r) were obtained in μm with h in kPa ($= 149/h$).

Evaluations of the air-filled porosity (φ) for different h values were also carried out. Air-filled porosity was obtained by relating the difference between the total porosity and θ for each h. These analyses made it possible to verify the influence of the weed control management on the available water at both the low and high pressure heads.

Relative differences (RDs) were also obtained among the SWRCs to evaluate the impact of the different weed control managements in relation to the reference area (FRT). RD was determined using the following equation:

$$RD(\%) = \left(\frac{SWRC_{wcm} - SWRC_{ref}}{SWRC_{ref}}\right).100 \qquad (3)$$

where the subscripts *wcm* and *ref* represent the weed control management and reference values, respectively.

The influence of the weed control managements on the structure of the soil was also assessed through soil pore classification systems based on functional characteristics. The system proposed by Greenland (1977) was used for this purpose, in which pores with equivalent cylindrical radii <0.25 μm were considered bonding + residual pores; those from 0.25 to 25 μm were considered storage pores; those from 25–250 μm were considered transmission pores; and those >250 μm were considered fissures.

Results and discussion

In the upper surface layer (0–0.05 m), the weed control methods used in the areas between the coffee rows changed the SWRC behavior in comparison to the reference samples (FRT) (Figure 1). The soil under the native forest was characterized by a high number of transmission and fissure pores, which do not contribute to water retention. On the other hand, they probably increase recharge and minimize the amount of surface water in the watershed (Hillel, 1998).

Changes in the soil structure induced by the weed control methods between the coffee rows affected the distribution of pores, with an increase in the pore contribution related to the storage of water in relation to the FRT samples (Lal & Shukla, 2004).

In the upper surface layer, the MOW weed control method and the chemical POSH method induced modifications in the soil porous system in relation to FRT, resulting in an increase in the water volume retained in the region of small pore sizes (Figures 1a and 1g). The same behavior was observed for CHECK, although with less intensity. This result is an indication of the increase in the volume of storage and residual pores, especially for the former methods.

The CHECK samples (-0.50 kPa) presented a slight difference in the peak position in relation to FRT (-0.42 kPa), as observed in the C_θ results (Figure 1c). On the other hand, for the weed control method with soil disturbance such as hand-hoe weeding, the mechanical action of the hoe caused changes in the soil structure while maintaining the soil without cover on the soil surface. The bare soil contributed negatively to the soil structure and compromised the organic carbon content and protection against water erosion (Faria, Schaefer, Ruiz, & Costa, 1998; Alcântara & Ferreira, 2000; Yang, Wang, Tang, & Chen et al., 2007; Araujo-Junior et al., 2011a; 2011b; Watanabe et al., 2007; Martins et al., 2015). Information on the organic carbon content for the same weed control methods and soil depths analyzed in this study can be found in Araujo-Junior et al. (2011b).

POSH and MOW seemed to cause the same effects with respect to the soil structure (Figure 1g), mainly under high h. The C_θ peak for these two management types was found at -0.30 kPa (MOW) and -0.26 kPa (POSH). A shift toward higher h was observed for these two management types in relation to FRT. It is important to stress that a large number of pores in the region of transmission pores contributes positively to water and air fluxes through the soil profile and directly affects plant growth by facilitating root penetration (Pagliai, Vignozzi, & Pellegrini, 2004).

These changes in the position of the C_θ peak help demonstrate the modifications in the soil structure induced by different management options (Leij, Ghezzehei, & Or, 2002). The C_θ results also demonstrate the differences in the frequency of pores for CHECK, POSH and MOW in relation to FRT (Figure 1c), which indicates important changes in the soil structure induced by the weed control managements (Lal & Shukla, 2004).

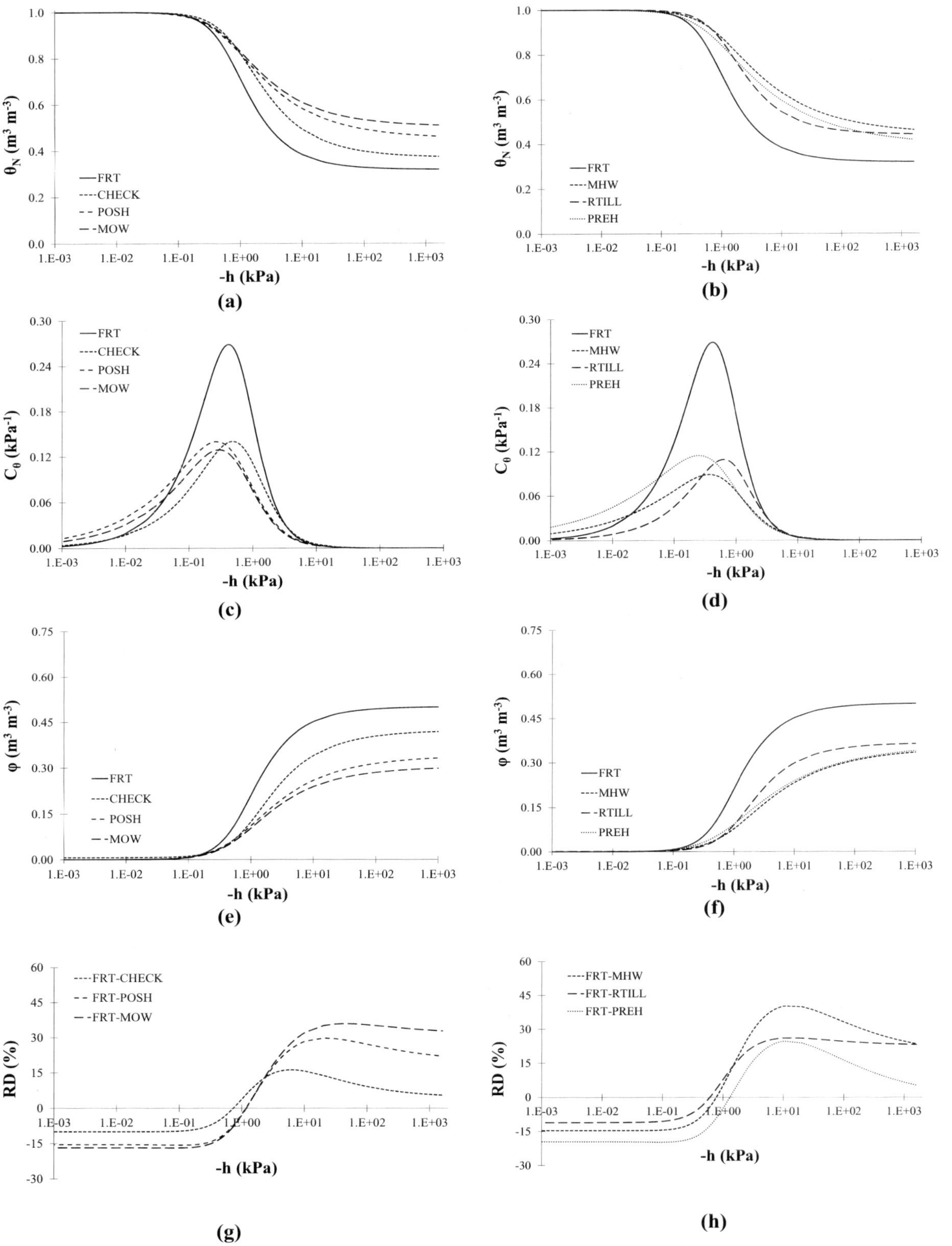

Figure 1. Normalized soil water retention curve (SWRC), volumetric water capacity (C_θ), air-filled porosity (φ) variation and SWRC relative differences (RD) for samples without soil disturbance and with ground cover (a, c, e, g) and with soil disturbance and without ground cover (b, d, f, h) for the 0 to 0.05 m layer. FRT=forest; CHECK=weed check; POSH=post-emergence herbicide; MOW=mechanical mower; MHW=hand-hoe weeding; RTILL=rotary tiller; PREH=pre-emergence herbicide.

By analyzing the SWRC data, it is possible to observe high values of mean soil porosity for FRT (0.733 m^3 m^{-3}) followed by CHECK (0.667 m^3 m^{-3}), POSH (0.620 m^3 m^{-3}) and MOW (0.610 m^3 m^{-3}). FRT retained a large amount of water (~29%) up to an r value of approximately 149 μm in comparison to the other managements (~18% – CHECK and POSH; ~17% – MOW) (Figure 1a). An inversion in the water retention pattern (not shown in the normalized SWRC) was observed for r values smaller than 149 μm among the managements (FRT<CHECK<POSH <MOW) (Figure 1a). This observation is an indication of a large amount of small pores (storage and residual+bonding pores) for MOW and POSH in relation to FRT (Greenland, 1977). The variations in φ with h provide insight about this result (Figure 1e). The smallest value of φ, in the region of small pores (small h), was found for MOW followed by POSH, CHECK and FRT.

In summary, the weed control methods MOW and POSH caused practically the same effects on the structure of the soil (Figures 1a and 1g), and the CHECK method resulted in water retention similar to that of FRT (Figure 1e and g).

Considering the managements with soil disturbance (Figure 1b), MHW and PREH presented a water retention behavior similar to FRT (Figure 1h). RTILL had closer agreement with FRT mainly regarding the region of large macropores. However, this method is not recommended for weed control in tropical regions because it results in damage caused to the soil structure due to aggregate breakage in small fractions as observed in the field. Additionally, the long-term use of this method resulted in the selection of the weed *Cyperus rotundus* ("tiririca"). There are negative aspects associated with the use of certain mechanical weed control methods, e.g., conventional tillage systems, such as the formation of surface crusts and plough pans at the lower cultivation limit (Pagliai, Vignozzi, & Pellegrini, 2004).

Differences in the SWRCs of these managements resulted mainly as a consequence of soil disturbance (Lipiec, Kus, Slowinska-Jurkiewicz, & Nosalewicz, 2006; Ugarte Nano, Nicolardot, Quinche, Munier-Jolain, & Ubertosi, 2016) (Figures 1b and 1h). The C_θ results demonstrate these differences among the weed crop methods (Figure 1d). The peak C_θ value for the different managements was found at -0.62 kPa (RTILL), -0.38 kPa (MHW) and -0.25 kPa (PREH); a shift of the peak toward large h values was observed in relation to FRT (-0.42 kPa), with the exception of RTILL. The C_θ values for MHW and PREH indicated a slightly broad distribution for these managements. FRT had the largest frequency of pores in relation to the other methods (Figure 1d), which is mainly related to the highest total pore volume (Lu, Malik, Chen, & Wu, 2014).

High values of mean soil porosity were found for RTILL (0.653 m^3 m^{-3}) followed by MHW (0.627 m^3 m^{-3}) and PREH (0.590 m^3 m^{-3}). Regarding pores smaller than 15 μm, the soil under MHW retained more water (~63%) than the other managements and the forest (FRT: ~39% < RTILL: ~55% < PREH: ~60%) (Figure 1b). This result indicates a large amount of small sized pores (storage and residual+bonding pores) for this management in relation to the others. Due to this fact, smaller values of φ were found for MHW followed by PREH and RTILL (Figure 1f), which is a consequence of soil disturbance (Lipiec, Hajnos, & Swieboda, 2012).

Regarding the weed control methods that cause soil disturbance, MHW and PREH presented similarities with respect to water retention (Figures 1b and 1f). The former had the smallest volume of pores in relation to the other managements (Figure 1d). RTILL was characterized by the narrowest C_θ distribution, which indicates a concentration of pores of similar sizes (Figure 1d). The air-filled porosity results showed similarities in their values among managements for large h values. On the other hand, MHW presented the smallest φ values and RTILL the largest values for small values of h (small pore sizes).

For the lower layer analyzed (0.10–0.15 m), the results showed similarities in the SWRCs among managements for soil without disturbance, mainly for h < -10 kPa (Figure 2a). Regarding pores larger than 300 μm, only slight differences in the volume of water retained between CHECK (~9% of water retained between 0 and -0.5 kPa) and POSH (~9%), and between MOW (~16%) and FRT (~17%) were observed (Figure 2g).

A large mean soil porosity was found for samples under FRT (0.663 m^3 m^{-3}), and similar values were found for the other managements, which indicates that for this soil layer, CHECK (0.607 m^3 m^{-3}), POSH (0.603 m^3 m^{-3}) and MOW (0.610 m^3 m^{-3}) led to the same type of modification in the soil structure if its porosity is considered as a reference attribute (Moraes et al., 2016).

However, the C_θ results indicated differences in the frequency of pores in the following sequence: FRT>MOW>POSH>CHECK (Figure 2c). The peak C_θ distribution presented a shift toward small h values among the managements in relation to FRT (-0.20 kPa), with MOW being an exception (-0.16 kPa). This result can be an indication that MOW maintained a soil structure more similar to FRT (Figure 2a and e), mainly for high h values (large pore sizes). POSH and CHECK were also characterized by narrow C_θ distributions.

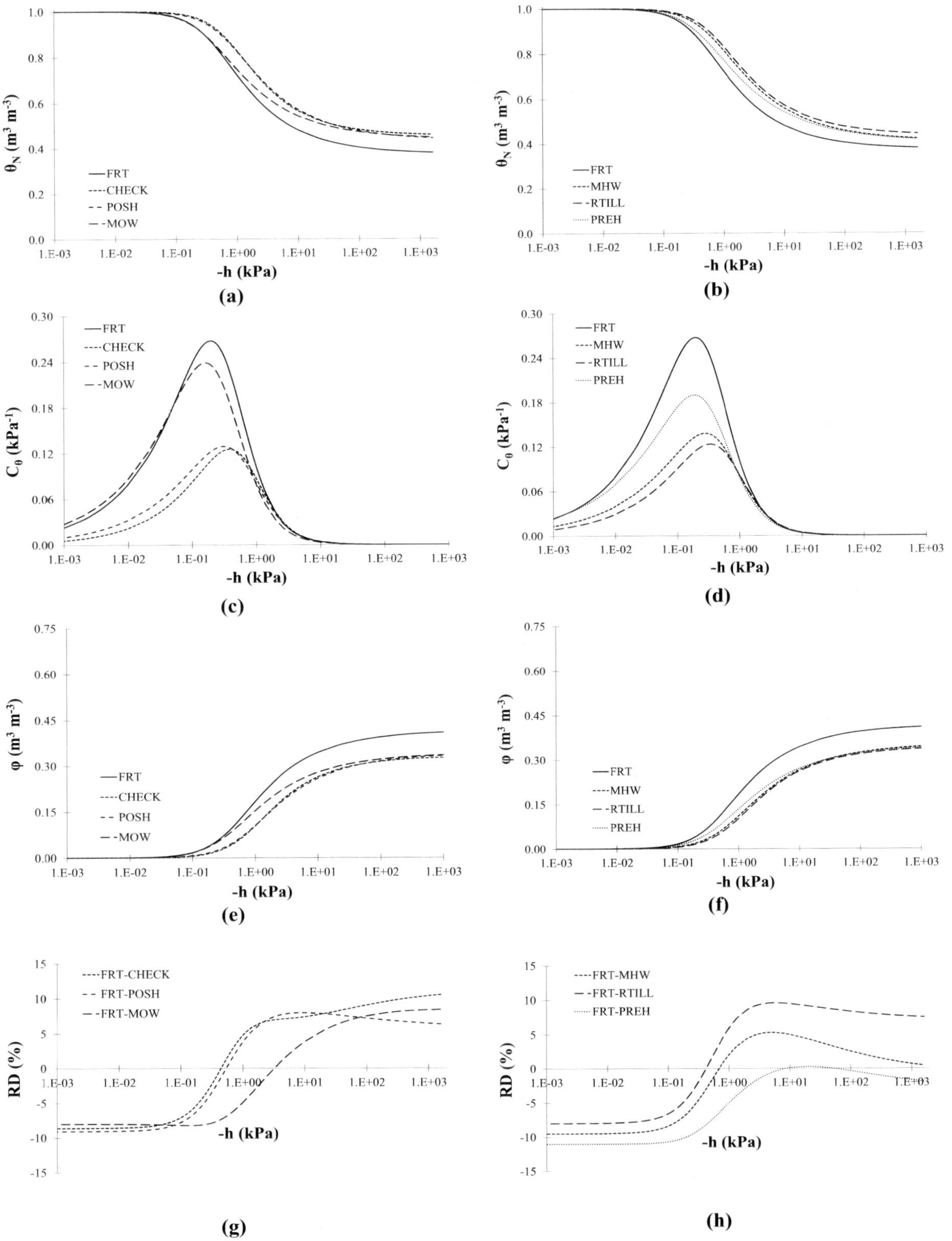

Figure 2. Normalized soil water retention curve (SWRC), volumetric water capacity (C_θ), air-filled porosity (φ) variation and SWRC relative differences (RD) for samples without soil disturbance and with ground cover (a, c, e, g) and with soil disturbance and without ground cover (b, d, f, h) for the 0.10 to 0.15 m layer. FRT=forest; CHECK=weed check; POSH=post-emergence herbicide; MOW= mechanical mower; MHW=hand-hoe weeding; RTILL=rotary tiller; PREH=pre-emergence herbicide.

With respect to the weed control methods that cause soil disturbance, the SWRC results showed similarities among managements for h ranging from 0 to -0.10 kPa (MHW: ~1.3% of water retained in this interval, RTILL: ~1.0%, PREH: ~2.2%) and from -20 to -1500 kPa (MHW: ~9.3%, RTILL: ~8.5%, PREH: ~8.4%) (Figure 2b and h). FRT presented the largest mean soil porosity (0.663 m^3 m^{-3}) followed by RTILL (0.610 m^3 m^{-3}), MHW (0.600 m^3 m^{-3}) and PREH (0.590 m^3 m^{-3}), which differed only slightly.

Regarding pores smaller than 10 μm, the managements that cause soil disturbance practically presented the same behavior in relation to water retention (MHW: ~52% of water retained, RTILL: ~53%, PREH: ~50%) (Figure 2b). This behavior is an indication that for the smallest pores, the effect of the changes in soil structure caused by these managements was similar in relation to FRT (Figure 2f).

The C_θ results showed distinct changes in the soil structure induced by the weed control methods for the lower layer studied (Figure 2d). A decrease in the frequency of pores was observed among managements in the following sequence: PREH>MHW>RTILL. The largest decrease in the frequency of pores found for RTILL is an indication of the damage caused to the soil structure by this weed control method (Lal & Shukla, 2004). The narrowest C_θ distribution was also found for this management. In addition, the C_θ peak presented a shift toward small h values among the managements in relation to FRT (-0.20 kPa).

An analysis of pore size distribution based on the Greenland classification was also carried out (Greenland, 1977). For the upper layer (0–0.05 m) studied, 57.1% of the FRT pores comprised transmission pores and fissures (Figure 3a), which are responsible for water transmission and redistribution throughout the soil profile (Hillel, 1998; Libardi, 2005; Lal & Shukla, 2004). Regarding these pore classes, FRT was followed by CHECK>POSH>MOW (managements without soil disturbance) and RTILL>PREH>MHW (managements with soil disturbance). The smallest proportion of transmission pores and fissures was found for MHW (32.1%). On the other hand, 67.9% of MHW pores belonged to residual (48.0%) and storage (19.9%) pores (Figure 3a). In the case of FRT, 42.9% of the pores were included in these two classes (33.2%: <0.25 μm and 9.7%: 0.5–0.25 μm).

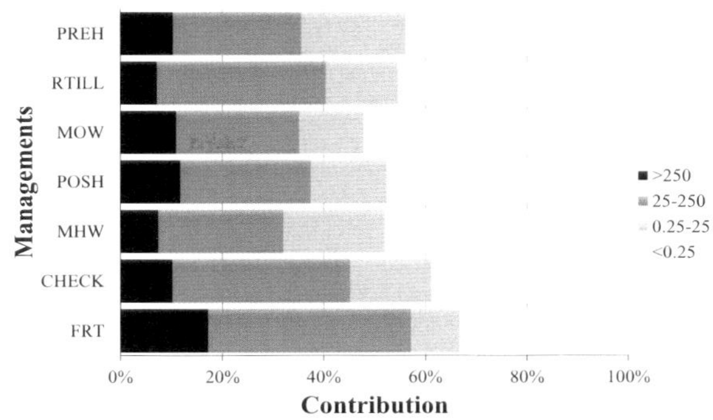

(a)

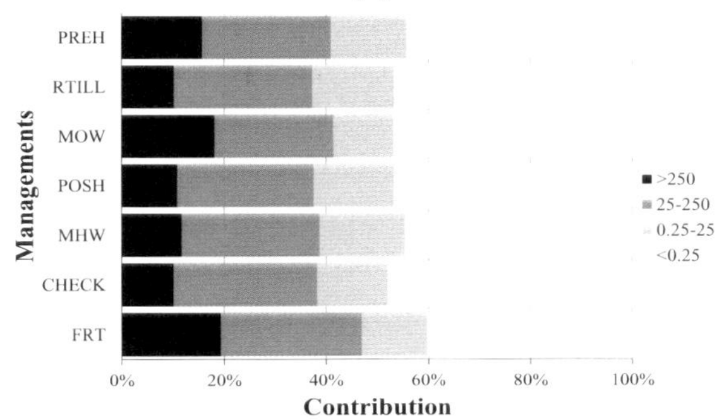

(b)

Figure 3. Pore size distribution according to the Greenland (1977) classification system for the different soil layers studied: 0–0.05 m (a) and 0.10–0.15 m (b). FRT=forest; CHECK=weed check; POSH=post-emergence herbicide; MOW=mechanical mower; MHW=hand-hoe weeding; RTILL=rotary tiller; PREH=pre-emergence herbicide.

Similarly, for the lower layer (0.10–0.15 m), FRT presented the largest contribution (47.0%) of transmission pores and fissures followed by MOW>CHECK>POSH (managements without soil disturbance) and PREH>MHW>RTILL (managements with soil disturbance) (Figure 3b). The smallest proportion of transmission pores and fissures was found for RTILL (37.3%). The largest proportion of residual pores was found for CHECK (48.0%) and that of storage pores for MHW (16.7%). In the case of FRT, 53.0% of the pores were included in these two classes (40.3%: <0.25 μm and 12.7%: 0.5–0.25 μm).

Conclusion

The application of different weed control methods produced changes in soil structure in relation to the native forest as observed by measurements of water retention, volumetric water capacity, pore size distribution and air-filled porosity. The upper layer (0–0.05 m) was more sensitive to changes than the lower layer (0.10–0.15 m).

In the interrow area (between coffee rows), the mechanical mower weed control method resulted in a lower layer soil structure that was very similar to the native forest. For the upper layer, the weed check and post-emergence herbicide methods exhibited the best agreements. The rotary tiller resulted in the selection of a weed plant (*Cyperus rotundus* – "tiririca") that is undesirable for coffee plantation management.

Acknowledgements

The authors would like to thank the National Council of Technological and Scientific Development (CNPq), Fapemig and the Araucária Foundation of the Support to Scientific and Technological Development of the State of Paraná for the research grants for Dr. Luiz F. Pires (303726/2015-6), Dr. Moacir de S. Dias Junior and Dr. Cezar F. Araujo-Junior and the Brazilian Consortium of Coffee Research and Development.

References

Alcântara, E. N., & Ferreira, M. M. (2000). Efeitos de métodos de controle de plantas daninhas na cultura do cafeeiro (*Coffea arabica* L.) sobre a qualidade física do solo. (In Portuguese, with English abstract). *Revista Brasileira de Ciência do Solo*, *24*(4), 711-721.

Araujo-Junior, C. F., Dias Junior, M. S., Guimarães, P. T. G., & Alcântara, E. N. (2011a). Sistema poroso e capacidade de retenção de água em Latossolo submetido a diferentes manejos de plantas invasoras em lavoura cafeeira. (In Portuguese, with English abstract). *Planta Daninha*, *29*(3), 499-513, 2011a.

Araujo-Junior, C. F., Dias Junior, M. S., Guimarães, P. T. G., & Alcântara, E. N. (2011b). Capacidade de suporte de carga e umidade crítica de um Latossolo induzida por diferentes manejos. (In Portuguese, with English abstract). *Revista Brasileira de Ciência do Solo*, *35*(1), 115-131.

Cássaro, F. A. M., Borkowski, A. K., Pires, L. F., Rosa, J. A., & Saab, S. C. (2011). Characterization of a Brazilian clayey soil submitted to conventional and no-tillage management practices using pore size distribution analysis. *Soil & Tillage Research*, *111*(2), 175-179.

Dane, J. H., & Hopmans, J. W. (2002). Pressure plate extractor. In J. H. Dane, & G. C. Topp (Eds.), *Methods of Soil Analysis: Part 4-Physical Methods* (p. 688-690). Madison, USA: Soil Science Society of America.

Faria, J. C., Schaefer, C. E. R., Ruiz, H. A., & Costa, L. M. (1998). Effects of weed control on physical and micropedological properties of a brazilian Ultisol. *Revista Brasileira de Ciência do Solo*, *22*(4), 731-741.

FAO. (1998). *Word reference base for soil resources: Word soil resources report*. Rome, IT: International Soil Reference and Information Centre ISRIC.

Greenland, D. J. (1977). Soil damage by intensive arable cultivation: temporary or permanent? *Philosophical Transactions of the Royal Society B*, *281*(1), 193-208.

Hillel, D. (1998). *Environmental soil physics*. San Diego, USA: Academic Press.

Klute, A. (1986). Water retention: laboratory methods. In C. A. Black (Ed.), *Methods of soil analysis: I. Physical and mineralogical methods* (p. 635-662). Madison, USA: Soil Science Society of America.

Lal, R., & Shukla, M. K. (2004). *Principles of soil physics*. New York, USA: Marcel Dekker.

Leij, F. J., Ghezzehei, T. A., & Or, D. (2002). Modeling the dynamics of the soil pore-size distribution. *Soil & Tillage Research*, *64*(1), 61-78.

Libardi, P. L. (2005). *Dinâmica da água no solo*. São Paulo, SP: EDUSP.

Lipiec, J., Kus, J., Slowinska-Jurkiewicz, A., & Nosalewicz, A. (2006). Soil porosity and water infiltration as influenced by tillage methods. *Soil & Tillage Research*, *89*(2), 210-220.

Lipiec, J., Hajnos, M., & Swieboda, R. (2012). Estimating effects of compaction on pore size distribution of soil aggregates by mercury porosimeter. *Geoderma*, *179-180*(1), 20-27.

Lu, S-G., Malik, Z., Chen, D-P., & Wu, C-F. (2014). Porosity and pore size distribution of Ultisols and correlations to soil iron oxides. *Catena*, *123*(1), 79-87.

Martins, B. H., Araujo-Junior, C. F., Miyazawa, M., Vieira, K. M., & Milori, D. M. B. P. (2015). Soil organic matter quality and weed diversity in coffee plantation area submitted to weed control and cover crops management. *Soil & Tillage Research*, *153*(1), 169-174.

Moraes, M. T., Debiasi, H., Carlesso, R., Franchini, J. C., da Silva, V. R., & da Luz, F. B. (2016). Soil physical quality on tillage and cropping systems after two decades in the subtropical region of Brazil. *Soil & Tillage Research*, *155*(1), 351-362.

Ogunwole, J. O., Pires, L. F., & Shehu, B. M. (2015). Changes in the structure of a Nigerian soil under different land management practices. *Revista Brasileira de Ciência do Solo*, *39*(3), 830-840.

Pagliai, M., Vignozzi, N., & Pellegrini, S. (2004). Soil structure and the effect of management practices. *Soil & Tillage Research*, 79(2), 131-143.

Radcliffe, D. E., & Simùnek, J. (2010). *Soil physics with Hydrus: Modelling and applications*. Boca Raton, USA: CRC Press.

Romano, N., Hopmans, J. W., & Dane, J. H. (2002) Suction table. In J. H. Dane, & G. C. Topp (Eds.), *Methods of Soil Analysis: Part 4-Physical Methods* (p. 692-698). Madison, USA: Soil Science Society of America.

Santos, H. G., Jacomine, P. K. T., Anjos, L. H. C., Oliveira, V. A., Lumbreras, J. F., Coelho, M. R., ... Oliveira, J. B. (2013). *Sistema brasileiro de classificação de solos*. Brasília, DF: Embrapa.

Seki, K. (2007). SWRC fit - a nonlinear fitting program with a water retention curve for soils having unimodal and bimodal pore structure. *Hydrology and Earth System Sciences*, *4*(1), 407-437.

Silva, M. M., Alves, M. C., Sousa, A. P., & Fernandes, F. C. S. (2006). Impacto do manejo nos atributos físico-hídricos de um Latossolo Vermelho sob cerrado, no município de Selvíria, Estado do Mato Grosso. (In Portuguese, with English abstract). *Acta Scientiarum. Agronomy*, *28*(1), 13-22.

Soil Survey Staff. (1998). *Keys to soil taxonomy*. Washington, DC: US Department of Agriculture, Natural Resource Conservation Service.

Stange, C. F., & Horn, R. (2005). Modeling the soil water retention curve for conditions of variable porosity. *Vadose Zone Journal*, *4*(3), 602-613.

Trintinalio, J., Tormena, C. A., Oliveira Júnior, R. S., Machado, J. L., & Constantin, J. (2005). Alterações nas propriedades físicas e químicas de um Latossolo Vermelho distrófico por diferentes manejos na entrelinha da cultura da pupunha (*Bactris gasipaes* Kunth). (In Portuguese, with English abstract). *Acta Scientiarum. Agronomy*, *27*(4), 753-759.

Ugarte Nano, C. C., Nicolardot, B., Quinche, M., Munier-Jolain, N., & Ubertosi, M. (2016). Effects of integrated weed management based cropping systems on the water retention of a silty clay loam soil. *Soil & Tillage Research*, *156*(1), 74-82.

van Genuchten, M. Th. (1980). A closed-form equation for predicting the hydraulic conductivity of unsaturated soils. *Soil Science Society of America Journal*, *44*(5), 892-898.

Victoria Filho, R. (2000). Estratégias de manejo de plantas daninhas. In L. Zambolin (Ed.), *Manejo integrado de doenças, pragas e plantas daninhas* (p. 349-363). Viçosa, MG: UFV.

Watanabe, A., Takada, H., Afandi, S., Adachi, T., Oki, Y., & Senge, M. (2007). Assessment of degradation and generation of humus in a coffee soil affected by weed cover by means of a stable carbon isotopic ratio. *European Journal of Soil Science*, *58*(5), 1042-1049.

Yang, Y., Wang, H., Tang, J., & Chen, X. (2007). Effects of weed management practices on orchard soil biological and fertility properties in southeastern China. *Soil & Tillage Research*, *93*(1), 179-185.

10

Liming and fertilisation in *Pinus taeda* plantations with severe nutrient deficiency in savanna soils

Araína Hulmann Batista[1*], Antônio Carlos Vargas Motta[1], Carlos Bruno Reissmann[1], Thays Schneider[2], Iara Lang Martins[1] and Maurício Hashimoto[1]

*[1]Programa de Pós-graduação em Ciência do Solo, Universidade Federal do Paraná, Rua dos Funcionários, 1540, 80035-050, Curitiba, Paraná, Brazil. [2]Universidade Estadual de Ponta Grossa, Ponta Grossa, Paraná, Brazil. *Author for correspondence. E-mail: arainahulmann@gmail.com*

ABSTRACT. Soils with high acidity and low exchangeable bases may be responsible for low yields of *Pinus taeda* in a forest plantation at Jaguariaíva, Paraná State, Brazil. The aim of this study was to evaluate the effect of liming and fertilisation, applied over litter, on two selected areas with *Pinus taeda* plantations. Soil, litter and pine needles were evaluated for K, Ca and Mg concentrations and soil acidity parameters. Seven treatments were applied: (i) complete (N, P, K, Zn, Cu, B, Mo, and lime); (ii) without N, P, and K; (iii) without Zn, Cu, B, and Mo; (iv) without K; (v) without Zn; (vi) without lime; and (vii) control (without nutrients and lime). Soil samples were collected at five soil depths (0-5, 5-10, 10-20, 20-40 and 40-60 cm) simultaneously with litter samples. Needles were also collected from the first and second pine flushes. Liming induced soil pH, Ca^{2+}, and Mg^{2+} increases, and the opposite was observed for Al^{3+} and Al saturation. Fertilisation increased soil exchangeable K^{+} concentrations and needle and litter K concentrations. The low Ca and Mg concentrations found in the plant needles might be attributable to their low mobility.

Keywords: nutrient deficiency, reforestation, litter, soil reaction, acidification.

Introduction

Commercial pine plantations in southern Brazil generally show fast growth and no visual nutrient deficiency symptoms, which has led to the idea that these species could grow without liming and fertilisation (FERREIRA et al., 2001; REISSMANN; ZÖTTL, 1987; REISSMANN; WISNIEWSKI, 2004). This assumption has continued amongst most growers, as the majority of pine plantations do not receiving any fertiliser or lime application.

Since the commercial plantations were introduced, the planted areas have been spread over a wide range of vegetation types, climates and geological and soil classes within a characteristic subtropical pattern. The expansion over the highly weathered soils of the Cerrado (Brazilian savannah) has tested the ability of commercial pine species to grow under very low levels of exchangeable bases and highly acidic soils. Under this natural condition, *Pinus caribaea* showed nutritional imbalances in some places, probably associated with Ca, Mg and Mn deficiencies, which may have caused plant death (CHAVES; CORREA, 2005). The above authors' findings suggest the need for lime or fertiliser application in very poor soils at the first planting time. Several regions are now in the second or third pine planting cycles without the replacement of exported nutrients. Therefore, nutrient deficiency problems are expected to increase with time, especially in the low-nutrient soils of the Cerrado.

In the last few decades, the negative effects of inadequate soil management in Brazilian areas of forest plantations have been investigated and highlighted (CHAVES; CORREA, 2005; LILIENFEIN et al., 2000; REISSMANN; ZÖTTL, 1987) As reported in more recent trials, significant increases in productivity due to lime and fertiliser applications have been observed in coniferous forests (FERREIRA et al., 2001). However, according to Brito et al. (1986) *Pinus caribeae* forests planted in soils treated with liming and macronutrients showed increasing growth-ring mass production up to the third year; but opposite results were obtained with older trees. Reduced productivity and nutrient deficiencies in forest plantations in Germany, France and Norway have been investigated since the 1950's. Positive responses in wood production to liming and fertilisation (N, P, K) with reduced plant deficiency symptoms have been observed (HÜTTL, 1989). However, according to a survey done by Hüttl and Zöttl (1993), several factors must be taken into consideration before the recommendation of liming and fertilisation, as the negative effects may outweigh the benefits, especially the leaching of large amounts of NO_3^-.

Litter plays a key role in forest ecosystems, affecting soil protection and nutrient cycling. Several studies in southern Brazil have attempted to find relationships between litter quality and quantity and pinus yield (SCHUMACHER et al., 2008; VIERA; SCHUMACHER, 2010). Also, plant needle chemical analysis may be an interesting tool to evaluate pine nutrition and its demand for nutrients (ZÖTTL; TSCHINKEL, 1971). However, Chaves and Correa (2005) could not find clear evidence of nutrient imbalances using needle chemical analysis.

The aim of this study was to characterise the effect of superficial liming and mineral fertilisation on soil acidity, on soil K^+, Ca^{2+} and Mg^{2+} dynamics and on K, Ca and Mg concentrations in litter and pine needles in two five-year-old *Pinus taeda* plantations with severe nutrient deficiency visual symptoms.

Material and methods

Characterisation of the site

The experiment was performed in two selected areas of *Pinus taeda* commercial plantations of the Forest Value Company, located at Jaguariaíva county (area I) (24° 13' 29.13" S latitude and 49° 36' 21.12" W longitude) and Arapoti county (area II) (24° 18' 56.72" S latitude and 49° 44' 13.16" W longitude). The areas are situated in the second high plain of Paraná State, Brazil (960 to 1,320 m above sea level), where the soil parent material is derived from the geological sandstone formation of Itararé and Furnas. The vegetation was predominantly Cerrado (savannah) and was gradually replaced by forest plantations, agriculture and livestock (MELLO et al., 2003). Soil samples were collected and analysed in each area near the pinus plantation for the main physical and chemical characteristics (Table 1).

Table 1. Soil chemical analysis and particle size determination in samples from areas I (Jaguariaíva) and II (Arapoti), performed prior to treatments.

	Depth	pH $CaCl_2$	pH SMP	Al^{3+}	H + Al^{3+}	Ca^{2+}	Mg^{2+}	K^+	SB[2]	P
	cm					$cmol_c\ kg^{-1}$				$mg\ dm^{-3}$
	0-20	3.83	4.98	2.70	10.88	0.38	0.20	0.05	0.63	6.38
Area I	20-40	4.08	5.40	1.70	7.85	0.28	0.10	0.03	0.41	3.23
	40-60	4.10	5.53	1.60	7.13	0.25	0.10	0.03	0.38	1.93
	0-20	3.85	5.65	1.25	6.45	0.15	0.06	0.03	0.24	0.38
Area II	20-40	4.00	5.88	0.88	5.50	0.15	0.10	0.02	0.27	0.18
	40-60	4.05	5.93	0.75	5.30	0.14	0.08	0.02	0.25	0.10
	Depth	CEC[1] effective	CEC potential	O.M.[3]	m[4]	V[5]	Sand	Silt	Clay	Textural class
	cm	$cmol\ kg^{-1}$			%			$g\ kg^{-1}$		
	0-20	3.33	11.5	5	81	6	833	80	87	sandy
Area I	20-40	2.11	8.26	4	81	5	827	83	90	sandy
	40-60	1.98	7.50	3	81	5	812	108	80	sandy
	0-20	1.49	6.69	3	84	4	734	26	240	medium
Area II	20-40	1.14	5.77	2	77	5	713	50	237	medium
	40-60	1.00	5.55	2	75	4	684	66	250	medium

[1]Cation Exchange capacity. [2]Sum of Bases. [3]Organic matter. [4]Al saturation. [5]Base saturation.

The regional climate is classified as subtropical (Cfa) transitioning to tropical (Cfb), according to Köppen. The average annual rainfall and temperature vary between 1,440 and 1,600 mm and 17 and 19°C, respectively. The winter is usually dry with low rainfall and occurrence of frost.

Experimental design

Two areas with commercial pine plantations were selected in 2008, both five years of age. The Jaguariaíva area (area I) showed an undulating topography and a thick litter layer. The Arapoti area (area II) is on uneven land and presents an irregular plant distribution due to the lower pine tree growth. The main disorders typically occur in the upper mid-portion of tree, with the ends of needles yellowing or necrotic. There is more premature plant death and slower growth compared with plants of the same age in other areas, which may be symptoms of severe nutrient deficiency.

Prior to this planting of *Pinus taeda*, currently at the second rotation, there was a *Eucalyptus* sp. plantation that, in turn, replaced the native cerrado vegetation. The two tree species were cultivated without liming and fertilisation. Before the current rotation, both areas were burnt after harvest. In the spring of 2008, both pinus plantations were pruned to a 2.0 m height using a manual saw, and the pruned material was left on the ground.

The experimental design consisted of randomised complete blocks with seven treatments and four replications. The experimental unit (plot) consisted of eight lines, each line with eight plants, totalling 64 plants in an area of 384 m^2. All areas together represent a total area of 10,752 m^2 (approximately 1 ha). The central 16 plants of each plot were considered for measurement and analysis.

The first fertiliser application occurred in October 2008 and the second in January 2010. The fertiliser inputs per 1 ha were: 40 kg nitrogen (89 kg urea, 45% N), 60 kg P_2O_5 (146 kg triple superphosphate, 41% P_2O_5), 80 kg K_2O (133 kg KCl, 60% K_2O), 3 kg zinc (20 kg $ZnSO_4$, 15% Zn), 2 kg boron (20 kg ulexite, 10% of B), 2 kg copper (8.3 kg $CuSO_4$, 24% Cu), and 20 g molybdenum (61 g Na_2MoO_4, 39% Mo).

Lime (1.3 Mg ha^{-1}) was manually broadcasted over litter after fertiliser application. The applied lime quantity was calculated to supply approximately 500 kg Ca, which is the amount exported by *Pinus* sp. in two cycles. The limestone used presented the following characteristics: CaO = 28.9%; MgO = 19.9%; NP (neutralisation power) = 99.4%; RE (reactivity efficiency) = 90%; and RPTN (relative power of total neutralisation) = 89.4%.

To diagnose the nutrient disorders, the following treatments with the omission of nutrients were adopted: (i) complete - application of all nutrients [N, P, K, Zn, Cu, B, Mo, Ca + Mg (dolomitic lime)]; (ii) without macro: application of nutrients without the macronutrients N, P, K; (iii) without micro: application of nutrients without the micronutrients Zn, Cu, B and Mo; (iv) without K: application of nutrients without K; (v) without Zn: application of nutrients without Zn; (vi) without lime: application of all nutrients except for lime (Ca + Mg); and (vii) control: no fertiliser or lime addition.

Sampling and soil analysis

The soil and litter sampling were performed in the same place at the same time. The soil samples were collected at the following depths: 0-5, 5-10, 10-20, 20-40, and 40-60 cm. The soil samples were air dried at room temperature, ground and sieved. Chemical analyses were performed according to the method of Embrapa (1997) for the determination of the following: active acidity (pH in 0.01 mol L^{-1} $CaCl_2$ solution); potential exchangeable acidity (Al^{3+}), extracted with a 1 mol L^{-1} KCl solution and determined by titration with a 0.2 mol L^{-1} NaOH solution; exchangeable Ca^{2+} and Mg^{2+} extracted with a 1 mol L^{-1} KCl solution and determined by atomic absorption spectrometry; and exchangeable K^+ extracted with Mehlich I and determined by flame photometry.

Sampling and analysis of litter and plants

Litter samples were collected from each plot and separated in Ln and La subsamples, according to Babel's (1972) definition: Ln (new litter) is composed of freshly fallen needles, and La (aged litter) consists of needles at the beginning of decomposition. Four points were randomly chosen within the plots. The litter was then marked by a wooden frame of 0.25 x 0.25 m, cut using a blade and removed. The material was collected in August 2010.

The needle samplings were conducted in November 2009 and December 2010. In each plot, five dominant trees were selected, which is equivalent to one hundred trees per hectare with a larger diameter at breast height (DBH). The needle samples were collected from the first and second releases (the second whorl) of the top at the north face.

The samples were dried in a forced-air oven at 60°C up to constant weight and were weighed and ground in a Willey type mill. Then, 1 g was submitted to dry digestion in a muffle furnace at 500°C, and the ashes were suspended in 10 mL of a 3 mol L^{-1} HCl solution, according to method described by Martins and Reissmann (2007). The concentrations of Ca and Mg were determined by atomic absorption spectrophotometry, while the concentration of K was determined by flame photometry.

Statistical analysis

The experiment was arranged in a completely randomised design. All the data reported are average values of four replicates. The data were statistically evaluated using ANOVA, and significant differences between means were compared with the Tukey test normal distribution ($p < 0.05$). The statistical analyses were performed using Assistat 7.9 (SILVA, 2012).

Results and discussion

Influence of liming and fertilisation on pH and levels of Al^{3+} in soil

The soil analysis results for the control treatment revealed that areas I and II were both very acidic, with high levels of exchangeable Al^{3+}. The pH values at a 0-5 cm depth were 3.03 and 3.18 for areas I and II, respectively (Table 2).

Table 2. Values of pH (in 0.01 mol dm^{-3} $CaCl_2$ solution), exchangeable Al^{3+} and Al saturation in soil samples from Jaguariaíva (Area I) and Arapoti (Area II), collected from five soil layer depths, in response to application of lime and fertilizer.

		pH		Al^{3+}		m^1	
Soil depth	Treatments	Area I	Area II	Area I	Area II	Area I	Area II
cm				cmol$_c$ kg^{-1}		%	
	Complete	4.07^{ab}	3.76^{ab}	0.65^{b}	1.35^{b}	11^{b}	42^{b}
	Without macro	4.07^{ab}	3.67^{ab}	0.63^{b}	1.48^{b}	14^{b}	57^{ab}
	Without micro	3.82^{bc}	4.09^{a}	0.73^{b}	0.98^{b}	16^{b}	34^{b}
0-5	Without K	4.15^{ab}	3.75^{ab}	0.50^{b}	1.20^{b}	9^{b}	49^{b}
	Without Zn	4.74^{a}	3.73^{ab}	0.15^{b}	1.38^{b}	2^{b}	43^{b}
	Without lime	3.20^{cd}	2.95^{b}	2.38^{a}	2.85^{a}	73^{a}	78^{a}
	Control	3.03^{d}	3.18^{ab}	2.55^{a}	2.43^{a}	86^{a}	92^{a}
	Complete	3.74^{ab}	3.83^{a}	1.35^{ab}	1.05^{b}	44^{bc}	51^{ab}
	Without macro	4.36^{a}	3.74^{ab}	1.18^{b}	1.30^{ab}	39^{c}	65^{ab}
	Without micro	3.85^{ab}	3.96^{a}	1.40^{ab}	0.93^{b}	48^{abc}	37^{b}
5-10	Without K	3.80^{ab}	3.78^{ab}	1.38^{ab}	1.33^{ab}	45^{bc}	58^{ab}
	Without Zn	4.05^{ab}	3.76^{ab}	1.18^{b}	1.20^{ab}	41^{c}	53^{ab}
	Without lime	3.43^{b}	3.14^{b}	2.18^{a}	2.28^{a}	77^{ab}	88^{a}
	Control	3.41^{b}	3.44^{ab}	2.00^{ab}	1.88^{ab}	81^{a}	92^{a}
	Complete	3.70^{a}	3.84^{a}	1.68^{a}	1.05^{a}	78^{a}	77^{a}
	Without macro	3.72^{a}	3.90^{a}	1.55^{a}	0.93^{a}	72^{a}	86^{a}
	Without micro	3.75^{a}	3.94^{a}	1.68^{a}	0.88^{a}	77^{a}	78^{a}
10-20	Without K	3.67^{a}	3.88^{a}	1.63^{a}	1.00^{a}	81^{a}	86^{a}
	Without Zn	3.89^{a}	3.86^{a}	1.63^{a}	0.93^{a}	71^{a}	84^{a}
	Without lime	3.71^{a}	3.67^{a}	1.93^{a}	1.23^{a}	79^{a}	90^{a}
	Control	3.66^{a}	3.77^{a}	1.73^{a}	0.95^{a}	81^{a}	94^{a}
	Complete	3.57^{a}	3.91^{a}	1.70^{a}	0.93^{a}	79^{a}	75^{a}
	Without macro	3.73^{a}	3.93^{a}	1.65^{a}	0.83^{a}	80^{a}	80^{a}
	Without micro	3.77^{a}	3.91^{a}	1.78^{a}	0.90^{a}	80^{a}	79^{a}
20-40	Without K	3.75^{a}	3.89^{ab}	1.50^{a}	0.88^{a}	79^{a}	83^{a}
	Without Zn	3.90^{a}	3.89^{ab}	1.53^{a}	0.83^{a}	75^{a}	78^{a}
	Without lime	3.78^{a}	3.79^{b}	1.80^{a}	0.98^{a}	79^{a}	79^{a}
	Control	3.73^{a}	3.84^{ab}	1.55^{a}	0.90^{a}	80^{a}	87^{a}
	Complete	3.76^{a}	3.93^{a}	1.73^{a}	0.88^{a}	82^{a}	78^{a}
	Without macro	3.80^{a}	3.94^{a}	1.50^{a}	0.83^{a}	80^{a}	82^{a}
	Without micro	3.78^{a}	3.78^{a}	1.78^{a}	0.93^{a}	81^{a}	81^{a}
40-60	Without K	3.78^{a}	3.80^{a}	2.00^{a}	0.80^{a}	83^{a}	83^{a}
	Without Zn	3.93^{a}	3.93^{a}	1.65^{a}	0.83^{a}	79^{a}	81^{a}
	Without lime	3.81^{a}	3.85^{a}	1.70^{a}	0.95^{a}	80^{a}	83^{a}
	Control	3.75^{a}	3.92^{a}	1.60^{a}	1.00^{a}	81^{a}	85^{a}

[1]Al saturation. Stand parameters with no letters in common represent means that are significantly different ($p < 0.05$).

Soil samples taken close to the experimental area and representing the same soil formation and supporting native vegetation showed similar pH values (Table 1). However, it is well known that pine plantations may acidify the soil (ALEXANDER; CRESSER, 1995; BRANDÃO; LIMA, 2002; INGERSLEV, 1997; LILIENTFEIN et al., 2000; SOUZA; SOUZA, 1981; PADUA et al., 2006).

The lime effect on soil pH was observed in both areas up to the 10 cm depth (Table 2). The results were highly satisfactory, considering the amount of low lime applied, the short period between applications and samplings and the high soil buffer capacity. Supporting this fast alteration under forestry conditions in Norway, Ingerslev (1997) showed changes in pH, Al, CEC and the sum of bases up to a 35 cm depth after eight years of liming. Maeda et al. (2008), studying sandy soils in the state of Paraná, verified decreased soil acidity after one year of surface application of plant biomass ashes up to a 20 cm depth.

More significant changes in pH due to liming were found in area I (pH increased from 3.03 to 4.75) compared with area II (3.18 to 3.76). This result was not expected, as the highest soil potential acidity levels (H + Al) expressed by the buffering power were found in area I up to a 20 cm depth. Other differences between the two soils are the texture and organic matter content (Table 1). Such pH variation may be explained by the changes in exchangeable Al and Al saturation in the first 10 cm of the soil (Table 2).

Although pine species are considered tolerant to Al^{3+} (ALEXANDER; CRESSER, 1995; GONÇALVES, 1995; HÜTTL; ZÖTTL, 1993) some studies have clearly shown negative effects of the element on tree development (NOWAK; FRIEND, 2006).

Liming and fertilisation effects on the soil-exchangeable K^+, Ca^{2+} and Mg^{2+} concentrations

Lime application resulted in higher sum of bases (data not shown) and higher base saturation values (Table 3) within the 0-10 cm layer. The fertiliser alone did not significantly change SB compared with the control treatment. In addition to the direct effect of lime on neutralising exchangeable Al, the consequent pH rise also promotes deprotonation of the organic colloid and mineral surfaces, which are essential to increasing the soil CEC and base saturation.

The base availability under native conditions (Table 3) was far below the minimum recommended for agricultural crops (IAC, 1997) for both areas. The low available K concentration in the control treatment evidenced the poor primary mineral content in the source material (Furnas Sandstone and Itararé). The K application provided significant soil K increments up to the 20 and 40 cm depths for areas I and II, respectively, confirming the high mobility of the soil K. Although the applied amount of K fertiliser was not high, the combined effect of sandy texture, low effective CEC and high precipitation may explain the high K mobility. The present results are corroborated by other studies that showed close relationships between soil characteristics and nutrient movement. Frank and Stuanes (2003) obtained an increase in K^+ levels up to a 22 cm depth in sandy soils with high Al saturation under the cultivation of *Pinus* sp. in Norway.

Table 3. Values of soil K^+, Ca^{2+} and Mg^{2+} concentrations, effective CEC and base saturation (V%) in samples from Jaguariaíva (area I) and Arapoti (area II), at five soil layer depths in response to application of lime and fertilizer.

		K^+		Ca^{2+}		Mg^{2+}		$CEC_{effective}$		V^1	
Soil depth	Treatment	Area I	Area II	Area I	Area II	Area I	Area II	Area I	Area II	Area I	Area II
cm				$cmol_c\ kg^{-1}$				$cmol_c\ kg^{-1}$		%	
	Complete	0.17^{ab}	0.12^{ab}	2.19^{a}	1.29^{a}	3.33^{a}	1.24^{ab}	6.34^{a}	4.00^{a}	51^{ab}	24^{ab}
	Without macro	0.03^{c}	0.04^{b}	1.41^{ab}	0.45^{b}	2.92^{a}	0.69^{ab}	4.98^{ab}	2.50^{a}	45^{ab}	17^{ab}
	Without micro	0.13^{abc}	0.10^{ab}	1.44^{ab}	1.26^{a}	2.55^{a}	1.38^{a}	4.85^{ab}	3.75^{a}	42^{b}	34^{a}
0-5	Without K	0.07^{bc}	0.04^{b}	1.78^{ab}	0.85^{ab}	2.86^{a}	0.86^{ab}	5.22^{ab}	3.25^{a}	47^{ab}	27^{ab}
	Without Zn	0.19^{a}	0.16^{a}	2.38^{a}	0.88^{ab}	3.96^{a}	0.94^{ab}	6.67^{a}	3.25^{a}	61^{a}	26^{a}
	Without lime	0.14^{abc}	0.13^{ab}	0.47^{bc}	0.65^{b}	0.21^{b}	0.51^{ab}	3.19^{b}	4.14^{a}	11^{c}	13^{ab}
	Control	0.03^{c}	0.04^{b}	0.22^{c}	0.09^{c}	0.16^{b}	0.05^{b}	2.96^{b}	2.61^{a}	4^{c}	2^{b}
	Complete	0.15^{a}	0.11^{a}	0.59^{a}	0.52^{ab}	1.04^{ab}	0.53^{ab}	3.13^{a}	2.21^{a}	25^{ab}	16^{ab}
	Without macro	0.03^{b}	0.03^{b}	0.63^{a}	0.28^{ab}	1.46^{a}	0.40^{ab}	3.29^{a}	2.02^{a}	29^{a}	12^{ab}
	Without micro	0.08^{ab}	0.09^{ab}	0.53^{ab}	0.85^{a}	1.04^{ab}	0.95^{a}	3.05^{a}	2.81^{a}	25^{ab}	27^{a}
5-10	Without K	0.03^{b}	0.03^{b}	0.63^{a}	0.50^{ab}	1.15^{ab}	0.50^{ab}	3.17^{a}	2.35^{a}	26^{ab}	15^{ab}
	Without Zn	0.15^{a}	0.13^{a}	0.75^{a}	0.49^{ab}	1.25^{ab}	0.51^{ab}	3.33^{a}	2.32^{a}	31^{a}	18^{ab}
	Without lime	0.11^{ab}	0.12^{a}	0.31^{b}	0.18^{b}	0.21^{b}	0.06^{b}	2.80^{b}	2.63^{a}	10^{ab}	4^{b}
	Control	0.03^{b}	0.03^{b}	0.22^{b}	0.08^{b}	0.21^{b}	0.05^{b}	2.45^{b}	2.03^{a}	7^{b}	2^{b}
	Complete	0.09^{a}	0.09^{a}	0.22^{a}	0.27^{a}	0.16^{a}	0.21^{a}	2.14^{a}	1.61^{a}	25^{a}	12^{a}
	Without macro	0.02^{b}	0.07^{ab}	0.25^{a}	0.16^{ab}	0.31^{a}	0.11^{ab}	2.14^{a}	1.25^{ab}	29^{a}	9^{a}
	Without micro	0.07^{ab}	0.07^{ab}	0.22^{a}	0.22^{ab}	0.21^{a}	0.19^{ab}	2.17^{a}	1.35^{ab}	25^{a}	12^{a}
10-20	Without K	0.01^{b}	0.02^{b}	0.16^{a}	0.17^{ab}	0.21^{a}	0.13^{ab}	2.00^{a}	1.31^{ab}	26^{a}	7^{a}
	Without Zn	0.10^{a}	0.09^{a}	0.22^{a}	0.17^{ab}	0.31^{a}	0.10^{ab}	2.26^{a}	1.27^{ab}	31^{a}	8^{a}
	Without lime	0.11^{a}	0.08^{a}	0.19^{a}	0.10^{bc}	0.21^{a}	0.05^{b}	2.43^{a}	1.45^{ab}	10^{a}	5^{a}
	Control	0.02^{b}	0.02^{b}	0.19^{a}	0.09^{c}	0.21^{a}	0.05^{b}	2.14^{a}	1.10^{b}	7^{a}	6^{a}
		K^+		Ca^{2+}		Mg^{2+}		$CEC_{effective}$		$V^{(1)}$	
Soil depth	Treatment	Area I	Area II	Area I	Area II	Area I	Area II	Area I	Area II	Area I	Area II
cm				$cmol_c\ kg^{-1}$				$cmol_c\ kg^{-1}$		%	
	Complete	0.08^{a}	0.07^{ab}	0.16^{a}	0.16^{a}	0.21^{a}	0.08^{a}	2.14^{a}	1.24^{a}	11^{a}	10^{a}
	Without macro	0.01^{ab}	0.01^{b}	0.22^{a}	0.14^{a}	0.21^{a}	0.06^{ab}	2.08^{a}	1.03^{a}	8^{a}	6^{a}
	Without micro	0.05^{ab}	0.05^{ab}	0.19^{a}	0.13^{a}	0.21^{a}	0.07^{ab}	2.22^{a}	1.14^{a}	8^{a}	6^{a}
20-40	Without K	0.01^{ab}	0.01^{b}	0.19^{a}	0.12^{a}	0.21^{a}	0.06^{ab}	1.90^{a}	1.05^{a}	8^{a}	5^{a}
	Without Zn	0.07^{ab}	0.07^{ab}	0.22^{a}	0.12^{a}	0.21^{a}	0.05^{ab}	2.02^{a}	1.07^{a}	10^{a}	6^{a}
	Without lime	0.05^{ab}	0.11^{a}	0.22^{a}	0.09^{a}	0.21^{a}	0.05^{ab}	2.27^{a}	1.22^{a}	10^{a}	6^{a}
	Control	0.00^{b}	0.01^{ab}	0.19^{a}	0.09^{a}	0.21^{a}	0.03^{b}	1.95^{a}	1.03^{a}	7^{a}	4^{a}
	Complete	0.02^{ab}	0.05^{ab}	0.16^{a}	0.14^{a}	0.21^{a}	0.06^{a}	2.11^{ab}	1.13^{a}	9^{a}	7^{a}
	Without macro	0.01^{ab}	0.05^{a}	0.16^{a}	0.13^{a}	0.21^{a}	0.06^{a}	1.87^{b}	1.01^{a}	7^{a}	6^{a}
	Without micro	0.04^{a}	0.06^{a}	0.16^{a}	0.12^{a}	0.21^{a}	0.06^{a}	2.17^{ab}	1.16^{a}	8^{a}	6^{a}
40-60	Without K	0.01^{ab}	0.05^{a}	0.19^{a}	0.11^{a}	0.21^{a}	0.05^{a}	2.40^{a}	0.96^{a}	9^{a}	5^{a}
	Without Zn	0.03^{ab}	0.06^{a}	0.19^{a}	0.11^{a}	0.21^{a}	0.04^{a}	2.07^{ab}	1.03^{a}	7^{a}	6^{a}
	Without lime	0.02^{ab}	0.07^{a}	0.19^{a}	0.08^{a}	0.21^{a}	0.04^{a}	2.12^{ab}	1.14^{a}	9^{a}	5^{a}
	Control	0.00^{b}	0.07^{a}	0.16^{a}	0.07^{a}	0.21^{a}	0.04^{a}	1.96^{b}	1.18^{a}	6^{a}	6^{a}

[1]V = Base saturation. Stand parameters with no letters in common represent means that are significantly different ($p < 0.05$).

Available Ca concentration increments were observed up to the 5 cm (area I) and 20 cm (area II) layers (Table 3). The Ca availability increased below a 5 cm depth for area II, suggesting a higher mobility under this condition than area I. The low Ca backgrounds indicated the influence of broadcast lime application, the observed Ca increases can be considered low. By comparing the complete treatment to that without macronutrients and without lime to control, it was possible to verify that fertiliser enhanced Ca availability in the 0-5 cm layers. This result was not expected, as the only extra source of Ca was triple superphosphate, with approximately 10% of Ca [$Ca(H_2PO_4)_2 \cdot 2H_2O$]; that is, the Ca input was much smaller than that of lime. The Ca increment difference between soil areas was not expected either, as both received the same amount of lime, but a higher available Ca concentration was found in area I than area II.

Ingerslev (1997), after eight years of limestone application in an area with conifers, found differences in the soil Ca^{2+} and Mg^{2+} concentrations up to a 35 cm depth. Frank and Stuanes (2003), in a four-year study, observed increased Ca^{2+} concentrations up to a 22 cm depth in sandy soils under coniferous planting. This result can be compared with this research, as the differences were significant up to the 20 cm depth, in less than two years.

Significant effects of lime application on available Mg^{2+} concentrations were also observed up to the 10 cm depth in area I and the 40 cm depth in area II. Similarly to the Ca responses, differences in Mg concentrations were observed between the complete treatment and the other treatments with lime. Also, higher available Mg concentrations were found in area I than in area II (Tables 1 and 3). Again, there is no explanation for the differences found in available Mg concentrations between treatments that received the same amount of Mg from lime.

It is important that there was a greater increase in Mg^{2+} than in Ca^{2+} concentrations with soil profile depth, despite the two nutrients being derived from the same corrective. According to Hüttl and Zöttl (1993), when both nutrients come from dolomitic lime, as in the present study, the Mg^{2+} is generally more available than Ca^{2+}, which may be explained by the higher $MgCO_3$ (0.106 g L^{-1}) solubility than $CaCO_3$ (0.014 g L^{-1}) in the limestone (the first is 7.6 times more soluble).

The symptoms of Mg^{2+}deficiency in forest plantations and the study of appropriate rates of Mg fertiliser have been investigated since the 1980s in various regions of Germany (EVERS; HÜTTL, 1990; HÜTTL, 1989; HÜTTL; ZÖTTL, 1993; UEBEL; HEINSDORF, 1997). In a review paper on conifer plantations in Germany, Hüttl and Zöttl (1993) concluded that the application of Mg^{2+} through liming was effective at increasing the Mg level in the soil and reversing the nutrient deficiency symptoms. However, Uebel and Heinsdorf (1997), studying the application of fertilisers containing Mg^{2+} and K^{+} in sandy soils under *Pinus* sp., observed no significant improvement in the two nutrient availability levels in the soil, despite the increased productivity. Nevertheless, the Mg^{2+} dynamics in soils under forest plantations are still unknown, and there are few answers about its addition to the soil, especially in soils under coniferous plantations in regions with warmer climates.

The responses achieved in the present study allow inferring that changes in the levels of K^{+}, Ca^{2+} and Mg^{2+} observed at even the most superficial depths (less than 10 cm) are important. The finest roots, responsible for most nutrient absorption, occupy the first 20 cm, including litter or the organic horizon. Thus, the surface application of fertilisers in forest areas can be considered an excellent procedure to correct low nutrient levels in the soil, even in established plantings.

Distribution of K^{+}, Ca^{2+} and Mg^{2+} in the new litter (Ln) and aged litter (La) in response to liming and fertilisation

The low K^{+}, Ca^{2+}, and Mg^{2+} concentrations found in the litter from the control treatment (Table 4) confirmed their low availability in the biogeochemical cycle of the studied areas. The obtained values were lower than those observed by Brum (1980) in *Pinus elliottii* plantations in northern Santa Catarina. Brum (1980) reported K^{+} and Ca^{2+} levels of approximately 0.6 and 2.5 g kg^{-1}, respectively. Even more expressive differences are observed in comparison with the data reported by Viera and Schumacher (2010), who found Ca^{2+} concentrations of 7.88 g kg^{-1} in the litter of *Pinus taeda* grown in basaltic soils of Rio Grande do Sul State.

Table 4. Concentrations of K^{+}, Ca^{2+}, and Mg^{2+} in the new litter (Ln) and aged litter (La) from Jaguariaíva (area I) and Arapoti (area II), in response to application of lime and fertilizers.

		K		Ca		Mg	
	Treatment	Ln	La	Ln	La	Ln	La
				g kg^{-1}			
Area I	Complete	1.18 ᵃ	0.53 ᵃᵇ	1.95 ᵃ	20.46 ᵃ	1.23 ᵃᵇ	8.90 ᵃ
	Without Macro	0.38 ᵇ	0.25 ᶜ	2.45 ᵃ	17.82 ᵃ	1.64 ᵃ	9.97 ᵃ
	Without micro	0.83 ᵃᵇ	0.45 ᵃᵇᶜ	2.25 ᵃ	18.51 ᵃ	1.23 ᵃᵇ	10.13 ᵃ
	Without K	0.50 ᵇ	0.40 ᵃᵇᶜ	2.39 ᵃ	13.38 ᵃ	1.50 ᵃᵇ	6.68 ᵇ
	Without Zn	1.13 ᵃ	0.43 ᵃᵇᶜ	1.80 ᵃ	19.81 ᵃ	1.35 ᵃᵇ	7.43 ᵃ
	Without lime	1.08 ᵃ	0.58 ᵃ	1.23 ᵃ	2.79 ᵇ	0.55 ᵇ	0.30 ᵇ
	Control	0.35 ᵇ	0.33 ᵇᶜ	1.33 ᵃ	1.28 ᵇ	0.76 ᵃ	0.40 ᵇ
Área II	Complete	1.20 ᵃ	0.60 ᵃᵇᶜ	2.45 ᵃᵇ	18.29 ᵃ	1.08 ᵃᵇ	8.81 ᵃ
	Without Macro	0.30 ᵇ	0.28 ᶜᵈ	2.41 ᵃᵇ	13.35 ᵃᵇᶜ	1.18 ᵃ	7.83 ᵃ
	Without micro	1.23 ᵃ	0.60 ᵃᵇᶜ	1.58 ᵃᵇᶜ	16.97 ᵃᵇ	0.78 ᵃᵇᶜ	8.72 ᵃ
	Without K	0.45 ᵇ	0.40 ᵇᶜᵈ	2.84 ᵃ	8.86 ᵃᵇᶜ	1.19 ᵃ	5.43 ᵃᵇ
	Without Zn	1.33 ᵃ	0.83 ᵃ	1.55 ᵃᵇᶜ	6.93 ᵃᵇᶜ	0.69 ᵃᵇᶜ	4.46 ᵃᵇ
	Without lime	1.33 ᵃ	0.70 ᵃᵇ	0.83 ᵇᶜ	1.19 ᶜ	0.41 ᵇᶜ	0.34 ᵇ
	Control	0.35 ᵇ	0.25 ᵈ	0.48 ᶜ	3.01 ᵇᶜ	0.33 ᶜ	1.75 ᵇ

Stand parameters with no letters in common represent means that are significantly different ($p < 0.05$).

The results (Table 4) indicate a distinct behaviour between the K^{+} and Ca^{2+} plus Mg^{2+} concentrations in the litter from the same treatment. Thus, the litter samples from the complete treatment showed higher K^{+} levels in more recently fallen needles (Ln) than aged needles (La), suggesting K leaching by rainfall. However, the low-litter K concentrations were found in the litter samples from treatments without the addition of macronutrients or K and control, and the expected decrease in aged-needle K concentrations was either unclear or did not happen.

In the complete treatment, however, significantly lower concentrations of Ca^{2+} were observed in the new litter, Ln (1.95 and 2.45 g kg^{-1}, in areas I and II, respectively), than in the aged litter, La (20.46 and 18.29 g kg^{-1}, in areas I and II, respectively). Similar results were obtained for Mg^{2+}: 1.23 and 1.08 g kg^{-1} for Ln in areas I and II, respectively, and 8.90 and 8.81 for g kg^{-1} for La in areas I and II, respectively. The increase in Ca and Mg concentrations during the aging process may be due to these elements not being easily lixiviated. Nevertheless, different limestone effects on Ca concentrations were observed between the two areas. In area I, significantly higher Ca concentrations with lime addition were only observed in the aged litter (La). In area II, significantly higher Ca concentrations with lime addition were found in both the La and Ln litter samples. Similar behaviour was observed for Mg. From these results, it might be inferred that the thick litter layer retained non-reacted lime in the organic layer and restricted the lime contact with soil, with consequent lower Ca and Mg availability to the soil-plant system. This hypothesis was supported by visual observation during the litter sampling. Another hypothesis is that part of lime reacted but that the released Ca and Mg remained bound to the litter exchange sites, becoming unavailable.

Effect of treatments on the needle K^+, Ca^{2+}, and Mg^{2+} concentrations in the first and second flushes during the years 2009 and 2010

A clear positive effect of K fertiliser on the needle K concentration was observed for both flushes and years at area I (Table 5). At area II, this effect was significant for the year 2010. These results confirm the soil K availability enhancement (Table 2). Also, lime omission resulted in higher needle K concentration in the first flush at area I due to the antagonic effect of increased Ca and Mg on decreasing K uptake by pine plants.

Table 5. Concentrations of K^+, Ca^{2+} and Mg^{2+} in the first and second needle flushes, at Jaguariaíva (area I) and Arapoti (area II), during 2009 and 2010, in response to lime and fertilizer application.

		K		Ca		Mg	
	Treatment	2009	2010	2009	2010	2009	2010
		g kg^{-1}					
Area I 1° L[1]	Complete	3.74^{a}	3.61^{a}	0.82^{a}	2.02^{a}	0.61^{ab}	0.50^{a}
	Without Macro	2.74^{b}	2.35^{bc}	1.06^{a}	2.38^{a}	1.01^{a}	0.65^{a}
	Without micro	3.84^{a}	3.77^{a}	0.58^{a}	1.91^{a}	0.52^{b}	0.42^{a}
	Without K	2.54^{b}	2.17^{c}	0.91^{a}	2.18^{a}	0.87^{ab}	0.57^{a}
	Without Zn	3.60^{a}	3.59^{ab}	0.67^{a}	1.81^{a}	0.55^{b}	0.38^{a}
	Without lime	3.89^{a}	4.39^{a}	0.85^{a}	1.52^{a}	0.59^{ab}	0.30^{a}
	Control	2.42^{b}	2.24^{c}	0.66^{a}	2.23^{a}	0.62^{ab}	0.55^{a}
Area I 2⁰ L[2]	Complete	3.74^{ab}	3.94^{a}	1.00^{abc}	2.12^{a}	0.92^{ab}	0.64^{ab}
	Without Macro	2.69^{bc}	2.55^{b}	1.37^{a}	2.78^{a}	1.16^{a}	0.75^{a}
	Without micro	4.09^{a}	3.87^{a}	0.64^{bc}	2.08^{a}	0.56^{b}	0.56^{ab}
	Without K	2.64^{bc}	2.49^{b}	1.31^{ab}	2.58^{a}	1.17^{a}	0.84^{a}
	Without Zn	3.27^{abc}	4.13^{a}	0.87^{abc}	2.53^{a}	0.67^{ab}	0.64^{ab}
	Without lime	3.66^{ab}	4.01^{a}	0.50^{c}	2.53^{a}	0.52^{b}	0.31^{b}
	Control	2.34^{c}	2.49^{b}	0.77^{abc}	2.80^{a}	0.72^{ab}	0.49^{ab}
Area II 1° L[1]	Complete	3.04^{a}	3.34^{ab}	2.21^{a}	1.49^{a}	0.60^{a}	0.52^{a}
	Without Macro	2.37^{a}	2.07^{c}	2.40^{a}	1.09^{a}	0.76^{a}	0.49^{a}
	Without micro	3.37^{a}	3.62^{a}	1.86^{a}	0.70^{a}	0.45^{a}	0.39^{a}
	Without K	2.52^{a}	1.97^{c}	1.95^{a}	1.21^{a}	0.61^{a}	0.47^{a}
	Without Zn	3.14^{a}	2.85^{b}	2.14^{a}	1.29^{a}	0.54^{a}	0.53^{a}
	Without lime	3.39^{a}	3.11^{ab}	1.86^{a}	0.46^{a}	0.39^{a}	0.27^{a}
	Control	2.54^{a}	1.90^{c}	2.15^{a}	0.91^{a}	0.57^{a}	0.42^{a}
Area II 2⁰ L[2]	Complete	2.81^{a}	3.58^{a}	2.18^{a}	1.66^{a}	0.50^{a}	0.57^{a}
	Without Macro	2.37^{a}	2.27^{b}	2.31^{a}	1.41^{a}	0.77^{a}	0.76^{a}
	Without micro	2.94^{a}	3.47^{a}	2.04^{a}	1.51^{a}	0.49^{a}	0.52^{a}
	Without K	2.27^{a}	2.19^{b}	2.24^{a}	1.32^{a}	0.67^{a}	0.55^{a}
	Without Zn	2.69^{a}	3.26^{a}	2.14^{a}	1.15^{a}	0.45^{a}	0.38^{a}
	Without lime	2.94^{a}	3.61^{a}	2.05^{a}	0.92^{a}	0.48^{a}	0.31^{a}
	Control	2.32^{a}	2.21^{b}	2.07^{a}	0.87^{a}	0.59^{a}	0.40^{a}

[1]First flush. [2]Second flush. Stand parameters with no letters in common represent means that are significantly different ($p < 0.05$).

However the observed K values were lower than the average K concentration (6 g kg^{-1}) suggested as adequate for pine trees by several authors (REISSMANN; WISNIEWSKI, 2004; VOGEL, 2003). Apparently, K fertilisation was not enough to reach the satisfactory value (6.0 g kg^{-1}), despite the high soil K concentration observed in the 0-20 cm layer. Nevertheless, the abovementioned authors indicated an adequate K concentration for pine trees without taking into consideration the sampling time and species. Chaves and Correa (2005) found no K deficiency symptoms in *Pinus caribeae* with needle K of 3.8 mg kg^{-1}.

Although many authors did not relate K^+ with yield (CHAVES; CORRÊA, 2005; HÜTTL; ZÖTTL, 1993), studies with *Pinus* sp. in the second plateau of Paraná, indicated that K is one of the most growth-limiting nutrients (REISSMANN; WISNIEWSKI, 2004). The present results indicate that more research is needed to better evaluate the effects of K on pinus nutrition and development. However, a low needle K concentration in the control treatment and a consistent increasing needle K concentration in fertilised treatments were observed. These results suggest that there must be concern over pinus tree demand for K and that this nutrient may be responsible for the deficiency symptoms observed in the field.

Reissmann and Wisniewski (2004) observed higher demand for K^+ than for Ca^{2+} in this species. The authors concluded that plots with poor quality soil showed K Ca^{-1} ratios close to 1, while the best areas showed K Ca^{-1} ratios above 3. In this study, the K Ca^{-1} ratios were observed to be close to 1, except for some results where the ratios approached 3 in response to the treatments. Low ratios confirmed plant K deficiency.

In contrast, significant differences in the Ca concentrations appeared only in the second flush in 2009 (Table 5), despite their increase in the superficial soil layer (Table 3).

In some samples, the needle Ca concentration was below 1.0 or 1.5 g kg^{-1}, values that are considered adequate for pine trees by Zöttl and Tschinkel (1971) and Allen (1987), respectively. However, the needle Ca concentration was above 1 g kg^{-1} in the control treatment, where the natural soil Ca^{2+} levels were very low, indicating the influence of other factors on Ca uptake by pine trees. Furthermore, Chaves and Corrêa (2005), studying plantations of *Pinus caribeae* in the Cerrado of Minas Gerais State, Brazil, found visual symptoms of mineral deficiency followed by early death in plants with Ca^{2+} ranging from 0.1 to 0.2 g kg^{-1}. Meanwhile normal plants showed needle Ca concentrations ranging from 0.6 to 2.0 g kg^{-1}. Therefore, the results obtained in the present research do not allow certainty that the Ca deficiency occurred under this experimental condition.

The needle Mg concentrations obtained in this study were consistently lower than 0.8 and 0.6 g kg^{-1}, values indicated as adequate for pine trees by Zöttl and Tschinkel (1971) and Reismann and Wisniewski (2004), respectively. Significant differences for this nutrient were only observed in area I. The highest needle Mg^{2+} concentration was observed in the first and second flushes in the plots without macro in 2009 and in the second flush in 2010 (Table 5). The K omission in this treatment may have contributed to the increased Mg^{2+} absorption. The needle chemical analysis confirmed the low Mg concentrations previously observed in the soil and litter samples, indicating that Mg deficiency was much more plausible than Ca in this study.

In general, there was close agreement between the first and second needle flushes for predicting Ca, Mg and K nutritional status. The large and irregular span between years for the Ca results did not allow a conclusion to be made, and further studies are required. The distinct movements of nutrients in plant tissues evidenced the well-known high K and Mg mobility in the plants compared with Ca (REISSMANN; WISNIEWSKI, 2004), which explains the high K/Ca ratio in young tissues.

The present results indicate that low levels of soil exchangeable bases, high acidity and Al^{3+} saturation were partially corrected in a short-term experiment with *Pinus taeda* plantations. However, more research is needed and long-term experiments must be performed to better evaluate the effects of increasing fertilisation and liming rates on both soil and plant nutrient concentrations. Specific lime and fertiliser calibration curves for pine forest plantations should be investigated to ensure the maximum positive effects of this type of crop management in pine tree productivity.

Conclusion

Potassium concentration increments in soils, plants and litter were more evident than those of other macronutrients in response to fertilisation. These responses suggest that surface fertilisation may be effective to correct K disorders in pine trees. However, the same cannot be said for Ca and Mg, as their needle concentrations showed irregular responses to the treatments.

There is no doubt about the need to provide nutrients via liming and fertilisation, even in forest plantations. This management is essential to maintaining the quality of soils and plants. However, more studies such as analysis of height increments and diameter are essential to evaluate the efficiency of liming and fertilisation on reversing the nutritional deficiency and improving the low growth rates in sites with these kinds of problems. A long period between application and sampling is needed for a more accurate conclusion.

Acknowledgements

We thank Valor Florestal, and its technicians Renato Teixeira Lima, Antônio Esteves Lemes and Fabiana Vargas Reis for their support. We also thank to professor Angela Furlani for helping us in translating the material, the researcher Edilson Batista de Oliveira for all the help in this search and even the Federal University of Paraná for all support to achieve work.

References

ALEXANDER, C. E.; CRESSER, M. S. An assessment of the possible impact of expansion of native woodland on the chemistry of Scottish freshwater. **Forest Ecology Management**, v. 73, n. 1, p. 1-27, 1995.

ALLEN, H. L. Forest fertilizers: nutrient amendment, stand productivity and environmental impact. **Journal of Forestry**, v. 85, n. 2, p. 37- 46, 1987.

BABEL, U. **Moderprofile in waldern**. Ulmen: Stuttgart, 1972.

BRANDÃO, S. L.; LIMA, S. C. pH e condutividade elétrica de *Pinus* e cerrado na Chapada, em Uberlândia (MG). **Caminhos de Geografia**, v. 3, n. 1, p. 46-54, 2002.

BRITO, J. O.; FERRAZ, E. S. B.; BARRICHELO, L. E.; COUTO, H. T. Z. A adubação mineral e seus efeitos sobre os anéis de crescimento da madeira de *Pinus caribaea* VAR. *bahamensis*. **IPEF**, v. 32, n. 1, p. 5-17, 1986.

BRUM, E. Composição química de acículas verdes e da litteira de *Pinus elliottii* Engelm. **Revista Floresta**, v. 11, n. 1, p. 17-23, 1980.

CHAVES, R. Q.; CORRÊA, G. F. Macronutrientes no sistema solo - *Pinus caribaea* Morelet em plantios apresentando amarelecimento das acículas e morte das plantas. **Revista Árvore**, v. 29, n. 5, p. 691-700, 2005.

EMBRAPA-Empresa Brasileira de Pesquisa Agropecuária. **Manual de análises químicas de solos, plantas e fertilizantes**. Brasília: Embrapa Comunicação para Transferência de Tecnologia, 1997.

EVERS, F. H.; HÜTTL, R. F. A new fertilization strategy in declining forests. **Water, Air and Soil Pollution**, v. 54, n. 1, p. 495-508, 1990.

FERREIRA, C. A.; SILVA, H. D.; REISSMANN, C. B.; BELLOTE, A. F. J.; MARQUES, R. **Nutrição de *Pinus* no sul do Brasil, diagnóstico e prioridades de pesquisa**. Colombo: Embrapa Florestas, 2001.

FRANK, J.; STUANES, A. O. Short-term effects of liming and nutrient leaching in a Scots pine ecosystem in

Norway. **Forest Ecology Management**, v. 176, n. 1-3, p. 371-386, 2003.

GONÇALVES, J. L. M. Recomendações de adubação para *Eucalyptus*, *Pinus* e espécies típicas da Mata Atlântica. **Documentos Florestais**, v. 15, n. 1, p. 1-23, 1995.

HÜTTL, R. F. Liming and fertilization as mitigation tools in declining forest ecosystems. **Water, Air and Soil Pollution**, v. 44, n. 1, p. 93-118, 1989.

HÜTTL, R. F.; ZÖTTL, H. W. Liming as a mitigation tool in Germany's declining forests- reviewing results from former and recent trials. **Forest Ecology Management**, v. 61, n. 3-4, p. 325-338, 1993.

IAC-Instituto Agronômico de Campinas. **Recomendação de adubação e calagem para o Estado de São Paulo**. Campinas: Instituto Agronômico; Fundação IAC, 1997.

INGERSLEV, M. Effects of liming and fertilization on growth, soil chemistry and soil water chemistry in a Norway spruce plantation on a nutrient-poor soil in Denmark. **Forest Ecology Management**, v. 92, n. 1, p. 55-66, 1997.

LILIENFEIN, W. W.; AYARZA, M. A.; VILELA, L.; LIMA, S. C.; ZECH, W. Soil acidification in *Pinus caribaea* forests on Brazilian savanna Oxisols. **Forest Ecology Management**, v. 73, n. 1, p. 1-27, 2000.

MAEDA, S.; SILVA, H. D.; CARDOSO, C. Resposta de *Pinus taeda* à aplicação de cinza de biomassa vegetal em Cambissolo Húmico, em vaso. **Pesquisa Florestal Brasileira**, v. 56, n. 1, p. 43-52. 2008.

MARTINS, A. P. L.; REISSMANN, C. B. Material vegetal e as rotinas laboratoriais nos procedimentos químico-analíticos. **Scientia Agraria**, v. 8, n. 1, p. 1-17, 2007.

MELLO, M. S.; MORO, R. S.; GUIMARÃES, G. B. **Patrimônio natural dos campos gerais**. Ponta Grossa: UEPG, 2003.

NOWAK, J.; FRIEND, A. Loblolly pine and slash pine responses to acute aluminum and acid exposures. **Tree Physiology**, v. 26, n. 9, p. 1207-1215, 2006.

PADUA, T. R. P.; SILVA, C. A.; MELO, C. A. Calagem em latossolo sob influência de coberturas vegetais: neutralização da acidez. **Revista Brasileira de Ciência do Solo**, v. 30, n. 5, p. 869-878, 2006.

REISSMANN, C. B.; WISNIEWSKI, C. Nutricional aspects of pine plantation. In: GONÇALVES, J. L. M.; BENEDETTI, V. (Org). **Forest nutrition and fertilization**. Piracicaba: IPEF, 2004. p. 141-170.

REISSMANN, C. B.; ZÖTTL, H. W. Problemas nutricionais em povoamentos de *Pinus taeda* em áreas do arenito da formação Rio Bonito-Grupo Guatá. **Revista Ciência Agrária**, v. 9, n. 1, p. 75-80, 1987.

SCHUMACHER, M. V.; VIERA, M.; WITSCHORECK, R. Produção de serapilheira e tranferência de nutrientes em área de segunda rotação com floresta de *Pinus taeda* no município de Cambará do Sul, RS. **Ciência Florestal**, v. 18, n. 4, p. 471- 480, 2008.

SILVA, F. A. **ASSISTAT versão beta**. Grande: Assistência Estatística, Departamento de Engenharia Agrícola do CTRN – Universidade Federal de Campina Grande, Campus de Campina. Available from: <http://www.assistat.com>. Access on: Aug. 5 2012.

SOUZA, D. M. P.; SOUZA, M. L. P. Alterações provocadas pelo florestamento de *Pinus sp.* na fertilidade de solos da região da Lapa-PR. **Revista Floresta**, v. 12, n. 1, p. 36-50, 1981.

UEBEL, E.; HEINSDORF, D. Results of long-term K and Mg fertilizer experiments in afforestation. **Forest Ecology Management**, v. 91, n. 1, p. 47-52, 1997.

VIERA, M.; SCHUMACHER, M. V. Teores e aporte de nutrientes na serapilheira de *Pinus taeda* L., e sua relação com a temperatura do ar e pluviosidade. **Revista Árvore**, v. 34, n. 1, p. 85-94, 2010.

VOGEL, H. L. M. Crescimento de *Pinus taeda*, relacionado a doses de N, P, K e sua diagnose nutricional pelo DRIS. R. **Ciência Florestal**, v. 13, n. 1, p. 182-188, 2003.

ZÖTTL, H. M.; TSCHINKEL, H. **Nutrición y fertilización forestall**: una guía práctica. Medellin: Universidade Nacional de Colombia, 1971.

Modeling of soil penetration resistance using statistical analyses and artificial neural networks

Fábio Lúcio Santos[1*], Valquíria Aparecida Mendes de Jesus[2] and Domingos Sárvio Magalhães Valente[1]

[1]Departamento de Engenharia Agrícola, Universidade Federal de Viçosa, Av. P. H. Rolfs, s/n, 36570-000, Viçosa, Minas Gerais, Brazil.
*[2]Programa de Pós-graduação em Fitotecnia, Universidade Federal de Viçosa, Viçosa, Minas Gerais, Brazil. *Author for correspondence.*
E-mail: fabio.ls@ufv.br

ABSTRACT. An important factor for the evaluation of an agricultural system's sustainability is the monitoring of soil quality via its physical attributes. The physical attributes of soil, such as soil penetration resistance, can be used to monitor and evaluate the soil's quality. Artificial Neural Networks (ANN) have been employed to solve many problems in agriculture, and the use of this technique can be considered an alternative approach for predicting the penetration resistance produced by the soil's basic properties, such as bulk density and water content. The aim of this work is to perform an analysis of the soil penetration resistance behavior measured from the cone index under different levels of bulk density and water content using statistical analyses, specifically regression analysis and ANN modeling. Both techniques show that soil penetration resistance is associated with soil bulk density and water content. The regression analysis presented a determination coefficient of 0.92 and an RMSE of 0.951, and the ANN modeling presented a determination coefficient of 0.98 and an RMSE of 0.084. The results show that the ANN modeling presented better results than the mathematical model obtained from regression analysis.

Keywords: modeling, soil physical properties, neural networks.

Introduction

An important factor for the evaluation of an agricultural system's sustainability is the monitoring of soil quality via its physical attributes. The monitoring of these attributes can result in better quality agricultural products, the promotion of more efficient mechanization processes and the establishment of the reasonable use of raw materials and natural resources (BEUTLER et al., 2001). Penetration resistance is a physical attribute of soil that can be used to monitor and evaluate soil quality (ISLAM; WEIL, 2000).

Penetration resistance influences the growth of roots, and it can be used as a parameter for evaluating the effects of tillage systems on the roots' environment, the detection of compacted layers, the prediction of the traction force needed to perform mechanized processes and the prevention of the appearance of a physical barrier that can be reduce the development of the plants

(CAMPANHARO et al., 2009; CUNHA et al., 2002).

The determination of the soil penetration resistance is performed a device called penetrometer, which allows the soil resistance to be measured quickly (TAVARES FILHO; RIBON, 2008). According to Dexter et al. (2007), the resistance to penetration is governed by fundamental properties of the soil, such as shear strength, compressibility and the friction force from the soil-metal interaction during the trial using the penetrometer. Hence, soil penetration resistance can be estimated as a quantity called cone index. This quantity can be expressed as the ratio of force per unit area of the base of the cone at a determined depth (CAMPANHARO et al., 2009; CUNHA et al., 2002).

Studies have been carried out to evaluate the influence of water content on the behavior of soil penetration resistance (CUNHA et al., 2002). Mathematical models have also been developed to predict the penetration resistance from basic soil properties, such as soil composition, bulk density and water content (CUNHA et al., 2002; DEXTER et al., 2007; SINGH; KAY, 1997). In this context, the use of Artificial Neural Networks (ANN) can be considered an alternative approach for predicting soil penetration resistance from soil bulk density and water content.

ANN have been employed to solve many problems in agriculture (ERZIN et al., 2008, 2010; KIM; GILLEY, 2008). Varella et al. (2002) used ANN for the determination of land cover from digital images. Khazaei and Daneshmandi (2007) used ANN to model the drying kinetics of sesame seeds. They concluded that the ANN technique presented better results than traditional mathematical modeling. Sarmadian et al. (2009) used ANN to model soil properties, and the results were better than the multivariate regression analysis, showing the effectiveness of the ANN technique.

The objective of this work is to determine the effect of the soil bulk density and water content on soil penetration resistance behavior measured from the cone index, using statistical analyses, specifically regression analysis, and ANN modeling.

Material and methods

Experimental procedure

The study was conducted at Cidade Gaucha, which is located in northwest Paraná State, Brazil. Samples were collected at a location where the predominant soil is classified as Rhodic Acrustox (EMBRAPA, 1999). The samples were collected at a depth of 0.10 m using steel cylinders in three areas that had different levels of management and, therefore, presented different levels of soil compaction. Thus, 12 samples were collected at each point, totaling 36 samples.

First, preliminary tests were conducted in an oven for the samples related to each area to establish the average bulk density of the soil, the moisture saturation and the water loss gradient. Three replicates were used for the preliminary tests for each sampled area, totaling nine samples. The samples were saturated for 48 hours and then were placed in an oven at 105°C. The samples were removed from the oven at intervals of 30 minutes to check the weight loss.

A penetrometer was used to determine the soil penetration resistance. The penetrometer had a 4 mm diameter rod and a load cell of 200 kgf. Readings were taken at intervals of 1 second during the probe penetration into the soil sample. The penetration resistance was obtained from the average of the points obtained during the test. At the end of the tests, the soil densities of the samples were determined.

From the experimental procedure described, a dataset was obtained considering three replications for three average soil bulk densities (1.75, 1.90 and 2.05 kg dm^{-3}) and three average water content levels (0.04, 0.08 and 0.12 kg kg^{-1}).

Statistical analyses

From the experimental results, a model was chosen considering the soil penetration resistance as a function of the soil bulk density and water content. A completely randomized factorial design with three densities was chosen, in which the evaluated factors were composed of three levels of bulk density and three levels of water content. The soil penetration resistance data were submitted to an analysis of variance at a 5% significance level. The effects of the soil bulk density and water content were studied by regression analysis.

All statistical analyses were performed using the SAS program, version 8.0. The model was chosen based on the coefficient of determination, the significance of regression coefficients and the lack of adjustment of the model.

Artificial neural network modeling

According to Haykin (1999), Artificial Neural Networks (ANN) are massively parallel networks, are self-adaptive and are interconnected by basic structures called neurons. Neurons are processing units with limited learning capacity; however, their interactions allow the ANN to learn from a determined set of input data and their output patterns. Figure 1 illustrates an ANN architecture, which is composed of an input layer, a processing layer (also known as a hidden layer) and an output layer. This type of architecture is called a "Multilayer Perceptron Network".

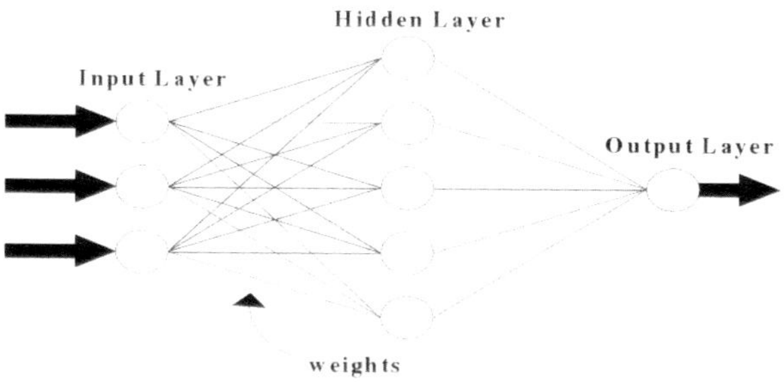

Figure 1. Multilayer Perceptron Network.

In this work, a model was developed based on the ANN technique to predict the soil penetration resistance using the soil bulk density and water content as the input data. The ANN modeling was composed of two stages: training and validation. In the training stage, architectures were considered that consisted of 2-n1-n2-1, where there were 2 elements in input vector and n1 and n2 represented the number of neurons in each hidden layer, with just one neuron in the output layer. Several configurations were tested in the ANN hidden layer, where the number of neurons in first hidden layer (n1) ranged from 1 to 15 and the number of neurons in second layer ranged (n2) from 0 to 15, totaling 240 architectures analyzed. Out of these architectures, 15 were composed of one hidden layer ANN when the number of neurons in layer n1 ranged from 1 to 15 and the number of neurons was equal to zero in layer n2.

An error backpropagation algorithm was used in the training stage. The data set was divided into a training set and a validation set. The training set consisted of 20 input and output patterns, where the input vector was composed of values of the soil bulk density and water content and the output consisted of the soil penetration resistance. To improve the ANN generalization capability, the output data were normalized, which allowed output values ranging from 0 to 1, according to Equation1.

$$PR_N(y) = 1 - \left(\frac{PR(y)}{PR_{max}} \right) \quad (1)$$

where:

$PR_N(y)$ = normalized penetration resistance;

$PR(y)$ = penetration resistance to be normalized;

PR_{max} = maximum value of the soil penetration resistance.

In the first step of the training stage, the ANN architectures with the best performance were determined during the training process. Thus, only architectures that reached a root mean square error (RMSE) of 0.001 were selected. However, to avoid over training, ANN models with minimal dimensions were selected. In the second step of the training stage, a study was developed to determine ANN parameters such as learning rate and momentum. The networks were trained so that these parameters could be determined properly. In this step, the RMSE and the number of training epochs were considered for the selection of the ANN architectures.

Once a given ANN was trained using the training data set, its performance must be evaluated using a validation set of data. The validation stage is essential to avoid ANN over-training. Thus, the performance of the ANNs selected were tested and compared using the determination coefficient (R^2) and the RMSE. The final ANN selection considered the lowest errors presented in the training and validation stages.

All programs used in the ANN training and validation stages were developed in the C programming language using a gcc-gnu compiler and the FANN library (Fast Artificial Neural Network Library) for the Linux operating system UBUNTU.

Results and discussion

Statistical Analyses

Table 1 presents the results of the analyses of variance obtained from the experimental data of the soil penetration resistance determined from the cone index while considering the different levels of soil bulk density and water content. The interaction between the soil bulk density and water content factors was significant at a 5% probability.

Table 1. Analyses of variance for soil mechanical penetration resistance considering different levels of density and water content.

SV	DF	SS	MS	F	P-value
Bulk Density (*D*)	2	306511347.90	153255673.90	--	--
Water Content (*WC*)	2	67906589.20	33953294.60	--	--
D x *WC*	4	9337733.30	2334433.30	692.45*	< 0.001
Residue	18	60682.70	3371.30		
Total	26	383816353.10			

*significant at 5% probability.

Equation 2 represents the selected model from the regression analyses. The model was chosen considering the determination coefficient (R^2), the significance of regression coefficients and the lack of adjustment of the model. It can be observed that the model presented a determination coefficient of 0.92. The selected model also presented an RMSE of 0.951.

$$(PR = -41821.28 + 26701.59D - 40100.79WC \; R^2 = 0.92) \quad (2)$$

where:

PR = soil penetration resistance, kPa;
D = soil bulk density, kg dm^{-3};
WC = soil water content, kg kg^{-1}.

Figure 2 presents the surface response relating the soil penetration resistance to the soil bulk density and water content. The model analysis shows that the highest penetration resistance tended to occur at a higher soil density and lower water content, as reported in the literature (CUNHA et al., 2002; DEXTER et al. 2007).

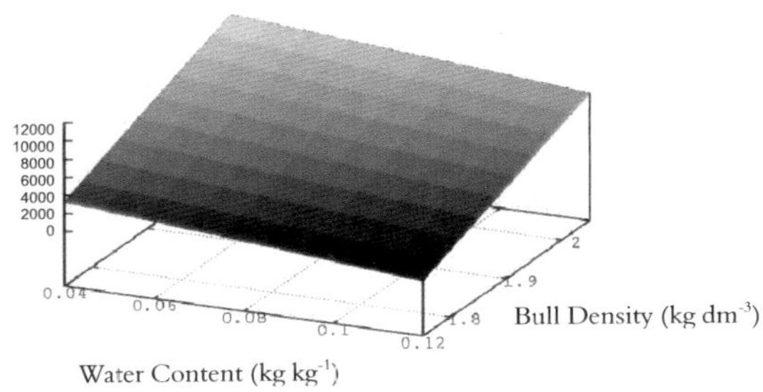

Figure 2. Surface response for penetration resistance (*PR*) related to the density (*D*) and water content (*WC*) of the soil.

To improve the visualization of these factors' influence on soil penetration resistance, the response surface is presented in Figure 3 for the different levels of soil bulk density and water content.

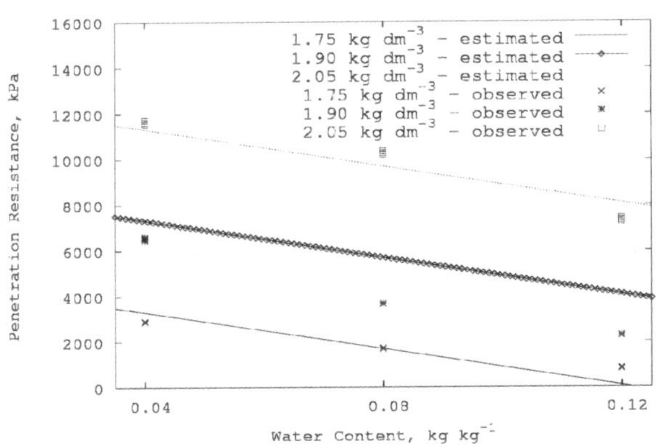

Figure 3. Response surface for the soil penetration resistance obtained from the different levels of soil bulk density and water content.

Figure 3 shows that the soil penetration resistance tended to decrease with higher values of soil water content, which can be explained by the reduction of the cohesion forces and the internal friction (KLEIN et al., 1998). Moreover, the soil penetration resistance tended to increase with higher values of soil bulk density. This effect can be explained by the reduction of the soil pore spaces, which resulted in an increase of the penetration resistance (CUNHA et al. 2002).

Artificial neural network modeling

In this study, an ANN model was employed to predict the soil penetration resistance from the soil bulk density and water content as input data. Figure 4 verifies the result of the study performed during the training stage, in which ANN architectures were evaluated. The study considered one and two hidden layer ANNs.

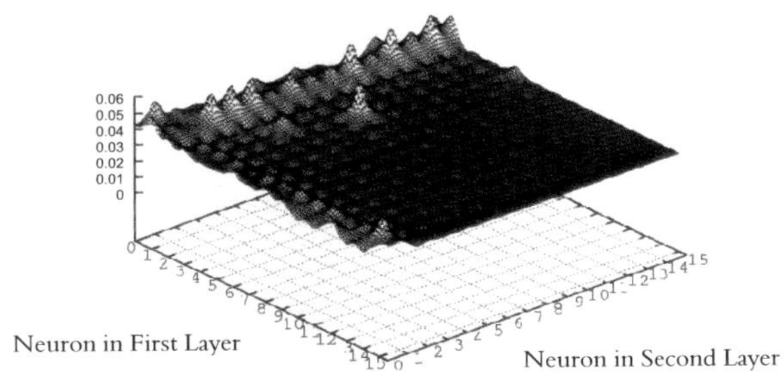

Figure 4. Artificial Neural Network learning capacity considering the number of neurons in the first hidden layers (n_1) and the second hidden layer (n_2).

Figure 4 shows that the learning capacity of the two hidden layers was significantly higher than one hidden layer. This feature indicates that when increasing the number of hidden layers, the ANN learning capacity increases. However, the number of neurons in each hidden layer can vary according to the complexity of the problem (ERZIN et al., 2010; KHAZAEI; DANESHMANDI, 2007; KIM; GILLEY, 2008).

Considering that several evaluated architectures reached the RMSE established during the training stage, for the validation stage, only the ANN architectures that presented minimal dimensions were selected. Among these, the ANN architecture composed by the 2-2-2-1 configuration presented the best results. The ANN results are presented in Table 2.

Table 2. Parameters and results of ANN architecture 2-2-2-1 selected after the training and validation stages.

ANN parameters			Transfer	Validation stage results			
architecture	learning rate	momentum	function	training RMSE	validation RMSE	R^2	global mean error (%)
2-2-2-1	0.4	0.5	Sigmoid	0.0320	0.0840	0.98	6.74

Table 2 shows that, during the training and validation, ANN architecture 2-2-2-1 presented a RMSE equal to 0.032 and 0.084, respectively. Moreover, it can be observed that there is a significant difference between the results obtained from the ANN modeling (RMSE equal to 0.084 and R^2 equal to 0.98) and the results obtained from the mathematical model performed by regression analysis (RMSE equal to 0.951 and R^2 equal to 0.92).

Figure 5 presents the results for the ANN validation stage (architecture 2-2-2-1) using a validation set composed of 9 patterns. In general, the global mean error between the ANN estimated values and the observed values was below 6.74%, which confirms the great prediction capability of the ANN for the proposed problem. Additionally, there are several works in the literature demonstrating ANN' capabilities when applied to systems modeling (AHAMAD et al., 2007; GUNAYDIN et al., 2010; KHAZAEI; DANESHMANDI, 2007; SANTOS et al., 2009; SARMADIAN et al., 2009; TURK et al., 2001).

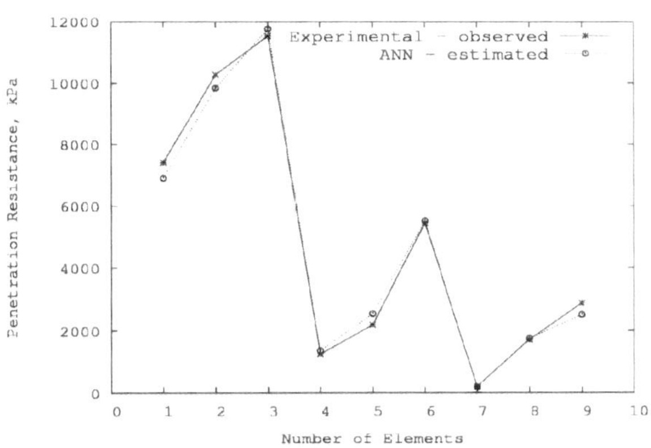

Figure 5. Validation stage results for 2-2-2-1 ANN architecture.

Conclusion

Penetration resistance is associated with the soil bulk density and water content. The highest penetration resistance values tended to occur at a higher density and lower water content, whereas the lowest penetration resistance values tended to occur at lower soil bulk density and higher water content.

The ANN trained by the backpropagation algorithm was able to learn the correlation between the penetration resistance with the soil bulk density and the water content. ANN modeling can be used to predict the soil penetration resistance from soil bulk density and water content as the input data.

ANN architecture 2-2-2-1 presented an RMSE less than 0.085, an R^2 equal to 0.98 and a global mean error of approximately 6.75%, whereas the model obtained from statistical analyses presented an RMSE of 0.951 and an R^2 of 0.92. These results show that the ANN model presented better results than the statistical model obtained from regression analysis.

References

AHAMAD, I.; NAGGAR, H. E.; KHAN, A. N. Artificial neural network application to estimate kinematic soil pile interaction response parameters. **Soil Dynamics and Earthquake Engineering**, v. 27, n. 9, p. 892-905, 2007.

BEUTLER, A. N.; SILVA, M. L. N.; CURI, N.; FERREIRA, M. M.; CRUZ, J. C.; PEREIRA FILHO, I. A. Resistência à penetração e permeabilidade de latossolo vermelho distrófico típico sob sistemas de manejo na região dos cerrados. **Revista Brasileira de Ciência do Solo**, v. 25, n. 1, p. 167-177, 2001.

CAMPANHARO, W. A.; SPERANDIO, H. V.; CECÍLIO, R. A.; HOLLANDA, M. P.; GUARIZ, H. R. Variabilidade espacial da resistência a penetração do solo a penetração em áreas sob cultivos puros e consorciados de café e eucalipto. **Revista Brasileira de Agroecologia**, v. 4, n. 2, p. 2721-2724, 2009.

CUNHA, J. P. A. R.; VIEIRA, L. B.; MAGALHÃES, A. C. Resistência mecânica do solo à penetração sob diferentes densidades e teores de água. **Engenharia na Agricultura**, v. 10, n. 1, p. 1-7, 2002.

DEXTER, A. R.; CZYŹ, E. A.; GATE, O. P. A method for prediction of soil penetration resistance. **Soil and Tillage Research**, v. 1, n. 93, p. 412-419, 2007.

EMBRAPA-Empresa Brasileira de Pesquisa Agropecuária. **Sistema brasileiro de classificação de solos**. Rio de Janeiro: Centro Nacional de Pesquisa de Solos, 1999.

ERZIN, Y.; RAO, B. H.; SINGH, D. N. Artificial neural network models for predicting soil resistivity. **International Journal of Thermal Sciences**, v. 47, n. 10, p. 1347-1358, 2008.

ERZIN, Y.; RAO, B. H.; PATEL, A.; GUMASTE, S. D.; SINGH, D. N. Artificial neural network models for predicting electrical resistivity of soils from thermal resistivity. **International Journal of Thermal Sciences**, v. 49, n. 1, p. 118-130, 2010.

GUNAYDIN, O.; GOKOGLU, A.; FENER, M. Prediction of artificial soil's unconfined compression strength test using statistical analyses and artificial neural networks. **Advances in Engineering Software**, v. 41, n. 9, p. 1115-1123, 2010.

HAYKIN, S. **Neural networks**: a comprehensive foundation. New Jersey: Prentice Hall, 1999.

ISLAM, K. R.; WEIL, R. R. Soil quality indicator propertiers in mid-Atlantic soils as influenced by conservation management. **Journal of Soil and Water Conservation**, v. 55, n. 1, p. 69-78, 2000.

KHAZAEI, J.; DANESHMANDI, S. Modeling of thin-layer drying kinetics of sesame seeds: mathematical and neural networks modeling. **International Agrophysics**, v. 21, n. 1, p. 335-348, 2007.

KIM, M.; GILLEY, J. E. Artificial neural network estimation of soil erosion and nutrient concentration in runoff from land application areas. **Computers and Electronics in Agriculture**, v. 64, n. 2, p. 268-275, 2008.

KLEIN, V. A.; LIBARDI, P. L.; SILVA, A. P. Resistência mecânica do solo à penetração sob diferentes condições de densidade e teor de água. **Engenharia Agrícola**, v. 18, n. 2, p. 45-54, 1998.

SANTOS, F. L.; DUARTE, M. L. M.; FARIA, M. T. C.; EDUARDO, A. C. Balancing of a rigid rotor using artificial neural network to predict the correction masses. **Acta Scientiarum. Technology**, v. 31, n. 2, p. 151-157, 2009.

SARMADIAN, F.; MEHRJARDI, T.; AKBARZADEH, A. Modeling of some soil properties using artificial neural network and multivariate regression in Gorgan province, north of Iran. **Australian Journal of Basic and Applied Sciences**, v. 3, n. 1, p. 323-329, 2009.

SINGH, A. P.; KAY, B. D. Estimating the least limiting water range of soils from properties and management. **Soil Science Society of America Journal**, v. 61, n. 1, p. 877-883, 1997.

TAVARES FILHO, J.; RIBON, A. A. Resistência do solo à penetração em resposta ao número de amostras e tipo de amostragem. **Revista Brasileira de Ciência do Solo**, v. 32, n. 1, p. 487-494, 2008.

TURK, G.; LOGAR, J.; MAJES, B. Modelling soil behaviour in uniaxial strain conditions by neural networks. **Advances in Engineering Software**, v. 32, n. 10, p. 805-812, 2001.

VARELLA, C. A. A.; PINTO, F. A. C.; QUEIROZ, D. M.; SENA JÚNIOR, D. G. Determinação da cobertura do solo por análise de imagens digitais e redes neurais. **Revista Brasileira de Engenharia Agrícola e Ambiental**, v. 6, n. 2, p. 225-229, 2002.

12

Land cover changes affect soil chemical attributes in the Brazilian Amazon

Murilo Rezende Machado[1*], Rodrigo Camara[2], Paulo de Tarso Barbosa Sampaio[3], Marcos Gervasio Pereira[2] and João Baptista Silva Ferraz[3]

*[1]Instituto Chico Mendes de Conservação da Biodiversidade, Floresta Nacional do Rio Preto, Zona Rural, 29960-000, Conceição da Barra, Espírito Santo, Brazil. [2]Universidade Federal Rural do Rio de Janeiro, Seropédica, Rio de Janeiro, Brazil. [3]Instituto Nacional de Pesquisas da Amazônia, Manaus, Amazonas, Brazil. *Author for correspondence. E-mail: murilo.machado@icmbio.gov.br*

ABSTRACT. Forest plantations may minimize the effects of deforestation in the Amazon. However, there are differences among species in terms of their influences on soil recovery. The effects of monospecific plantations of *Acacia mangium*, *Dipteryx odorata*, *Jacaranda copaia*, *Parkia decussata*,and *Swietenia macrophylla,* and areas of pasture and native forest on the chemical soil attributes of the Brazilian Amazon were evaluated. One bulked soil sample was collected per plot (0.00-0.05, 0.05-0.10, and 0.10-0.30 m; three plots of 128 m^2) in each area. No significant differences in most of the soil attributes were observed among the forest plantations. However, soil K^+ and P were higher in the *Swietenia macrophylla* plantations, while higher values of Ca^{2+}, sum of bases, and pH occurred in *Jacaranda copaia* plantations. In the native forest, the pH, and P content were lower, whereas the soil organic matter (SOM) content, soil organic carbon (SOC) content, cation exchange capacity (CEC), N content, H+Al content, and Al^{3+} content were higher than in the plantations. The lowest values of SOM, SOC, CEC, K^+, Mg^{2+}, N, H+Al, and Al^{3+} occurred in the pasture. None of the forest species led to the return of the original soil chemical attributes of the native forest. However, *S. macrophylla* and *J. copaia* plantations presented the highest positive edaphic influences.

Keywords: Amazon deforestation, edaphic attributes, environmental reclamation, forest plantations.

Introduction

The Amazon region encompasses the world's highest level of biodiversity. However, the loss of diversity is affecting millions of individuals of different species of plants, animals and microbial life-forms (Foley et al., 2007) due to annual deforestation of over 20,000 km^2 (Arraes, Mariano, & Simonassi, 2012). Thus, approximately 20% of its 3.6 million km^2 have already been lost, resulting in "savannization", which negatively affects biodiversity and water availability across other biomes (Sawyer, 2009). This accelerated deforestation is driven by the trees being cut for commercial purposes, with subsequent burn before initiating agricultural activities (Arraes et al., 2012). Consequently, this region is an important Brazilian source of carbon emissions for the atmosphere, which significantly contribute to the greenhouse effect (Machado, 2009).

Therefore, the scientific community has agreed that fighting against Amazon deforestation is a matter of national security (Machado, 2009; Sawyer, 2009). Thus, promoting the conservation of natural resources combined with sustainable socio-economic and environmental development will contribute to minimizing this situation (Foley et al., 2007). The establishment of forest plantations enhances forest regeneration (Sansevero, Prieto, Moraes, & Rodrigues, 2011). This fact is a consequence of the attraction of birds, which are seed dispersal agents, and the amelioration of severe microclimate conditions by the recovery of plants in the soil, which facilitates the establishment of seedlings (Meli & Dirzo, 2013). Moreover, the presence of a dense root system contributes to the control of soil erosion and to the input of soil organic matter and nutrients through litterfall (Chada, Campello, & Faria, 2004).

The high amount of biomass in the Amazon forest is due to efficient nutrient cycling between plants and soil (Ferreira, Luizão, Miranda, Silva, & Vital, 2006). Thus, it is necessary to select suitable species that can be effectively established on degraded soils and contribute to its nutritional enrichment (Meli, Martinez-Ramos, Rey-Benayas, & Carabias, 2014). Studies focused on this information should include native and well-adapted exotic species, pioneer and non-pioneer species, and nitrogen-fixing species, which may facilitate the reforestation and long-term recovery of ecosystem functions (Chada et al., 2004; Macedo et al., 2008).

This study aimed to evaluate the impacts of monospecific plantations established in pasture areas on the chemical properties of topsoil in the Brazilian Amazon. Accordingly, we tested the hypothesis that there is no difference among five forest species with respect to their effects on improving soil chemical properties.

Material and methods

The experimental area was located at the coordinates 2° 56' 13 "S and 58° 55' 55" W, 250 km from the AM-010 State Road in Itacoatiara, Amazonas State, Brazil. According to the National Institute of Meteorology (INMET), the local climate is characterized by high total annual rainfall (2,551 mm year^{-1}) and a short period with less rain from August to October, and an annual mean temperature of 25.9°C. The drier month presents total rainfall less than 60 mm, and the mean temperature of the colder month is always equal to or greater than 18°C. The climate was classified as "Am", which means that an excessive amount of rainfall occurs (≥ 2,500 mm per year) and the winter is dry (Köppen, 1948).

The local relief varies from flat to wavy, and the soils are predominantly Oxisols. For the establishment of pasture areas, tree species with high economical value were removed from the original vegetation, Ombrophilous Dense Forest, with subsequent burning of the remaining vegetation. Thereafter, areas with *Bracharia humidicula* (Rendle) Schweickerdt pastures were established. In December 2003, part of this area was selected on a farm called Nova Vida for the establishment of monospecific plantations of five species (Table 1). The planting spacing was 2 x 2 m. More information about the preparation of the area for the installation of the plantations was provided by Machado, Sampaio, Ferraz, Camara, and Pereira (2016).

In addition to the forest stands, two contiguous areas with representative types of land use in the region were considered: a *Brachiaria humidicola* pasture area and a native forest area. The soil chemical attributes in the pasture were considered as the soil conditions prior to the installation of the forest stands, in the experimental area. On the other hand, the native forest was considered as a reference for the original conditions of the soil.

The soil chemical attributes were evaluated in three plots of 128 m^2 in each of the seven land use types. In each of the plots, one soil sample was randomly collected at three depths (0.00-0.05 m, 0.05-0.10 m, and 0.10-0.30 m), each consisting of six subsamples, using a thread auger in December 2007. The experimental design was completely randomized, consisting of seven treatments and three replicates within each treatment.

After air-drying on the lab bench, the soil samples were manually ground, passed through a 2-mm sieve, and the plant residues were removed. Values of the following variables were determined according to Embrapa (2011): pH (H_2O); potential acidity (H+Al); Al^{3+}, K^+, Ca^{2+}, and Mg^{2+} content; sum of exchangeable bases (SB); potential cation-exchange capacity (CEC at pH 7.0); available P content; N content; soil organic carbon (SOC) content; and soil organic matter (SOM) content.

Table 1. General information about the forest species and the respective planted area in Itacoatiara, Amazonas, Brazil.

Scientificname	Popular name	Native	Ecologicalgroup	Area of forest plantation (ha)
Acacia mangium Willd.	Acácia	No	Pioneer	7.50
Dipteryx odorata (Aubl.) Willd.	Cumaru	Yes	Non-pioneer	0.80
Jacaranda copaia (Aubl.) D Don	Caroba	Yes	Pioneer	5.20
Parkia decussata Ducke	Faveira	Yes	Non-pioneer	2.76
Swietenia macrophylla King.	Mogno	Yes	Non-pioneer	1.60

The results were subjected to analysis of variance using the F test, and the mean values were compared using the LSD test ($p < 0.05$). These analyseswere performedusing Systat software, version 8.0 (Wilkinson, 1998). We also performed multivariate analysis of hierarchical clustering using Ward's method to identify possible patterns in the effects of the different land use on soil chemical attributes. Therefore, we considered the mean values of the soil attributes calculated among the soil depths (0.00-0.05 m, 0.05-0.10 m, and 0.10-0.30 m) in each ecosystem. A dendrogram was constructedusing PAST software, version 2.17c (Hammer, Harper, & Ryan, 2001).

Results and discussion

In general, the land use significantly influenced all of the chemical attributes at the three sampling depths (Table 2). Comparing only the monospecific plantations, some patterns of this effect were observed. Higher K^+ and P soil values occurredin *Swietenia macrophylla* plantations. On the other hand, the higher soil values of Ca^{2+}, which influenced higher SB values, occurred in soils with*Jacarandacopaia*. As a result, higher pH values were observedin this area, but these pH values were not significantly different from the pH values verifiedin the soil in the pasture area.

Considering all seven types of land use, the highest values of SOM, CEC, H+Al, Al^{3+}, SOC, and N occurred in the native forest soil (Table 3). In the contrast, the lower values of SOM, CEC, H+Al, Al^{3+}, SOC, K^+, Mg^{2+}, and N were verified in the pasture soil, while the lowest values of pH and P were found in the native forest soil.

The relationships among SOM and the other soil chemical attributes were previously found by other authors. The direct relationship between SOM and CEC is a function of the high reactivity power of SOM, in which the diversified organic radicals influence the CEC (Iwata et al., 2012). Thus, the higher content of SOM increases the low fertility of Oxisols, in which the organic matter is responsible for a substantial proportion of the CEC (Effgen et al., 2012). In this way, the Oxisolshave a high content of very low-charged 1:1 clays and Fe and Al oxide-hydroxides (Santos et al., 2013). Thus, the Oxisols have high Al^{3+}contents and low CEC values, which are highly dependent upon the organic matter content.

Organic carbon is the main constituent of the SOM (Effgen et al., 2012). Thus, SOC and SOM quickly respond to changes in land-use and soil management (Smith, 2008). The soil N content is also directly related to the SOM content, which reflects the input of the deciduous plant material into the soil (Barreto & Lima, 2006); this input is significantly higher in forests than in pasture areas (Machado, Piña-Rodrigues, & Pereira, 2008).

Table 2. The mean values of pH, N, P, K^+, Ca^{2+}, and Mg^{2+} in soil (0.00-0.05 m, 0.05-0.10 m, and 0.10-0.30 m) in monospecific plantations of *Acacia mangium*, *Dipteryx odorata*, *Jacaranda copaia*, *Parkia decussate*,and *Swietenia macrophylla*, and in areas of native forest and pasture in Itacoatiara, Amazonas, Brazil.

Area	pH	N	P	K^+	Ca^{2+}	Mg^{2+}
	(H_2O)	(g kg^{-1})	----(mg dm^{-3})----		----($cmol_c$ dm^{-3})----	
0.00-0.05 m						
A. mangium	4.17±0.27b	1.93±0.27bc	3.67±0.58a	24.00±5.57b	0.51±0.15ab	0.29±0.11ab
D. odorata	4.50±0.19b	1.88±0.09c	5.00±1.00a	30.33±5.51ab	0.65±0.25a	0.36±0.07a
J. copaia	4.66±0.14ab	2.28±0.15ab	4.33±0.58a	28.00±1.00ab	1.00±0.37a	0.41±0.21a
P. decussata	4.34±0.04b	1.69±0.21c	4.33±0.58a	21.33±3.21b	0.19±0.07b	0.25±0.05ab
S. macrophylla	4.38±0.04b	1.96±0.20bc	5.00±1.00a	35.00±4.00a	0.32±0.09b	0.31±0.09ab
Nativeforest	3.66±0.01c	2.37±0.19a	3.67±2.08a	24.67±4.04b	0.06±0.03b	0.11±0.03b
Pasturearea	5.00±0.60a	0.91±0.25d	5.33±4.08a	11.67±6.35c	0.73±0.64ab	0.19±0.18b
0.05-0.10 m						
A. mangium	4.14±0.23b	1.26±0.23ab	3.33±0.58ab	16.67±4.04b	0.24±0.10b	0.18±0.07ab
D. odorata	4.45±0.25b	1.30±0.27ab	3.67±0.58ab	21.00±1.00ab	0.44±0.24ab	0.28±0.11a
J. copaia	4.51±0.16ab	1.59±0.18a	4.00±0.00ab	22.67±2.52a	0.57±0.18a	0.29±0.15ab
P. decussata	4.26±0.07b	1.07±0.25b	3.67±1.15ab	17.33±4.04b	0.11±0.01b	0.16±0.04ab
S. macrophylla	4.29±0.06b	1.63±0.26a	4.33±0.58a	23.33±3.21a	0.21±0.12b	0.21±0.11ab
Nativeforest	3.73±0.11c	1.65±0.34a	2.33±0.58b	15.00±1.00b	0.04±0.01b	0.07±0.01b
Pasturearea	4.85±0.42a	0.81±0.15c	3.00±0.00b	8.67±2.08c	0.36±0.17b	0.12±0.09ab
0.10-0.30 m						
A. mangium	4.33±0.11bc	0.90±0.06b	1.33±0.58ab	6.67±0.58bc	0.06±0.04bc	0.08±0.03b
D. odorata	4.50±0.22ab	0.97±0.08ab	1.67±0.58ab	9.00±1.00b	0.20±0.14ab	0.16±0.07ab
J. copaia	4.48±0.16ab	1.05±0.13ab	1.67±0.58ab	12.67±4.51ab	0.26±0.15a	0.17±0.10a
P. decussata	4.37±0.08bc	0.85±0.09b	1.67±0.58ab	11.33±4.73ab	0.08±0.01bc	0.11±0.03ab
S. macrophylla	4.31±0.08bc	0.96±0.10ab	2.00±0.00a	15.33±3.06a	0.09±0.03bc	0.09±0.03ab
Nativeforest	4.13±0.05c	1.10±0.06a	1.00±0.00b	7.33±0.58bc	0.03±0.02c	0.04±0.00b
Pasturearea	4.68±0.27a	0.61±0.03c	2.00±0.00a	4.00±0.00c	0.07±0.04bc	0.05±0.02b

*Mean values ± standard deviation followed by the same letter in the same column, in the same depth, are not statistically different (LSD test; $p < 0.05$).

Table3. The mean values of SOC (soil organic carbon), SOM (soil organic matter), $H^{+}+Al^{3+}$, Al^{3+}, potential cation-exchange capacity (CEC at pH 7.0) and sum of bases (SB) in soil (0.00-0.05 m, 0.05-0.10 m, and 0.10-0.30 m) in monospecific plantations of *Acacia mangium*, *Dipteryx odorata*, *Jacaranda copaia*, *Parkia decussate*,and *Swietenia macrophylla*, and in areas of native forest and pasture in Itacoatiara, Amazonas, Brazil.

Area	SOC	SOM	H+Al	Al^{3+}	CEC (pH 7)	SB
	-----(g kg^{-1})-----		----------------------($cmol_c$ dm^{-3})----------------------			
			0.00-0.05 m			
A. mangium	27.22±5.05b	46.82±8.68b	9.32±1.33b	1.35±0.38b	10.20±1.30b	0.88±0.19abc
D. odorata	27.59±1.25b	47.45±2.15b	9.15±0.57b	1.13±0.33bc	10.24±0.28b	1.10±0.30ab
J. copaia	31.35±1.21b	53.92±2.07b	9.16±0.63b	0.87±0.21bc	10.66±0.57b	1.50±0.58a
P. decussata	23.39±2.16b	40.22±3.71b	9.04±0.73b	1.35±0.16b	9.53±0.78bc	0.50±0.08bc
S. macrophylla	27.63±2.43b	47.52±4.18b	8.72±0.63b	1.21±0.14b	9.45±0.73bc	0.73±0.17bc
NativeForest	42.02±11.05a	72.27±19.01a	12.99±2.91a	2.42±0.42a	13.27±2.96a	0.28±0.08c
Pasturearea	13.21±2.25b	22.72±3.87c	6.13±1.38c	0.58±0.51c	7.09±0.79c	0.96±0.85abc
			0.05-0.10 m			
A. mangium	19.61±2.33b	33.73±4.00b	7.46±0.37b	1.30±0.25b	7.94±0.37b	0.48±0.12b
D. odorata	19.29±0.60b	33.18±1.03b	7.37±0.65b	1.00±0.38bc	8.16±0.50b	0.78±0.35ab
J. copaia	23.53±1.54a	40.47±2.64a	7.46±0.47b	0.95±0.19bc	8.39±0.35ab	0.93±0.35a
P. decussata	17.63±1.37bc	30.32±2.34b	7.00±0.34bc	1.27±0.10b	7.32±0.35bc	0.32±0.05b
S. macrophylla	21.89±2.85ab	37.65±4.90ab	7.77±0.43b	1.26±0.10b	8.26±0.67b	0.49±0.24b
NativeForest	25.40±2.75a	43.67±4.72a	9.29±1.04a	1.89±0.13a	9.46±1.05a	0.17±0.03b
Pasturearea	11.16±2.21c	19.20±3.80c	5.79±1.19c	0.66±0.32c	6.30±0.98c	0.51±0.26b
			0.10-0.30 m			
A. mangium	9.62±0.39bc	16.64±0.68bc	5.04±0.16ab	0.94±0.06ab	5.20±0.17ab	0.16±0.06b
D. odorata	10.29±0.24b	17.69±0.41b	5.08±0.23ab	0.80±0.26b	5.47±0.15ab	0.39±0.20ab
J. copaia	12.12±1.80a	20.85±3.10a	4.62±1.03b	0.79±0.15b	5.09±0.88b	0.47±0.27a
P. decussata	10.16±0.84b	17.48±1.44b	5.31±0.38ab	0.97±0.10ab	5.54±0.41ab	0.23±0.03b
S. macrophylla	11.73±0.47a	20.18±0.81a	5.59±0.37a	1.02±0.06ab	5.81±0.37ab	0.22±0.05b
Nativeforest	12.67±0.30a	21.80±0.52a	5.81±0.42a	1.16±0.08a	5.92±0.43a	0.11±0.02b
Pasturearea	8.52±0.21c	14.65±0.36c	5.19±0.47ab	0.83±0.12b	5.32±0.49ab	0.13±0.06b

*Mean values ± standard deviation followed by the same letter in the same column in the same depth are not statistically different (LSD test; $p< 0.05$).

The anionic organic compounds in the SOM are capable of complexing the H^{+} and Al^{+3}cationsthat are free in the soil solution and add base cations (Ca^{2+}, Mg^{2+}, and K^{+}), reducing the soil acidity and increasing the soil pH (Pavinato & Rosolem, 2008). Therefore, the increase of SOM influenced the low soil values of H+Al and Al^{3+} in the native forest. Although we expected higher soil K^{+}, Ca^{2+}, and Mg^{2+} contents and pH in this ecosystem, this result did not occur. The higher pH values that were observedin the monospecific forest plantations and pasture area were likely due to liming and periodic burning, respectively.

Periodic burning in pasture areas leads to the release of bases to the soil from plant material (Sousa, Miranda, & Oliveira, 2007). As a result, the H^{+} and Al^{3+}contents in the soil solution decrease and the pH increases. This effect was verifiedin the present study, while the opposite effect was observed for Mg^{2+}, whichdecreased in the pasture area. This pattern was probably caused by nutrient losses through leaching under high rates of rainfall (Heringer, Jacques, Bissani, & Tedesco, 2002), which naturally occur in the Amazon.

Although the lower SOM content may have influenced the low K^{+} and Mg^{2+}contents in the soil in the pasture area, the higher SOM content did not increase the concentrations of these bases in the native forest. The probable explanation for this divergence are the differences between these ecosystems, with respect to the quality of the organic matter (C/N ratio and lignin content, for example), because this characteristic and the amount of plant organic matter into the soil influence the SOM content (Costa et al., 2009).

Acacia mangium, whichbelongs to the Fabaceae (Leguminosae) family, has the ability to form a mutualistic symbiosis with N-fixing bacteria (Perrineau et al., 2011). Therefore, both green and senescent leaves of this species have higher N contents compared to the N contents in *Dipteryx odorata*, *Parkia decussata* and *S. macrophylla* leaves, although the values for *J. copaia*leaves were not significantly different from those observed in*A. mangium* (Machado, Sampaio, Ferraz, Camara, & Pereira, 2016). Silva et al. (2015) verified that the total soil N content in a monospecific plantation of *A. mangium*was higherwhen compared to a monospecific plantation of *Eucalyptus camaldulensis* Dehn, in the north of Rio de Janeiro State.

However, Silva et al. (2015) noticed that the total N content was significantly lower in the soil in both the mixed stands of *A. mangium/E.camaldulensis* and *A.mangium/Sesbaniavirgata* (Cav.) Pers., compared to the monospecific plantation of *Acacia mangium*. This result was observed even though *S. virgata* (Fabaceae) also may associate with N-fixing bacteria. Therefore, factors other than forest species also determine the soil chemical attributes, such as weather conditions and the characteristics of the decomposer biota community.

The soil under the native forest presented higher soil fertility, probably because the richness of the plant community was higher than in the monospecific plantations and pasture area. Some studies pointed out that the litter in native forests can have higher nutrient contents than the litter ofmonospecific plantations with native or exotic tree species (Cunha Neto, Leles, Pereira, Bellumath, & Alonso, 2013). Thus, more diverse litter is more suitable for decomposers (Balvanera et al., 2006), which leads to the favorable release of mineral elements enclosed in the litter and results in soil with higher fertility in native forest ecosystems.

The coexistence of different tree species that continuously produce heterogeneous litter in terms of chemical quality and that have different rates of nutrients uptake from soil results in increased soil fertility, compared to the soil fertility of monospecific plantations (Gama-Rodrigues, Barros, & Comerford, 2007). Thus, the authors of the study previously mentioned suggested that more diverse forest ecosystems present complementarity between various tree species, which results in more efficient and balanced nutrient cycling dynamics.

The hierarchical clustering dendrogram obtained from the multivariate analysis indicated that *J. copaia* and *S. macrophylla* formed a group with high similarity (lower distance between connections) with the native forest (group 1) compared to the other ecosystems (Figure 1). In contrast, *A. mangium* and *P.decussata* formed another group with high similarity to *D .odorata* (group 2). These groups (1 and 2) presented low similarity to each other because the distance of connection between them (approximately 18) was greater than half of the total connection distance (30) that linked all of the types of land use (Figure 1). Moreover, the forest ecosystems (monospecific plantations and native forest) formed a macro group (group 3) in which the distance of connection to the pasture area was equal to 27, i.e., next to the value of the connection that encompassed all of the seven ecosystems. Thus, the soil chemical attributes showed that the pasture presented the highest dissimilarity to the forest ecosystems.

The information obtained by the present study indicated that the plantations did not result in the return of the soil chemical attributes to their original conditions that occurred under native forest. However, when comparing the different monospecific plantations with each other, the plantations of *J. copaia* and *S. macrophylla*had major positive impacts on the soil chemical attributes to a depth of 0.03 m. In contrast, the pasture resulted in major negative edaphic impacts.

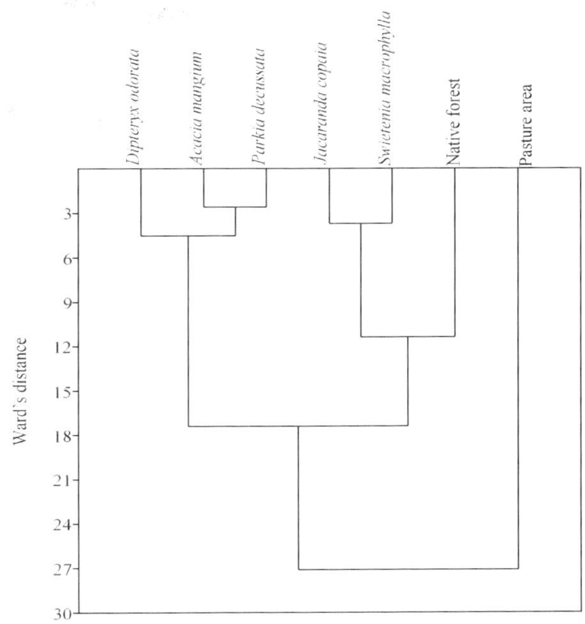

Figure 1. A hierarchical clustering dendrogram for the soil chemical attributes (mean values obtained among the three depths 0.00-0.05 m, 0.05-0.10 m, and 0.10-0.30 m) in monospecific plantations of *Acacia mangium*, *Dipteryx odorata*, *Jacaranda copaia*, *Parkia decussate*, and *Swietenia macrophylla*, and areas of native forest and pasture in Itacoatiara, Amazonas, Brazil.

The intensely weathered soils in Amazon present low natural fertility (Quesada et al., 2010). Thus, we may consider that the forest species strongly depend on the nutrient cycling to attend their nutritional needs. For this reason, there is a close association between the litter layer disposedon the soil surface and a dense net of fine roots, which decreases in quantity with an increasing soil depth in the Amazonian forest (Herrera, Jordan, Klinge, & Medina, 1978).

This process allows the direct uptake of nutrients by plants that are released during the decomposition and mineralization of litter, which is an important mechanism to minimize nutrient loss by soil leachingthat is favored by heavy rainfall in the region (Newbery, Alexander, & Rother, 1997). Moreover, another important mechanism of nutrient retention in the ecosystem is biochemical cycling, i.e., the internal translocation of mobile nutrients within plant organisms from senescent tissues to young tissues before leaf abscission (Almeida et al., 2014). This re-translocation process conserves approximately 60 to 85% of the total nutrient contents absorbed when the soil availability of nutrients is low (Malavolta, 2006). In this sense, the species that have higher rates of such a dynamic would be more adapted to the soil and climatic conditions that prevail in the Amazon.

The results obtained by Machado et al. (2016) indicated that higher rates of internal nutrient translocation occurred in *S. macrophylla* (for N), *A. mangium* (for P and S), *J. copaia* (for K) and *P.decussata* (for Mg), while *D.odorata* did not show high performance for the re-translocation of macronutrients. Because N and P are the most limiting nutrients in the soil for tropical forest productivity, *S. macrophylla* and *A. mangium* play important ecological roles. Thus, both species should be prioritized for recovering degraded Amazon areas compared to the other forest species studied.

Conclusion

The forest plantations did not return the soil chemical attributes to their original native forest conditions.

When comparing the forest plantations, *S. macrophylla* and *J. copaia*had the highest positive edaphic influence. Therefore, these forest species should be preferentially recommended in plantations for land reclamation in degraded areas with similar soil characteristics and climate conditions to those observed in this study.

Acknowledgements

The authors thank the Amazon Conservation Scholarship Program (Bolsas de Estudo para a Conservação da Amazônia - BECA)/International Institute of Education in Brazil (InstitutoInternacional de Educação do Brasil - IEB)/Moore Foundation for financial support of this project.

References

Almeida, C. X., Pita Junior, J. L., Rozane, D. E., Souza, H. A., Hernandes, A., Natale, W., & Ferraudo, A. S. (2014). Nutrient cycling in mango trees. *Semina: Ciências Agrárias, 35*(1), 259-266.

Arraes, R. A., Mariano, F. Z., &Simonassi, A. G. (2012). Causas do desmatamento no Brasil e seu ordenamento no contexto mundial. *Revista de Economia e Sociologia Rural, 50*(1), 119-140.

Balvanera, P., Pfisterer, A. B., Buchmann, N., He, J. S., Nakashizuka, T., Raffaelli, D., & Schmid, B. (2006). Quantifying the evidence for biodiversity effects on ecosystem functioning and services. *Ecology Letters, 9*(10), 1146-1156.

Barreto, A. C., & Lima, F. H. S. (2006). Características químicas e físicas de um solo sob floresta, sistema agroflorestal e pastagem no sul da Bahia. *Revista Caatinga, 19*(4), 415-425.

Chada, S. S., Campello, E. F. C., & Faria, S. M. (2004). Sucessão vegetal em uma encosta reflorestada com leguminosas arbóreas em Angra dos Reis, RJ. *Revista Árvore, 28*(6), 801-809.

Costa, O. V., Cantarutti, R. B., Fontes, L. E. F., Costa, L. M., Nacif, P. G. S., & Faria, J.C. (2009). Estoque de carbono do solo sob pastagem em área de tabuleiro costeiro no sul da Bahia. *Revista Brasileira de Ciência do Solo, 33*(5), 1137-1145.

Cunha Neto, F. V., Leles, P. S. S., Pereira, M. G., Bellumath, V. G. H., & Alonso, J. M. (2013). Acúmulo e decomposição da serapilheira em quatro formações florestais. *Ciência Florestal, 23*(3), 379-387.

Effgen, E. M., Nappo, M. E., Cecílio, R. A., Mendonça, A. R., Manzole, R., & Borcarte, M. (2012). Atributos químicos de um Latossolo Vermelho-Amarelo distrófico sob cultivo de eucalipto e pastagem no sul do Espírito Santo. *Scientia Forestalis, 40*(95), 375-381.

Empresa Brasileira de Pesquisa Agropecuária [Embrapa]. (2011). *Manual de métodos de análises de solos* (2a ed.). Rio de Janeiro, RJ: Embrapa Solos.

Ferreira, S. J. F., Luizão, F. J., Miranda, S. A. F., Silva, M. S. F., & Vital, A. R. T. (2006). Nutrientes na solução do solo em floresta de terra firme na Amazônia Central submetida à extração seletiva de madeira. *Acta Amazonica, 36*(1), 59-68.

Foley, J. A., Asner, G. P., Costa, M. H., Coe, M. T., De Fries, R., Gibbs, H.K., … Snyder, P. (2007). Amazonia revealed: forest degradation and loss of ecosystem goods and services in the Amazon Basin. *Frontiers in Ecology and the Environment, 5*(1), 25-32.

Gama-Rodrigues, A. C., Barros, N. F., & Comerford, N. B. (2007). Biomass and nutrient cycling in pure and mixed stands of native tree species in southeastern Bahia, Brazil. *Revista Brasileira de Ciência do Solo, 31*(2), 287-298.

Hammer, O., Harper, D. A. T., & Ryan, P. D. (2001). PAST: Paleontological statistics software package for education and data analysis. *Palaeontologia Electronica, 4*(1), 1-9.

Heringer, I., Jacques, A. V. A., Bissani, C. A., & Tedesco, M. (2002). Características de um Latossolo Vermelho sob pastagem natural sujeita à ação prolongada do fogo e de práticas alternativas de manejo. *Ciência Rural, 32*(2), 309-314.

Herrera, R., Jordan, C., Klinge, H., & Medina, E. (1978).Amazon ecosystems: their structure and functioning with particular emphasis on nutrients. *Interciencia, 3*(4), 223-231.

Iwata, B. F., Leite, L. F. C., Araújo, A. S. F., Nunes, L. A. P. L., Gehring, C., & Campos, L. P. (2012). Sistemas agroflorestais e seus efeitos sobre os atributos químicos em Argissolo Vermelho-Amarelo do Cerrado piauiense. *Revista Brasileira de Engenharia Agrícola e Ambiental, 16*(7), 730-738.

Köppen, W. (1948). *Climatologia*: con un estudio de los climas de la Tierra. Mexico: Fondo de Cultura Economica.

Macedo, M. O., Resende, A. S., Garcia, P. C., Boddey, R. M., Jantalia, C. P., Urquiaga, S., ... Franco, A. A. (2008). Changes in soil C and N stocks and nutrient dynamics 13 years after recovery of degraded land using leguminous nitrogen-fixing trees. *Forest Ecology and Management*, *255*(5-6), 1516-1524.

Machado, L. O. R. (2009). Desflorestamento na Amazônia brasileira: ação coletiva, governança e governabilidade em área de fronteira. *Sociedade e Estado*, *24*(1), 115-147.

Machado, M. R., Piña-Rodrigues, F. C. M., & Pereira, M. G. (2008). Produção de serapilheira como bioindicador de recuperação em plantio adensado e de revegetação. *Revista Árvore*, *32*(1), 143-151.

Machado, M. R., Sampaio, P. T. B., Ferraz, J., Camara, R., & Pereira, M. G. (2016). Nutrient retranslocation in forest species in the Brazilian Amazon. *Acta Scientiarum. Agronomy*, *38*(1), 93-101.

Malavolta, E. (2006). *Manual de nutrição de plantas*. São Paulo, SP: Agronômica Ceres.

Meli, P., & Dirzo, R. (2013). Effects of grasses on sapling establishment and the role of transplanted saplings on the light environment of pastures: implications for tropical forest restoration. *Applied Vegetation Science*, *16*(2), 296-304.

Meli, P., Martínez-Ramos, M., Rey-Benayas, J. M., & Carabias, J. (2014). Combining ecological, social and technical criteria to select species for forest restoration. *Applied Vegetation Science*, *17*(4), 744-753.

Newbery, D. Mc C., Alexander, I. J., & Rother, J. A. (1997). Phosphorus dynamics in a lowland African rain forest, the influence of ectomycorrhizal trees. *Ecological Monographs*, *67*(3), 367-409.

Pavinato, P. S., & Rosolem, C. A. (2008). Disponibilidade de nutrientes no solo - decomposição e liberação de compostos orgânicos de resíduos vegetais. *Revista Brasileira de Ciência do Solo*, *32*(3), 911-920.

Perrineau, M. M., Le Roux, C., Faria, S. M., Balieiro, F. C., Galiana, A., Prin, Y., & Béna, G. (2011). Genetic diversity of symbiotic *Bradyrhizobiumelkanii* populations recovered from inoculated and non-inoculated *Acaciamangium* field trials in Brazil. *Systematic and Applied Microbiology*, *34*(5), 376-384.

Quesada, C. A., Lloyd, J., Schwarz, M., Patiño, S., Baker, T. R., Czimczik, C., ... Piva, R. (2010). Variations in chemical and physical properties of Amazon forest soils in relation to their genesis. *Biogeosciences*, 7(5), 1515-1541.

Sansevero, J. B. B., Prieto, P. V., Moraes, L. F. D., & Rodrigues, P. J. F. P. (2011). Natural regeneration in plantations of native trees in Lowland Brazilian Atlantic Forest: community structure, diversity, and dispersal syndromes. *Restoration Ecology*, *19*(3), 379-389.

Santos, H. G., Jacomine, P. K. T., Anjos, L. H. C., Oliveira, V. A., Lumbreras, J. F., Coelho, M. R., ... Oliveira, J. B. (2013). *Sistema brasileiro de classificação de solos* (3a. ed.) Brasília, DF: Embrapa.

Sawyer, D. (2009). Fluxos de carbono na Amazônia e no Cerrado: um olhar socioecossistêmico. *Sociedade e Estado*, *24*(1), 149-171.

Silva, C. F., Loss, A., Carmo, E. R., Pereira, M. G., Silva, E. M. R., & Martins, M. A. (2015). Fertilidade do solo e substâncias húmicas em área de cava de extração de argila revegetada com eucalipto e leguminosas no Norte Fluminense. *Ciência Florestal*, *25*(3), 547-561.

Smith, P. (2008). Land use change and soil organic carbon dynamics. *Nutrient Cycling in Agroecosystems*, *81*(2), 169-178.

Sousa, D. M. G., Miranda, L. N., & Oliveira, S. A. (2007). Acidez do solo e sua correção. In R. F. Novais, V. H. Alvarez, N. F. Barros, R. L. F. Fontes, R. B. Cantarutti, & J. C. L. Neves (Eds.), *Fertilidade do solo* (p. 205-274). Viçosa, MG: Sociedade Brasileira de Ciência do Solo.

Wilkinson, L. (1998). *SYSTAT: The system for statistics. version 8.0*. Evanston, IL: SPSS. Inc.

13

Biological soil loosening by grasses from genus *Brachiaria* in crop-livestock integration

José Flávio Neto, Eduardo da Costa Severiano*, Kátia Aparecida de Pinho Costa, Wellingthon Silva Guimarães Junnyor, Wainer Gomes Gonçalves and Renata Andrade

Instituto Federal de Educação, Ciência e Tecnologia Goiano, Rod. Sul Goiana, Km 01, 75901-970, Campûs Rio Verde, Rio Verde, Goiás, Brazil.
**Author for correspondence. E-mail: severianoec@yahoo.com.br*

ABSTRACT. Soil compaction associated with pastures degradation can decrease animal productivity, forage longevity and compromise environmental sustainability. To confront this serious issue, the loosening potential of forages should be recognized. We evaluated the least limiting water range as indicator of biological loosening potential in relation to cultivation of grasses the genus *Brachiaria* in crop-livestock integration. We also evaluated the water availability to soybean crop that succeeded these grasses. Our studies were performed in two stages. In the first stage, we divided 32 plots into four randomized blocks in which we cultivated corn combined with following treatments: 1 - *Brachiaria brizantha* cultivar Marandu; 2 - Xaraes; 3 - Piata; 4 - MG4; 5 - *B. decumbens*; 6 - *B. ruziziensis*; 7 - Invasive plants; and 8 - uncovered soil. We evaluated soil, to quantify the biological soil loosening, and also forages. In the second stage, we cultivated soybean and added: 9 - conventional tillage as a control treatment, increasing number of plots to 36. Our results suggest that it is possible to cultivate *Brachiaria brizantha* or *Brachiaria decumbens* as management strategy to aid edaphic recovery. Xaraes and Piata grasses provide greater soil loosening while increasing water availability to successive soybean crop.

Keywords: pasture recovery, soil structural quality, least limiting water range.

Introduction

The extensive and continuous grazing activity of the Brazilian livestock industry is the leading cause of pasture degradation. Physical degradation of the soil occurs because of compaction caused by animal trampling and machinery traffic under unfavorable conditions with regards to soil water content (LANZANOVA et al., 2007). This degradation results in an accelerated decrease in pasture productivity and longevity (IMHOFF et al., 2000).

Pasture recovery and improved soil physical characteristics are achieved by reversing this process of degradation. By combining techniques to amortize operating costs and increase sustainability, as with Crop-Livestock Integration (CLI) (COSTA et al., 2010), the chemical, physical and biological soil properties can be improved. In addition, no-tillage systems can enable straw formation (PETTER et al., 2011).

Although many farmers are reluctant to adopt this system because of the possible negative effects of animal trampling (BAVOSO et al., 2010; FLORES et al., 2007), the use of forage plants provides an alternative for soil physical recovery. Unlike tillage performed by farm equipment, tillage by plants occurs uniformly throughout the soil layers by roots and forms biopores favorable to root growth, water infiltration and gas diffusion (SEVERIANO et al., 2010). Consequently, tillage plants mitigates deleterious effects and provides a favorable environment for succession crops (CHIODEROLI et al., 2012; GUIMARÃES et al., 2009).

Although some plants are known to penetrate compacted soil layers (LIMA et al., 2012), little is known about the tillage potential of Brachiaria (Trin.) Griseb. spp. (syn. Urochloa P. Beauv. spp.), especially its new cultivars. This knowledge is relevant when considering the significant area occupied by the genus (approximately 85% of the pastures cultivated in Brazil) because of both its nutritional value and its high stress resistance (ARROYAVE et al., 2011) and its increasing adoption into integrated production systems in tropical regions (PACHECO et al., 2008), given its feasibility for intercropping with annual crops (CALONEGO et al., 2011).

The aim of the present study was to evaluate the least limiting water range of the soil as an indicator of both its biological soil loosening potential for cultivating grass of the genus *Brachiaria* in CLI and its water availability for a successor soybean crop.

Material and methods

Our study was conducted in the experimental area of the Federal Institute of Education, Science and Technology Goiano, Rio Verde Câmpus, in the municipality of Rio Verde-Goiás State, 17° 48' 34.25" S and 50° 54' 05.36" W, at an altitude of 731 m and in an area covered with Dystroferric Red Latosol (EMBRAPA, 2006). A chemical and physical characterization of the soil is presented in Table 1.

According to Köppen's classification, the climate is megathermal or humid tropical (Aw) of the tropical savanna subtype, with dry winters and rainy summers. The average annual temperature in the region is 25°C, and the average annual rainfall is approximately 1,600 mm, with the maximum precipitation occurring in January and the minimum occurring in June, July and August (< 50 mm $month^{-1}$).

A 2,016 m^2 area was planted with corn (*Zea mays*) on November 19, 2010 and was intended for silage. The spacing between rows was 0.88 m (population of 55,000 plants ha^{-1}). Fertilization was performed at planting according to soil analysis results (Table 2) with 30 kg ha^{-1} N, 200 kg ha^{-1} P_2O_5, 60 kg ha^{-1} K_2O, 2 kg ha^{-1} B and 0.4 kg ha^{-1} Mo. The following fertilizer sources were used: simple superphosphate, potassium chloride, boric acid and sodium molybdate, following the recommendations of Sousa and Lobato (2004).

After crop emergence, plant thinning was conducted, and 36 plots of 5.4 × 6 m (32.4 m^2) were defined and randomly arranged into four blocks. Topdressing, which consisted of an application of 30 kg ha^{-1} N from ammonium sulfate and 90 kg ha^{-1} K_2O from potassium chloride, was performed thirty days after the corn emerged. Grass oversowing was performed with 9 kg ha^{-1} pure viable seeds of the genus *Brachiaria* to implement the CLI system according to the following treatments: 1. *Brachiaria brizantha* cv. Marandu; 2. *Brachiaria brizantha* cv. Xaraes; 3. *Brachiaria brizantha* cv. Piata; 4. *Brachiaria brizantha* cv. MG4; 5. *Brachiaria decumbens* and, 6. *Brachiaria ruziziensis*.

For comparison, the following control treatments were adopted: 1. mechanical recovery by tillage with subsoiling and delumping by rotary hoe (conventional tillage); 2. fallow soil in the absence of plant coverage; and 3. plant community (invasive plants) in terms of density and diversity predominated by a canopy of dicotyledonous plants of the following species: *Amaranthus spinosus* L. (caruru-de-espinho), *Nicandra physalodes* (L.) Gaertn. (joá-de-capote), *Crotalaria incana* L. (xique-xique), *Sida spinosa* L. (guanxuma), *Waltheria indica* L. (malva-veludo), (L.), *Conysa canadensis* (L.) Cronquist (buva), *Spermacoce latifolia* Aubl. (erva-quente), *Ageratum conyzoides* L. (mentrasto), *Stellaria media* (L.) Vill. (erva-de-passarinho). Monocotyledons, which were present to a lesser extent, were identified as follows: *Digitaria insularis* (L.) Fedde (capim-amargoso), *Eleosine indica* (L.) Gaertn., (capim-pé-de-galinha) and *Commelina benghalensis* L. (trapoeraba).

Table 1. Physical and chemical characterization of the Dystroferric Red Latosol used to cultivate crop-livestock integration systems, in the municipality of Rio Verde, Goiás State.

Depth	Pd	Bd	Particle Size							Sulfuric attack			Ki	Kr
			VCS	CS	MS	FS	VFS	Silt	Clay	SiO_2	Al_2O_3	Fe_2O_3		
(cm)	($kg\ dm^{-3}$)		($g\ kg^{-1}$)											
0–20	2.8	1.2	1	15	154	141	53	195	441	41	204	204	0.34	0.21

Pd: Particle density; Bd: Bulk density; VCS: very coarse sand; CS: coarse sand; MS: medium sand; FS: fine sand; VFS: very fine sand; Ki: molecular ratio (SiO_2/Al_2O_3); Kr: molecular ratio $SiO_2/(Al_2O_3 + Fe_2O_3)$.

Corn intended for silage production was mechanically harvested 90 days after sowing on February 17, 2011, with the dry matter content ranging from 30 to 35%. The development of *Brachiaria* grasses was subsequently monitored for 250 days during dry mass production, which varied with climatic seasonality.

We collected samples from areas measuring 1 m^2 to evaluate the forage by randomly placing a square within each plot and cutting 20 cm of height, the product of which was prepared and packaged for subsequent studies. The cuts were performed on 03/21/2011 (1st cut), 05/11/2011 (2nd cut), 07/04/2011 (3rd cut), 09/12/2011 (4th cut), 10/24/2011 (5th cut) and 11/25/2011 (6th cut).

Forage management was preceded by topdressing during the rainy season with 50 kg ha^{-1} N and 25 kg ha^{-1} K_2O applied in the form of urea and potassium chloride, respectively. After evaluation, a standardization cut was performed throughout the experimental area, and the resulting residue was removed.

After the mechanized collection of corn silage and after each forage cutting, we continued the undisturbed sampling of the soil in each plot, randomly sampling layers of 0 to 20 cm using metal rings with a 6.4 cm diameter and 5.0 cm height. After the samples were removed, they were wrapped in PVC film, embedded in paraffin and packaged in styrofoam boxes for transport and storage in the laboratory.

The undisturbed samples were initially saturated and subjected to a matrix potential of - 6 kPa to determine their microporosity and field capacity (EMBRAPA, 2011a; SEVERIANO et al., 2011). The water content of the soil was subsequently adjusted to 0.04 to 0.53 dm^3 dm^{-3}, and the samples were subjected to penetrometer testing according to the method of Tormena et al. (1998) using a MARCONI-MA 933 bench penetrometer equipped with an electronic speed variator and a data logging system.

Then, the samples were dried in an oven at 105°C for 48 hours to determine the bulk density (Bd) according to the method of Embrapa (2011a). The total porosity (TP) was determined by equation 1 as follows:

$$TP = [1 - (Bd/Pd)] \tag{1}$$

in which Pd is the particle density.

The penetration resistance curve (PRC) was obtained by adjusting the penetration resistance (PR) values as a function of the volumetric water content (θ) and Bd using the non-linear model proposed by Busscher (1990) according to equation 2:

$$PR = 0.34\theta^{-0.78} Bd^{5.65}; R^2 = 0.82^{**} \tag{2}$$

The LLWR was determined according to the procedures described by Silva et al. (1994). Either the upper bound of the soil water content retained at the matrix potential of - 6 kPa or the point at which the air-filled porosity (θ_{AP}) was 10% was considered the field capacity (θ_{FC}) (SEVERIANO et al., 2008). The θ_{AP} (GRABLE; SIEMER, 1968) was calculated for each sample using Eq. 3:

$$\theta_{AP} = TP - 0.1 \tag{3}$$

in which TP is the total porosity.

For the lower bound, we considered the water content retained at the matrix potential of - 1,500 kPa the permanent wilting point (θ_{PWP}) (RICHARDS; WEAVER, 1943) and/or the water content corresponding to a penetration resistance of 2.5 MPa (θ_{PR}) (SEVERIANO et al., 2011). These values were used in Equation 2. The LLWR was obtained by adjusting the bounds of soil water content as a function of Bd with upper bound set as the lowest value between θ_{FC} and θ_{AP} and lower bound set as greatest value between θ_{PWP} and θ_{PR}.

A soil sample collected 32 days after the last forage evaluation was cut to assess the current chemical condition of the soil. Then, the grasses were desiccated with a glyphosate herbicide at a dosage of 1,500 g ha^{-1}. The soybean crop (*Glycine max* L. Merril) was then sown at a rate of 400,000 plants ha^{-1} with row spacings of 0.47 m to begin the second portion of the study. Fertilization was then performed according to the results of the soil analysis (Table 2) following the recommendations of Sousa and Lobato (2004).

Table 2. The sorption complex of Dystroferric Red Latosol used to cultivate crop-livestock integration systems during the seeding of *Brachiaria* grasses intercropped with corn (harvested 2010/2011) and soybean (harvested 2011/2012) in the municipality of Rio Verde, Goiás State.(1)

Ca	Mg	Al	H + Al	P	K	V(2)	m(3)	O.M.(4)	pH ($CaCl_2$)
	---------- cmolc dm^{-3} ----------			----- mg dm^{-3} -----		----- % -----		g kg^{-1}	
				Harvest 2010/2011					
3.2	1.3	0.0	5.6	2.9	104.0	46.6	0.0	27.3	5.2
				Harvest 2011/2012					
2.76	1.41	0.03	5.34	2.12	138.33	45.8	0.74	45.49	5.21

(1)20 cm depth; (2)V: base saturation; (2)m: aluminum saturation; (4): O.M.: organic matter. P: determined by Mehlich extraction.

Daily monitoring of the soil water content (θ) was initiated at the time of soybean crop planting and was extended until physiological maturity, which occurred between 12/08/2011 and 03/23/2012 in the 0–20 cm soil layer. Sampling was performed using a Saci, model S-20, semi-automatic soil sampler, and most sampling took place in the morning. The samples were packaged and sent to the laboratory for humidity determination using a gravimetric technique according to Embrapa (2011a).

We established a frequency of θ within the available water amplitude during the soybean cycle (Fwithin) according to Silva and Kay (1997) to evaluate the water availability of biological soil loosening for each system, considering the vegetative and reproductive phases of the soybean crop.

The temperature and rainfall were monitored during the experiment, and the results are shown in Figure 1.

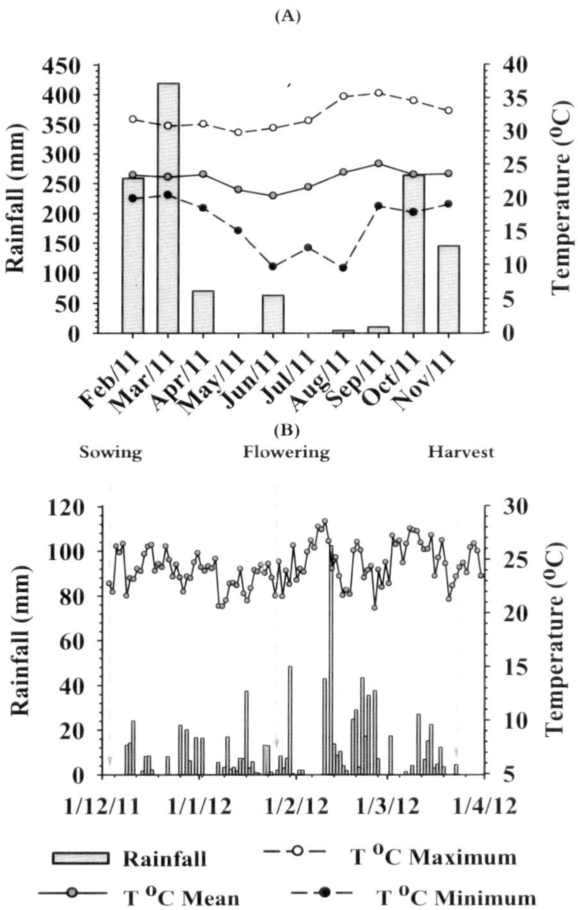

Figure 1. Rainfall (mm) and temperature (ºC): (A) monthly and (B) daily during the cultivation of *Brachiaria* grasses and soybean, respectively, in Rio Verde, Goiás State.

The dry matter production of *Brachiaria* grasses was subjected to an analysis of variance using a randomized block design, and, when significant, the mean values were compared using the Tukey test ($p < 0.05$).

Results and discussion

The variation in soil water content with increasing Bd and an emphasis on the LLWR are represented by the shaded area in Figure 2. The water retention increases with increasing Bd for both the field capacity (FC) and the permanent wilting point (PWP), which can be attributed to the change in pore size that arises from soil compaction (OLIVEIRA et al., 2007), especially the reduction in air-filled porosity with increasing density commonly observed in crop-livestock integration systems (SPERA et al., 2004).

If the increase in Bd provided a larger water adsorption surface for the solid particles (BLAINSKI et al., 2009) with a 0.15 $dm^3\ dm^{-3}$ maximum value for the LLWR at a Bd equal to 1.15 kg dm^{-3}, the mechanical impedance caused by the increase in PR prevented the soil water availability from reaching its maximum until it passed the critical density value (Bd_c) of 1.25 kg dm^{-3},at which the LLWR became null. This finding is consistent with the value obtained by Lima et al. (2012). Under these conditions, plant growth was limited at any humidity level, which suggests that the soil was physically degraded (SILVA et al., 1994). The amplitude of LLWR ranged from 0.01 to 0.15 $dm^3\ dm^{-3}$, which is common when latosols of this textural class are subjected to intensive handling (BLAINSKI et al., 2009), especially considering the critical penetration resistance (PR).

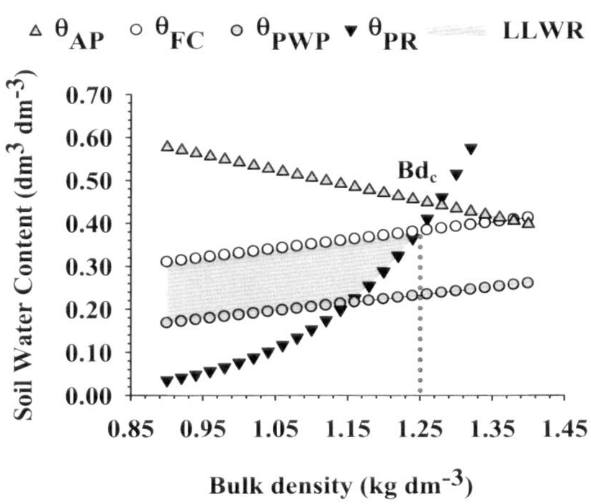

Figure 2. Variation in the soil water content (θ) with an increase in bulk density (Bd) within the critical bounds of the field capacity (θ_{FC}, - 6 kPa), the permanent wilting point (θ_{PWP}, - 1500 kPa), the air-filled porosity at 10% (θ_{AP}) and the soil penetration resistance of 2.5 MPa (θ_{PR}) for the Dystroferric Red Latosol used to cultivate crop-livestock integration systems in Rio Verde, Goiás State. The shaded area represents the least limiting water range (LLWR); Bd_c represents the critical density for plant development.

The decrease in soil porosity did not limit oxygenation (Figure 2) largely due to the development of structural pores in oxidic latosols (REATTO et al., 2007), which in turn favor their excessive aeration. According to Severiano et al. (2011), problems related to soil anoxia occur only when the structure is extremely degraded (Bd > Bd_c), or for a relatively short time after intense rainfall when the soil water content is above the field capacity because of the dynamic behavior of water in soil.

According to Blainski et al. (2008), the adoption of crop-livestock integration systems may provide an alternative for reestablishing soil physical quality characteristics. However, species and cultivars may vary in their capacity for biological loosening, as observed in Figure 3. The pastures of the evaluated genus *Brachiaria* exhibited variations in the degree to which the LLWR increased with the greater increase found in *B. brizantha*, particularly the cultivars Xaraes and Piata, whose forage root systems demonstrated high aggressiveness in the disruption of the compacted layers.

These results are in agreement with those of Bonelli et al. (2011), who, while assessing the effects of soil compaction on the yield and morphological characteristics of Piata and Mombaça grasses, found that soil compaction did not influence the production of Piata grass. This finding suggests that this plant has a more aggressive root system related to physiological processes and environmental interactions that can, in turn, enable these cultivars roots to break up the compacted layers.

Flores et al. (2008) reported that Xaraes grass has advantages over other cultivars of *Brachiaria*, such as faster regrowth rate and greater forage production, particularly during dry season. This greater metabolic activity appears to have contributed to loosening efficiency.

Marandu grass and *B. decumbens* exhibited an intermediate potential disruption of the compacted layers, which is consistent with the findings of Chioderoli et al. (2012) who observed higher yields in the successor crop when these two forages were sown at the time of the preceding corn topdressing. Calonego et al. (2011) also demonstrated that when *Brachiaria* was intercropped with corn for two consecutive years, it improved the structural conditions of soil, reducing its penetration resistance and consequently increasing the LLWR.

Moreover, MG4 grass and *B. ruziziensis* behaved as invading weed communities and exhibited limited potential for assisting in physical soil recovery. In addition, mechanical recovery via tillage provided the greatest recovery rates, as represented in Figure 3, which only covers last evaluation period because the tillage was performed while planting the summer crop.

When their nutritional demands were met, the cultivars Xaraes and Piata exhibited high yield potential for both the plant and animal components and adaptability to climatic challenges in production systems at high technological levels (COSTA et al., 2010; FLORES et al., 2008). This capacity mitigated the effects of seasonality in forage production caused by low rainfall and nocturnal temperatures during the fallow period (Figure 1).

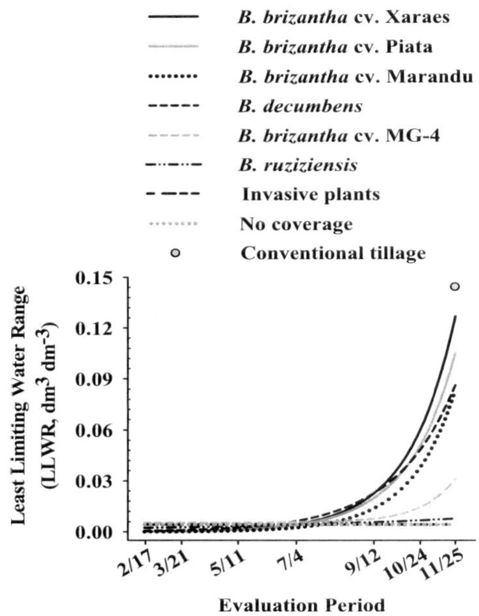

Figure 3. Changes in the least limiting water range (LLWR, dm^3 dm^{-3}) of Dystroferric Red Latosol cultivated in crop-livestock integration systems in Rio Verde, Goiás State after cutting the forage plants (1st evaluation of LLWR was performed upon harvest of previous corn crop).

Thus, even with the drastic reduction in forage production that occurs during the dry season, the highest yields for the cultivars were observed at the third, fourth and fifth grass cuttings compared with cuttings performed in the rainy season (1st, 2nd and 6th), as shown in Table 3. Thus, we suggest that accumulation of organic reserves in the stems of these grasses contributed to greater metabolic activity and regrowth in the resumption of rainy season. Consequently, greater root aggression led to improved biological loosening (Figure 3).

The increase in LLWR from the cultivation of *Brachiaria* grasses mitigated the effects of structural degradation on the soybean successor crop. In addition, Figure 4 suggests that the presence of straw on the soil surface from pasture drying enhanced the water availability to the crop. According to Andrade et al. (2011), water losses via evaporation in no-tillage systems, as compared with conventional tillage, are reduced in proportion to increases in waste, which preserves the water content of the soil by reducing thermal oscillation in the surface layer (BRAIDA et al., 2006; CHIODEROLI et al., 2012).

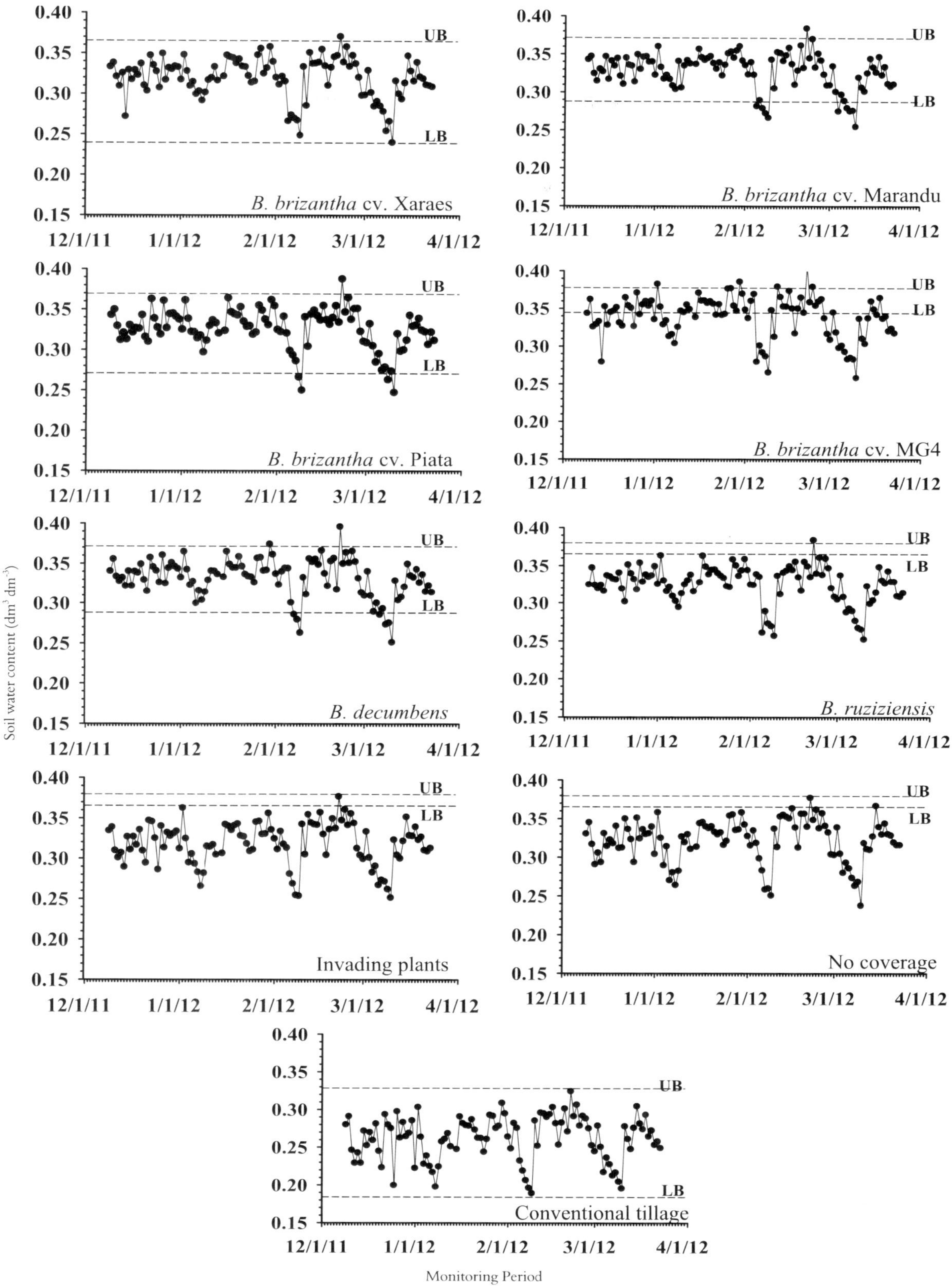

Figure 4. Temporal variation in the soil water content during the soybean crop cycle in relation to the critical bounds of the least limiting water range in a Dystroferric Red Latosol under crop-livestock integration systems. Ub: upper bound (θ_{FC}, - 6 kPa) and Lb: lower bound (θ_{PWP}, - 1500 kPa or θ_{PR} 2.5 MPa) of the LLWR for the monitored period.

Table 3. Production of forage grass dry mass of the genus *Brachiaria* in crop-livestock integration systems in Rio Verde, Goiás Stsate.[1]

Grass	Evaluation cut[2]						Total
	First	Second	Third	Fourth	Fifth	Sixth	
	kg ha^{-1}						
B. brizantha cv. Xaraes	5.433Ca	3.947Cb	1.446Ac	858Ad	1.119Ad	4.140Bb	16.943AB
B. brizantha cv. Piata	7.373Aa	3.847CDc	1.255Ad	883Ad	1.053Ad	4.915Ab	19.325A
B. brizantha cv. Marandu	5.302Ca	4.236BCb	976ABc	980Ac	786Ac	3.781BCc	16.060B
B. decumbens	6.136Ba	4.543Bb	1.283Ad	774Ae	1.053Ade	3.699Cc	17.487AB
B. brizantha cv. MG4	4.800Da	5,110Aa	1,268Ac	851Ac	839Ac	3.667Cb	16.536B
B. ruziziensis	5.528Ca	3.434Db	604Bd	286Bd	331Bd	2.217Dc	12.401C

[1]For each grass, mean values are followed by same uppercase letters in columns, and for each evaluation cut, mean values followed by same lower case letters in rows did not differ according to the Tukey's test at 5% probability (average of four replicates). 1st evaluation cut: 03/21/2011; 2nd evaluation cut: 05/11/2011; 3rd evaluation cut: 07/04/2011; 4th evaluation cut: 09/12/2011; 5th evaluation cut: 10/24/2011; 6th evaluation cut: 11/25/2011. [2]Sum of production from the six evaluation cuts.

Temporal variation in the soil water content during the subsequent crop cycle demonstrated that the upper limit of the LLWR had little influence on the water availability (Figure 4), even for systems in which the forages exhibited low tillage potential (Figure 3).

We observed occasional overall anoxia problems when monitoring was conducted following intense rainfall (e.g., 02/22/2012, Figure 1). As previously discussed, soil water content returned to LLWR bounds in subsequent evaluations due to predominance of structural pores and the dynamic behavior of water, as corroborated by Blainski et al. (2009).

The lower bound of the LLWR provided greater water restrictions on all crop-livestock integration systems (Figure 4) in accordance with Leão et al. (2006), which increases the likelihood of stress in the subsequent crop due to mechanical impediment to the root system (KLEIN; CÂMARA, 2007). However, because the soil was tilled by grass cultivation, the frequency with which θ remained within the LLWR bounds increased (Fwithin), as shown in Table 4.

Table 4. Least limiting water range (LLWR), physical soil recovery rate and analysis of θ frequency within the LLWR bounds (Fwithin) during the soybean cycle for crop-livestock integration systems using grasses of the genus *Brachiaria* on a Dystroferric Red Latosol in Rio Verde, Goiás State.

Treatment	LLWR	Physical Recovery	Fwithin (%)	
	dm^3 dm^{-3}	(%)	VP	RP
B. brizantha cv. Xaraes	0.13	88	100.0	98.3
B. brizantha cv. Piata	0.10	69	100.0	89.8
B. brizantha cv. Marandu	0.08	54	100.0	83.1
B. decumbens	0.08	54	100.0	84.7
B. brizantha cv. MG4	0.03	20	56.5	42.4
B. ruziziensis	0.01	7	0.0	0.0
Invading plants	0.01	7	0.0	1.7
No coverage	0.01	7	0.0	3.4
Conventional tillage	0.14	98	100.0	100.0

VP: Vegetative phase; RP: Reproductive phase.

Although conventional tillage systems allow the soil to remain uncovered and may lead to water loss via evaporation unlike a soil with straw on its surface (PERES et al., 2010), the mechanical action of tillage equipment was proven to provide greater physical recovery of the soil (Table 4) and consequently an increase in water available to the successor crop. Conventional tillage was the only treatment in which the PR was not within the lower bound, as corroborated by Serafim et al. (2008). Therefore, an absence of water stress is suggested and quantified by the occurrence of a θ value within the LLWR bounds (Fwithin) in 100% of the monitoring evaluations of this treatment. These data demonstrate that crop dependence on the regular rain distribution occurred during the crop cycle (Figure 1).

Table 4 also indicates that the biological recovery promoted by Xaraes grass resulted in a small reduction in Fwithin during the reproductive phase of the soybean ($LLWR_{100\%VP}$ and $LLWR_{98.3\%RP}$). We observed similar effects in the treatments with piata grass ($LLWR_{1.00\%VP}$ and $LLWR_{89.8\%RP}$), followed by *B. decumbens* ($LLWR_{100\%VP}$ and $LLWR_{84.7\%RP}$) and marandu grass ($LLWR_{100\%VP}$ and $LLWR_{83.1\%RP}$), which is agreement with data obtained by Bonelli et al. (2011), Calonego et al. (2011) and Chioderoli et al. (2012).

These results emphasize that the occurrence of water stress is coincident with the phenological stages of increased water demand by crop. According to Embrapa (2011b), the water requirements of soybean crop increase with plant development and peak at flowering and grain filling. Significant water deficits during these stages can cause physiological changes in the plants and lead to premature leaf and flower fall and pod abortion, which can also reduce grain yield. Because the soil is exposed in intensive agricultural production systems, the choice of grass cultivars as rotation crops is therefore critical as they provide a potentially effective form of biological loosening (Figure 3 and Table 4).

MG4 grass yielded an intermediate response ($LLWR_{56.5\%VP}$ and $LLWR_{42.4\%RP}$). An inferior response was observed in treatments that contained weeds ($LLWR_{nullVP}$ and $LLWR_{1.7\%RP}$) and were no coverage ($LLWR_{nullVP}$ and $LLWR_{3.4\%RP}$), suggesting that physical environment unfavorable to plant growth these conditions (SEVERIANO et al., 2011).

The species *B. ruziziensis* has been widely used as a cover crop for no-tillage systems in the Cerrados region because of its ease of desiccation during the formation of straw with its rapid death and clump reduced, which favor mechanized sowing. Moreover, the use of *B. ruziziensis* in intensive production systems is limited when the low potential for physical soil recovery is considered (LLWR = 0.01 dm^3 dm^{-3}, establishing an Fwithin of 0.0 and 0.0% for the vegetative and reproductive phases, respectively, as shown in Table 4).

We emphasize that the choice of tillage plants can be part of a management strategy for the recovery of soil structural quality in crop-livestock integration systems, in contrast with the mechanical recovery promoted by conventional tillage. Plant-based tillage can be used in addition to diversifying and verticalizing production, minimizing costs, diluting risks and aggregating the value of agricultural products. Plant-based tillage is also a strategy for environmental preservation because it is a low-carbon agricultural model.

Conclusion

Our results evidence that led to the following observations:

- The cultivation of *Brachiaria brizantha* or *Brachiaria decumbens* can be used as a management strategy in the edaphic recovery of crop-livestock integration systems; and
- The Xaraes and Piata grasses provided greater biological loosening and consequently increased water availability for the successor soybean crop.

Acknowledgements

The authors thanks the FAPEG (Fundação de Amparo à Pesquisa do Estado de Goiás) for partial funding of our research, Call 09/2010 - ProPós (Process number 201010267000906).

References

ANDRADE, J. G.; SEGUI, J. P.; CARLESSO, R.; TROIS, C.; KNIES, A. E. Perdidas de agua por evaporación en maíz con siembra convencional y directa para diferentes niveles de cobertura muerta. I. resultados experimentale. **Revista Ciencias Técnicas Agropecuárias**, v. 20, n. 2, p. 29-36, 2011.

ARROYAVE, C.; BARCELÓ, J.; POSCHENRIEDER, C.; TOLRÀ, R. Aluminium-induced changes in root epidermal cell patterning, a distinctive feature of hyperresistance to Al in *Brachiaria decumbens*. **Journal of Inorganic Biochemistry**, v. 105, n. 11, p. 1477-1483, 2011.

BAVOSO, M. A.; GIAROLA, N. F. B.; TORMENA, C. A.; PAULETTI, V. Preparo do solo em áreas de produção de grãos, silagem e pastejo: efeito na resistência tênsil e friabilidade de agregados. **Revista Brasileira de Ciência do Solo**, v. 34, n. 1, p. 227-234, 2010.

BLAINSKI, E.; TORMENA, C. A.; FIDALSKI, J.; GUIMARÃES, R. M. L. Quantificação da degradação física do solo por meio da curva de resistência do solo à penetração. **Revista Brasileira de Ciência do Solo**, v. 32, n. 3, p. 975-983, 2008.

BLAINSKI, E.; GONÇALVES, A. C. A.; TORMENA, C. A.; FOLEGATTI, M. V.; GUIMARÃES, R. M. L. Intervalo hídrico ótimo num Nitossolo Vermelho distroférrico irrigado. **Revista Brasileira de Ciência do Solo**, v. 33, n. 2, p. 273-281, 2009.

BONELLI, E. A.; BONFIM-SILVA, E. M.; CABRAL, C. E. A.; CAMPOS, J. J.; SCARAMUZZA, W. L. M. P.; POLIZEL, A. C. Compactação do solo: Efeitos nas características produtivas e morfológicas dos capins Piatã e Mombaça. **Revista Brasileira de Engenharia Agrícola e Ambiental**, v. 15, n. 3, p. 264-269, 2011.

BRAIDA, J. A.; REICHERT, J. M.; VEIGA, M.; REINERT, D. J. Resíduos vegetais na superfície e carbono orgânico do solo e suas relações com a densidade máxima obtida no ensaio proctor. **Revista Brasileira de Ciência do Solo**, v. 30, n. 4, p. 605-614, 2006.

BUSSCHER, W. J. Adjustment of flat-tipped penetrometer resistance data to common water content. **Transactions of the American Society of Agricultural Engineers**, v. 33, n. 2, p. 519-524, 1990.

CALONEGO, J. C.; BORGHI, E.; CRUSCIOL, C. A. C. Intervalo hídrico ótimo e compactação do solo com cultivo consorciado de milho e braquiária. **Revista Brasileira de Ciência do Solo**, v. 35, n. 6, p. 2183-2190, 2011.

CHIODEROLI, C. A.; MELLO, L. M. M.; GRIGOLLI, P. J.; FURLANI, C. E. A.; SILVA, J. O. R.; CESARIN, A. L. Atributos físicos do solo e produtividade de soja em sistema de consórcio milho e braquiária. **Revista Brasileira de Engenharia Agrícola e Ambiental**, v. 16, n. 1, p. 37-43, 2012.

COSTA, K. A. P.; OLIVEIRA, I. P.; SEVERIANO, E. C.; SAMPAIO, F. M. T.; CARRIJO, M. S.; RODRIGUES, C. R. Extração de nutrientes pela fitomassa de cultivares de Brachiaria brizantha sob doses de nitrogênio. **Ciência Animal Brasileira**, v. 11, n. 2, p. 307-314, 2010.

EMBRAPA-Empresa Brasileira de Pesquisa Agropecuária. **Sistema brasileiro de classificação de solos**. 2nd ed. Rio de Janeiro: Embrapa Solos, 2006.

EMBRAPA-Empresa Brasileira de Pesquisa Agropecuária. **Manual de métodos de análises de solo**. 2. ed. Rio de Janeiro: Embrapa Solos, 2011a.

EMBRAPA-Empresa Brasileira de Pesquisa Agropecuária. **Tecnologias de Produção de Soja** - Região Central do Brasil 2012 e 2013. Londrina: Embrapa Soja, 2011b.

FLORES, J. P. C.; ANGHINONI, I.; CASSOL, L. C.; CARVALHO, P. C. F.; LEITE, J. G. B.; FRAGA, T. I. Atributos físicos do solo e rendimento de soja em sistema plantio direto em integração lavoura-pecuária com diferentes pressões de pastejo. **Revista Brasileira de Ciência do Solo**, v. 31, n. 4, p. 771-780, 2007.

FLORES, R. S.; EUCLIDES, V. P. B.; ABRÃO, M. P. C.; GALBEIRO, S.; DIFANTE, G. S.; BARBOSA, R. A. Desempenho animal, produção de forragem e características estruturais dos capins marandu e xaraés submetidos a intensidades de pastejo. **Revista Brasileira de Zootecnia**, v. 37, n. 8, p. 1355-1365, 2008.

GRABLE, A. R.; SIEMER, E. G. Effects of bulk density, aggregate size, and soil water suction on oxygen diffusion, redox potential and elongation of corns roots. **Soil Science Society of America Journal**, v. 32, n. 2, p. 180-186, 1968.

GUIMARÃES, R. M. L.; TORMENA, C. A.; ALVES, S. J.; FIDALSK, J.; BLAINSK, E. Tensile strength, friability and organic carbon in an oxisol under a crop-livestock system. **Scientia Agricola**, v. 66, n. 4, p. 499-505, 2009.

IMHOFF, S.; SILVA, A. P.; TORMENA, C. A. Aplicação da curva de resistência no controle da qualidade física de um solo sob pastagem. **Pesquisa Agropecuária Brasileira**, v. 35, n. 7, p. 1493-1500, 2000.

KLEIN, V. A.; CAMARA, R. C. Rendimento da soja e Intervalo Hídrico Ótimo em Latossolo vermelho sob plantio direto escarificado. **Revista Brasileira de Ciência do Solo**, v. 31, n. 2, p. 221-227, 2007.

LANZANOVA, M. E.; NICOLOSO, R. S.; LOVATO, T.; ELTZ, F. L. F.; AMADO, T. J. C.; REINERT, D. J. Atributos físicos do solo em sistema de integração lavoura-pecuária sob plantio direto. **Revista Brasileira de Ciência do Solo**, v. 31, n. 5, p. 1131-1140, 2007.

LEÃO, T. P.; SILVA, A. P.; MACEDO, M. C. M.; IMHOFF, S.; EUCLIDES, V. P. B. Least limiting water range: a potential indicator of changes in near-surface soil physical quality after the conversion of Brazilian Savanna into pasture. **Soil and Tillage Research**, v. 88, n. 1-2, p. 279-285, 2006.

LIMA, V. M. P.; OLIVEIRA, G. C.; SERAFIM, M. E.; CURI, N.; EVANGELISTA, A. R. Intervalo Hídrico Ótimo como indicador de melhoria da qualidade estrutural de Latossolo degradado. **Revista Brasileira de Ciência do Solo**, v. 36, n. 1, p. 71-78, 2012.

OLIVEIRA, G. C.; SEVERIANO, E. C.; MELLO, C. R. Dinâmica da resistência à penetração de um Latossolo vermelho da Microregião de Goiânia, GO. **Revista Brasileira de Engenharia Agrícola e Ambiental**, v. 11, n. 13, p. 265-270, 2007.

PACHECO, L. P.; PIRES F. R.; MONTEIRO, F. P.; PROCÓPIO, S. O.; ASSIS, R. L.; CARMO, M. L.; PETTER, F. A. Desempenho de plantas de cobertura em sobressemeadura na cultura da soja. **Pesquisa Agropecuária Brasileira**, v. 43, n. 7, p. 815-823, 2008.

PERES, J. G.; SOUZA, C. F.; LAVORENTI, N. A. Avaliação dos efeitos da cobertura de palha de cana-de-açúcar na umidade e na perda de água do solo. **Engenharia Agrícola**, v. 30, n. 5, p. 875-886, 2010.

PETTER, F. A.; PACHECO, L. P.; PROCÓPIO, S. O.; CARGNELUTTI FILHO, A.; VOLF, M. R. Seletividade de herbicidas à cultura do milho e ao capim-braquiária cultivadas no sistema de integração lavoura-pecuária. **Semina: Ciências Agrárias**, v. 32, n. 3, p. 855-864, 2011.

REATTO, A.; BRUAND, A.; SILVA, E. M.; MARTINS, E. S.; BROSSARD, M. Hydraulic properties of the diagnostic horizon of Latosol of a regional toposequence across the Brazilian central plateau. **Geoderma**, v. 139, n. 1-2, p. 51-59, 2007.

RICHARDS, L. A.; WEAVER, L. R. Fifteen-atmosphere percentage as related to the permanent wilting point. **Soil Science**, v. 56, n. 5, p. 331-339, 1943.

SERAFIM, M. E.; VITORINO, A. C. T.; PEIXOTO, P. P. P.; SOUZA, C. M. A.; CARVALHO, D. F. Intervalo hídrico ótimo em um latossolo vermelho distroférrico sob diferentes sistemas de produção. **Engenharia Agrícola**, v. 28, n. 4, p. 654-665, 2008.

SEVERIANO, E. C.; OLIVEIRA, G. C.; DIAS JÚNIOR, M. S.; OLIVEIRA, L. F. C.; CASTRO, M. B. Pressão de preconsolidação e intervalo hídrico ótimo como indicadores de alterações estruturais do solo em decorrência das operações de colheita da cana-de-açúcar. **Revista Brasileira de Ciência do Solo**, v. 32, n. 4, p. 1419-1427, 2008.

SEVERIANO, E. C.; OLIVEIRA, G. C.; DIAS JUNIOR, M. S.; COSTA, K. A. P.; CASTRO, M. B.; MAGALHÃES, E. N. Potencial de descompactação de um Argissolo promovido pelo capim-tifton 85. **Revista Brasileira de Engenharia Agrícola e Ambiental**, v. 14, n. 1, p. 39-45, 2010.

SEVERIANO, E. C.; OLIVEIRA, G. C.; DIAS JÚNIOR, M. S.; COSTA, K. A. P.; SILVA, F. G.; FERREIRA FILHO, S. M. Structural changes in latosols of the cerrado region: I – relationships between soil physical properties and least limiting water range. **Revista Brasileira de Ciência do Solo**, v. 35, n. 3, p. 773-782, 2011.

SILVA, A. P.; KAY, B. D. Effect of Soil Water Content Variation on the Least Limiting Water Range. **Soil Science Society of America Journal**, v. 61, n. 3, p. 884-888, 1997.

SILVA, A. P.; KAY, B. D.; PERFECT, E. Characterization of the least limiting water range. **Soil Science Society of America Journal**, v. 58, n. 6, p. 1775-1781, 1994.

SOUSA, D. M.; LOBATO, E. **Cerrado**: correção do solo e adubação. 2. ed. Brasília: Embrapa Informação Tecnológica, 2004.

SPERA, S. T.; SANTOS, H. P.; FONTANELI, R. S.; TOMM, G. O. Efeitos de sistemas de produção de grãos envolvendo pastagens sob plantio direto nos atributos físicos de solo e na produtividade. **Revista Brasileira de Ciência do Solo**, v. 28, n. 3, p. 533-542, 2004.

TORMENA, C. A.; SILVA, A. P.; LIBARDI, P. L. Caracterização do intervalo hídrico ótimo de um Latossolo roxo sob plantio direto. **Revista Brasileira de Ciência do Solo**, v. 22, n. 4, p. 573-581, 1998.

14

Physical and hydraulic properties of aridisols as affected by nutrient and crop-residue management in a cotton-wheat system

Muhammad Mahmood-ul-Hassan*, Eajaz Rafique and Abdul Rashid

*Land Resources Research Institute, National Agricultural Research Centre, Islamabad 45500, Pakistan. *Author for correspondence. E-mail: mmh@comsats.net.pk*

ABSTRACT. A 5-year cotton–wheat rotation field experiment was conducted on two alkaline-calcareous soils to investigate the impact of integrated nutrient-management and crop-residue incorporation on soil physical and hydrological properties. The nutrient treatments were: T1–farmers' fertilizer use; T2–balanced nutrient management (recommended N, P, Zn. and B from mineral sources); T3–integrated nutrient management, same as T2, except that 75% N was applied from fertilizer and 25% as FYM; and T4 same as T2, except that every alternate year, wheat was substituted by *Berseem* green manure. All treatments were compared with and without crop residue incorporation. Soil organic matter (SOM) content in both the soils was significantly increased with T2, T3 and T4 as compared with T1, and was increased further where the nutrient-management treatments were applied in combination with crop-residue recycling. Higher increase in SOM content was observed in soil having relatively lower initial SOM (0.61%) than soil having initial 0.80% SOM content. Increased SOM content, in return, decreased soil bulk density, improved macro- and meso-porosity, and enhanced percent recovery of stable aggregates correspondingly. Infiltration rates were 20, 49 and 26% higher with T2, T3 and T4, respectively, over T1 and 64% higher with crop-residue incorporation over crop-residue removal. Positive impacts on soil physical properties were also observed in the sub-soil layers.

Keywords: cotton-wheat system, crop-residue recycling, integrated nutrient management, bulk density, water stable aggregate, hydraulic conductivity.

Introduction

Maintenance and improvement of soil productivity are among the most crucial concerns for sustaining agricultural productivity in arid and semiarid regions of the world. In these climates, desertification, overexploitation and low soil organic matter (SOM) are major constraints to soil productivity as they limit the rehabilitative physical, chemical and biological processes. The SOM is considered a key element for restoring these physical, chemical and biological processes for maintaining and enhancing soil productivity. Enhancement of SOM from the presently low levels can lead to soil productivity improvement on a sustainable basis by improving soil's biophysical properties.

Majority of soils, developed from alluvial deposits in South Asia, are low in organic matter and are weakly structured. These are susceptible to degradation as a result of wetting and drying, compaction caused by on-farm traffic and ill-timed tillage operations. The consequences of the degradation of structure include restricted water and air permeability and drainage, lowered total porosity, increased mechanical impedance to root proliferation and poor crop establishment (BLANCO-CANQUI; LAL, 2009).

Soil organic matter accumulation through management practices improves soil structure and resilience of infiltration and water-holding capacity, and reduces soil-temperature extremes and soil-water loss, all of which are necessary for sustainable agricultural systems (BLANCO-CANQUI; LAL, 2009). The amount of SOM accumulation is dependent on the annual addition of organic materials and the rate of decomposition, the latter being the highest in hot climatic regions (DE RIDDER; VAN KEULEN, 1990; ROWELL, 1994). In the cotton-belt of Pakistan, SOM is restored/maintained traditionally by recycling farm-animal waste organic materials. As adequate quantity of farmyard manure (FYM) is not available to farmers, crop-residue recycling constitutes an important management practice to restore SOM. In the cotton–wheat system, a major cropping system (3.07 million ha) in Pakistan, more than 10 million tons of cotton stalks are produced annually; which are totally removed, physically, from the fields and burn as fuel. Similarly, 15 million tons of wheat-straw, produced annually, is either removed from the field for dry silage/fodder usage or is burned in the fields. The physical removal of crop residues from cotton-wheat fields results in export of nutrients and organic matter from the soil system and, in turn, deteriorates the physical quality. Crop-residue burning causes not only a decline in soil quality but also air pollution (HOOKER et al., 1982; SINGH et al., 2005a).

Beneficial effects of SOM include increase cation-exchange capacity, facilitate water infiltration and conductance, enhance soil-water retention, and stabilize soil aggregates and reduce soil erosion. Contrarily, low SOM content adversely affects physical and hydrological properties of the soil. Despite high levels of fertilizer use, poor soil structure and low SOM threaten agricultural production in arid and semi-arid regions. Therefore, regeneration of soil structure and improvement in organic matter is an out-and-out requirement for sustainable soil health and crop productivity.

The study was conducted to investigate the effects of integrated nutrient and crop residue management on SOM content and soil physical and hydraulic properties in irrigated cotton–wheat cropping system practiced on Aridisols.

Material and methods

Site description

The 5-year field experiment was conducted on Aridisol - two predominant soil series, i.e., Shahpur – clay 29 and silt 56% (fine-loamy, mixed, hyperthermic Fluventic Camborthid) at Chah A. Rahim (29° 56′ N; 71° 23′ E) and Awagat – clay 12 and silt 46% (coarse silty, mixed, hyperthermic, Fluventic Camborthid) at Chak 5-Faiz (29° 58′ N; 71° 31′ E) in the Punjab province of Pakistan, where the cotton-wheat system is prevalent.

The Shahpur soil is deep, moderately well-drained, calcareous, weak coarse sub-angular blocky, whereas the Awagat soil is deep, well-drained, calcareous, massive and weak coarse sub-angular blocky. The soils of the experimental area were formed from alluvial deposits derived from the Himalayas, and the sediments range from recent to sub-recent and late Pleistocene (AKRAM et al., 1969).

The area has an arid subtropical continental climate with large seasonal fluctuations in temperature as well as rainfall. The summers are very hot, with a mean monthly maximum air temperature of 43°C and minimum of 6°C (Figure 1). The annual rainfall ranged from 80 to 300 mm year^{-1}. About two-thirds of the total rains were received during the monsoon season, from mid-July to mid-September, as heavy downpours. Prior to the initiation of the experiments, continuous cotton-wheat rotation had been practiced at both the field sites using traditional tillage (cultivation with multi-tine tractor driven cultivator and planking with a long flat piece of lumber) and cultural practices.

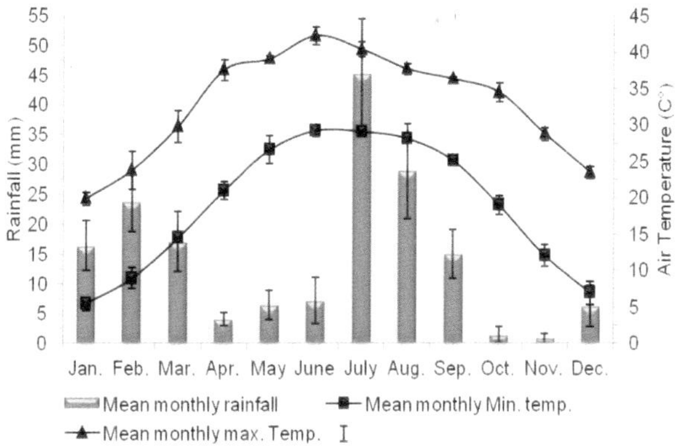

Figure 1. Mean monthly rainfall and air temperature during 2000–2005 at the experimental sites.

Experimental design and treatments

The permanent layout of cotton (*Gossypium hirsutum* L.)–wheat (*Triticum aestivum* L.) system field experiment was started in 2000, with four nutrient-management treatments: T1 – Farmers' fertilizer use (FFU), i.e., 110 kg N ha^{-1} in cotton and 80 kg N + 60 kg P_2O_5 ha^{-1} in wheat; T2 – Balanced nutrient management (BNM), i.e., 170 kg N + 60 kg P_2O_5 + 5 kg Zn + 1 kg B ha^{-1} in cotton and 140 kg N + 100 kg P_2O_5 + 5 kg Zn + 1 kg B ha^{-1} in wheat; T3 – Integrated nutrient management (INM), same as BNM, except that 75% N was applied as fertilizer and 25% N as FYM; T4 – same as BNM, except that wheat was substituted by *Berseem* green manure during years 2 and 4 of the experiment (BNM+GM). All the treatments were replicated four times and imposed with and without incorporation of cotton stalks and wheat-straw into the soil. In crop-residue-recycled plots, cotton stalks (after picking seed cotton) were incorporated into the soil with the help of a rotavator 14-21 days prior to wheat sowing, and wheat straw was incorporated into the soil 14-21 days prior to cotton sowing. In residue-removed plots, all the cotton stalks and wheat straw (harvested at ground level) were removed from the field.

Measurements

During May 2005, after harvesting the fifth wheat crop, soil samples from two depths, 0–15 and 15-30 cm, were taken randomly from all replicated plots. Four intact soil samples, 7.0 cm diameter and 5.0 cm long from each plot,, were taken using a double-cylinder steel core sampler, to measure bulk density and water retention characteristics. Relatively larger four intact samples, i.e., 10 cm diameter and 10 cm long, from each plot, were also collected for measuring saturated hydraulic conductivity. Also, disturbed composite soil samples from both depths were collected to measure aggregation status and organic carbon.

Soil physical and hydraulic properties were measured following the procedures described by Sparks et al. (1996) and Dane and Topp (2002). Organic carbon was determined via the rapid titration method (NELSON; SOMMERS, 1996). Aggregate stability was determined by wet-sieving method and water-stable aggregates were expressed as the sand-corrected weight, as described by Nimmo and Perkings (2002). The undisturbed soil cores were dried in an oven at 105°C for 24h and then weighed to calculate bulk density (GROSSMAN; REINSCH, 2002). Total porosity was calculated as 1-(bulk density/particle density), assuming that particle density = 2.65 mg m^{-3} (MARSHALL et al., 1996). Soil texture was determined using the hydrometer method (GEE; OR, 2002) and saturated hydraulic conductivity (K_s) with a constant-head method (REYNOLDS et al., 2002). Water-retention characteristics were measured using saturated (for 24 hours) undisturbed soil cores, successively at -1, -5, and -10 kPa using a fine-sand tension table with a hanging water column and at -30, -100, -500 and -1500 kPa with a pressure plate apparatus.

Pore-size distribution was calculated from water-retention data estimating equivalent mean pore diameter for a particular soil matric pressure using equation described by Danielson and Sutherland (1986). Unsaturated hydraulic conductivity, in relation to the water content, was calculated from the Campbell (1974) empirical relationship. Infiltration was measured, *in situ*, by double ring infiltrometers of 41– 52 cm inner diameter of rings according to Bertrand (1965).

Statistical Analysis

Analysis of variance of the measured parameters was performed using MSTAT-C and treatment means were compared using Duncan's multiple range test (DMRT) at 5% probability level.

Results and discussion

Soil organic matter

After 5 years of following the cotton-wheat rotation in the Aridisol SOM content in the surface soil (0–15 cm) was significantly increased by BNM, INM and BNM+GM treatments in comparison with the FFU treatment in both Awagat and Shahpur soils (Table 1). Significant increase in

SOM was also observed with crop-residue incorporation compared with crop-residue removal treatment in both depths ($p \leq 0.05$). Zeleke et al. (2004); Singh et al. (2007) also observed increases in organic carbon contents of soils following incorporation of crop residues compared with soils where residues were removed. The percent increases in SOM contents over FFU in Awagat soil were 4.7, 9.4, and 6.3 for the BNM, INM, and BNM+GM treatments, respectively; whereas the corresponding increases at Shahpur soil were 3.6, 7.1 and 3.6. The highest SOM content was recorded with INM, where 25% N was supplied through FYM. The magnitude of increase in SOM with INM over FFU was greater in Awagat soil (9.4%) compared with Shahpur soil (7.1%). Because the initial SOM in Shahpur soil was higher than that of Awagat soil. The observed 9% SOM increase after five continuous incorporations of wheat and cotton residues was rather low considering the 17% SOM increase after continuous incorporation of wheat and cotton residues for three years (GURSOY et al., 2011) and 63% in a three year field experiment under maize-based system (ZELEKE et al., 2004). Low SOM increase in this study could be because of high summer temperatures prevailing in the study area as well as because of the nature of the crop constituents. Thornley and Cannell (2001) reported that high temperature led to accelerated oxidation rates of the added crop residues.

The organic-matter buildup (up to ≈ 0.86%) is straightforward in aridisol (soils exist in study area) having low initial organic matter and afterwards increase is very slow (RAFIQUE et al., 2012). Higher increase in SOM (9%) in Awagat soil with relatively low initial SOM (0.61%) than in Shahpur soil having higher initial SOM content (0.80%) was in good agreement with the observation made by Zeleke et al. (2004). They reported lesser increase in SOM for Humbo site having relatively higher initial SOM than Alaba site having relatively lower initial SOM after continuous incorporation of maize residues for three years in a field experiment. In the 15-30 cm soil layer, SOM contents with FFU, BNM, and BNM+GM treatments were statistically similar and were significantly lower than in the INM treatment where 25% mineral N was substituted by organic N supplied (as FYM) in both the soils. A significant increase in SOM content was observed following incorporation of crop residue than when crop residue was removed. Increases in SOM content in the 15-30 cm layers suggested that a fraction of the added organic manure and crop residues was placed down to the 15-30 cm soil layers during incorporation operations or over the 5-year period of field experimentation.

Bulk density

Crop-residue incorporation across a period of five years significantly lowered bulk density of the surface 0-15 cm soil (Table 2). Bulk density was significantly less for balanced nutrient management treatment than for the farmers' fertilizer use treatment in both the soils. A bulk-density reduction trend was observed with the expected increase in biomass incorporated into the soils, i.e., 1.37, 1.34, 1.31 and 1.31 mg m^{-3} at Awagat and 1.34, 1.30, 1.27 and 1.29 mg m^{-3} bulk density at Shahpur corresponding to FFU, BNM, INM and BNM+GM treatments, respectively. Increasing biomass input to the soil through residue incorporation and enhanced crop growth have been reported to improve soil quality relative to organic carbon and biotic activity (LAL, 1989; KARLEN et al., 1994), and this might be the reason for the significantly lower bulk density following BNM, INM and BNM+GM treatments. This assumption was further substantiated by the strong correlations ($r = 0.87$ and 0.90) observed between bulk density and organic carbon content in Awagat soil and Shahpur soil ($p \leq 0.05$).

Table 1. Effect of nutrient managements and crop residue incorporation on soil organic matter content (%) after 5 years.

Nutrient Treatments	Awagat						Shahpur					
	0-15 cm			15-30 cm			0-15 cm			15-30 cm		
	CRR	CRI	Mean	CRR	CRI	Mean	CRR	CRI	Mean	CRR	CRI	Mean
FFU	0.62¶	0.66¶	0.64¶¶	0.42	0.45	0.44	0.81	0.87	0.84	0.53	0.58	0.56
BNM	0.64	0.69	0.67	0.45	0.49	0.47	0.84	0.90	0.87	0.57	0.61	0.59
INM	0.67	0.72	0.70	0.48	0.54	0.51	0.87	0.92	0.90	0.60	0.66	0.63
BNM+GM	0.66	0.70	0.68	0.44	0.50	0.47	0.84	0.90	0.87	0.58	0.64	0.61
Mean	0.65	0.69	0.67	0.45	0.50		0.84	0.90	0.87		0.57	0.62
LSD ($p \leq 0.05$)												
Treatments			0.02			0.02			0.02			0.02
Interaction			0.03			0.03			0.03			0.03

CRR: crop-residue removed; CRI: crop-residue incorporated; LSD:l least significance difference; ¶: average of four replicates; ¶¶: means of eight values.

Table 2. Effect of nutrient managements and crop residue incorporation on soil bulk density (mg m^{-3}) after 5 years.

Nutrient Treatments	Awagat						Shahpur					
	0-15 cm			15-30 cm			0-15 cm			15-30 cm		
	CRR	CRI	Mean	CRR	CRI	Mean	CRR	CRI	Mean	CRR	CRI	Mean
FFU	1.41¶	1.33¶	1.37¶¶	1.62	1.56	1.59	1.37	1.31	1.34	1.60	1.54	1.57
BNM	1.37	1.31	1.34	1.59	1.54	1.57	1.32	1.28	1.30	1.57	1.55	1.56
INM	1.34	1.28	1.31	1.55	1.50	1.53	1.30	1.25	1.27	1.54	1.49	1.51
BNM+GM	1.33	1.29	1.31	1.56	1.53	1.55	1.31	1.27	1.29	1.56	1.52	1.54
Mean	1.36	1.30		1.58	1.53		1.33	1.28		1.57	1.52	
LSD ($p \leq 0.05$)												
Treatments			0.0			0.02			0.02			0.02
Interaction			0.02			0.02			0.02			0.02

CRR: crop-residue removed; CRI: crop-residue incorporated; LSD: least significance difference; ¶: average of four replicates; ¶¶: means of eight values.

Mostly, the Aridisols with low organic matter under extensive cultivation and extreme climate conditions (high temperature and low rainfall), those exist in the cotton belt of Punjab, Pakistan, are even more susceptible to soil compaction. It is evident from the results reported by Rafique et al. (2012) that the bulk density of surface soil (0–15 cm) varied from 1.25 to 1.66 mg m^{-3} and sub-surface (15–30 cm) from 1.50 to 1.82 mg m^{-3} in the cotton-wheat system.

The compaction increases soil strength and adversely affects the water retention and availability to plant as well as nutrients supply and hence ultimately affects the crop productivity. Similar results have been reported by Zeleke et al. (2004); Ghuman and Sur (2001) in Ethiopia and by Singh et al. (2007) in Punjab, India; the latter observed significant decreases in bulk density after 3-5 years of crop-residue incorporation. The lowest bulk density of 1.25 mg m^{-3} of Shahpur soil and 1.28 mg m^{-3} of Awagat soil was observed following INM treatment in combination with crop-residue incorporation (Table 2). Our results reveal that FFU in combination with crop-residue incorporation had an effect on bulk density improvement (1.33 and 1.31 mg m^{-3} in Awagat and Shahpur soils, respectively) comparable to that of INM treatment alone (1.34 and 1.30 mg m^{-3} in Awagat and Shahpur soils, respectively). Thus, the addition of organic matter, either in form of FYM or crop residue after every crop, may be preferred to improve and sustain the soil physical conditions for optimum crop growth. This was consistent with the earlier reports of bulk density reduction as a result of inorganic fertilizer application and crop-residues incorporation (ZELEKE et al., 2004; BELLAKKI; BADANUR, 1997).

Aggregation

Percent water-stable aggregates (PWSA) in both the soil series were significantly ($p \leq 0.05$) higher with BNM, INM, and BNM+GM treatments than with FFU (Figure 2). The treatment effects on PWSA were larger in the surface soil layers (0-15 cm depth) than in the sub-soils (15-30 cm depth). The large effect on surface PWSA was most probably consequent of large increase in surface SOM that enhanced the microbial activities and help in stable aggregation (ROBERSON et al., 1995) The PWSA increased further with crop-residue incorporation. The order of treatment effects on aggregation improvement was in accordance with that of the SOM content (Table 1) resulting from BNM, INM and crop-residue incorporation practices.

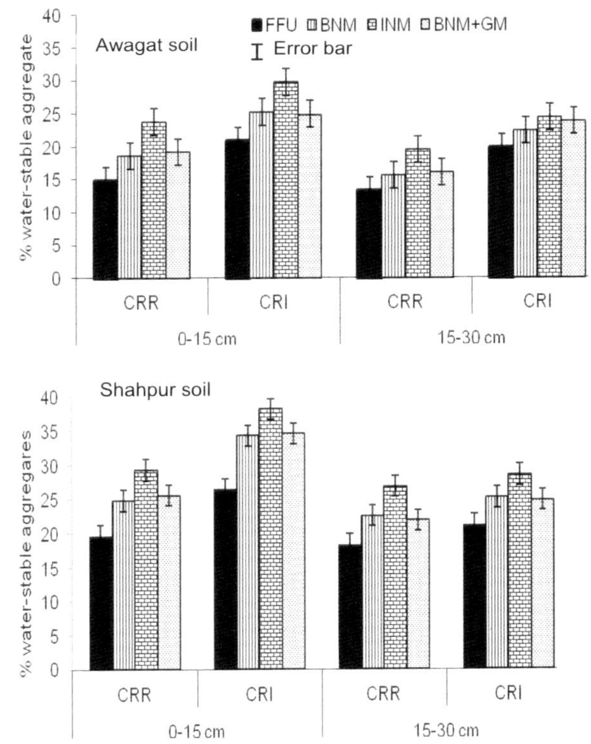

Figure 2. Effect of nutrient managements and crop residue incorporation on water stable aggregates after 5 years. CRR: crop residue removed; CRI: crop residue incorporated.

Different crop residues have different type of impacts on aggregation and aggregate stability. The impact depends upon the decomposition rates, chemical compositions, C:N ratio and quantity of crop residues. For example, residues of cereal and legumes crop are more frequently incorporated than cotton crop to improved SOM and soil structure. But these residues decomposed more rapidly than cotton crop because of lower C:N ratio (legume crop ≈ 23 and cereal crops ≈ 55). Crop residues that decomposed easily (cereal and legume) generate binding substances in the short term and even become even shorter in hot climate prevails in cotton areas of Pakistan. Contrarily, the cotton crop residues have both, short- and long-term impacts on aggregation and aggregate stability. Because its stalks fraction is rich in cellulose and lignin compounds decomposed slowly and provide binding agent over long-term. While, it leave fraction decompose rapidly and provide binding agent for short-term.

Balanced and integrated nutrient applications to crop not only increased the shoot biomass but also increased the root biomass. Crops that produce large amounts of residues result in greater soil aggregation than those with lower biomass production because of differences in decomposition rates. Large quantities of organic materials are supplied to surface and sub-surface soils by the shoot and root system, respectively, during their growth period. Exudates of added organic matter through shoot, roots and FYM resulted in the production of a large amount of active polysaccharide-binding agent in surface and sub-surface soil (DEGENS, 1997; HAYNES et al., 1991). Addition of organic matter to soils also enhances microbial activities and biomass, which would produce extracellular mucilaginous polysaccharide materials having the capacity to stabilize soil aggregates (ROBERSON et al., 1995). Increased aggregate stability in this experiment is consistent with the results of Singh et al. (2005b, 2007) who observed that, in rice-wheat cropping systems, soil aggregate stability was greater with straw incorporation than from where straw was removed. Similar results were also reported by Sarwad et al. (2005) under sorghum-chickpea crop sequence.

Contrary to earlier studies on cereal and legume crop residues incorporation (SINGH et al., 2005a; ZELEKE et al., 2004), significant positive impact on aggregate stability below 10 cm depth was observed. It is most likely that the deeper root system of cotton crop, compared with cereal crops and legumes, may have contributed to sub-surface SOM. Further, the addition of fresh carbon has a larger impact on aggregation of a soil with low organic carbon (e.g. soils under studied) compared with a high carbon content soil. The slow decomposition and long residence time of the cotton roots by-product in the soil may allow the byproduct to make a large and long-term contribution to SOM and hence on aggregation. The contribution of SOM by the cotton crop roots to sub-surface could directly be involved in stability of soil aggregates.

Total porosity and volumes of macro- and micro-pores

Total soil porosity of 0-15 cm depth was significantly higher for BNM, INM and BNM+GM treatments than for FFU treatment in both soil types (Table 3). The total soil porosity was highest with INM treatment in the Shahpur soil, which was statistically higher than with BNM and BNM+GM treatments; BNM and BNM+GM were statistically similar. However, in Awagat soil, INM and BNM+GM treatments were statistically similar but significantly higher than BNM. Further, the increase in total soil porosity after 5-year cotton and wheat crop-residue incorporation was higher in Awagat soil (coarse silty in texture and low in initial SOM) than that of Shahpur (fine loam in texture and high in initial SOM). The results showed that the relatively lighter textured soils with low initial SOM and nearly massive structure were more susceptible to addition of fresh carbon in improving soil structure (aggregation and porosity) than those otherwise.

Table 3. Effect of nutrient managements and crop residue incorporation on total porosity (%) after 5 years.

Nutrient Treatments	Awagat						Shahpur					
	0-15 cm			15-30 cm			0-15 cm			15-30 cm		
	CRR	CRI	Mean	CRR	CRI	Mean	CRR	CRI	Mean	CRR	CRI	Mean
FFU	46.8[¶]	49.8[¶]	48.3[¶¶]	38.9	41.3	40.1	48.4	50.7	49.6	39.8	41.9	40.8
BNM	48.2	50.6	49.4	39.9	41.9	40.9	50.1	51.6	50.8	40.9	41.6	41.3
INM	49.4	51.6	50.5	41.4	43.3	42.4	51.1	53.0	52.0	42.0	43.8	42.9
BNM+GM	49.7	51.2	50.4	41.1	42.3	41.7	50.4	52.2	51.3	41.0	42.6	41.8
Mean	48.5	50.8		40.3	42.2		50.0	51.9		40.9	42.5	
LSD ($p \leq 0.05$)												
Treatments			0.45			0.66			0.66			0.75
Interaction			0.53			0.69			0.65			0.78

CRR: crop-residue removed; CRI: crop-residue incorporated; LSD: least significance difference; ¶: average of four replicates; ¶¶: means of eight values.

Similarly, the increase in volume of the macropores (equivalent diameter > 60 μm) and mesophores (equivalent diameter 60 – 30 μm), described by Luxmoore (1981), determined from the water retention curves of the soils, by crop-residue incorporation was higher in Awagat soil than the Shahpur soil (Figure 3). However, this impact was more intense where the INM treatment was applied in combination with crop-residue incorporation. The results of this study revealed that the volumes of macro- and mesopores (as result of aggregation) depend on crop residue incorporation rather than clay content of the soils, seems to contradict the results of Wagner et al. (2007).

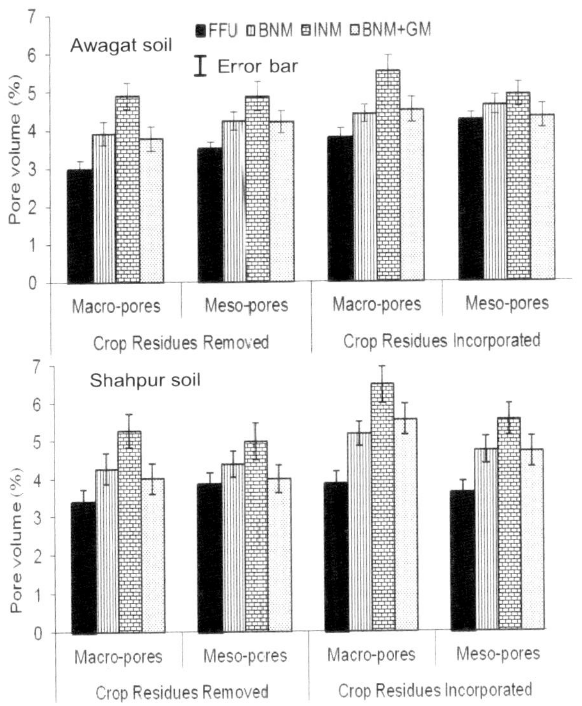

Figure 3. Effect of nutrient managements and crop residue incorporation on volume of macro-pores (equivalent diameter > 60 μm) and meso-pores (equivalent diameter 30-60 μm).

They suggested that macro-aggregation largely depends on soil clay content rather than straw incorporation. According to Tisdall and Oades (1982), inorganic and relatively persistent organic binding agents are important to the development of microaggregates, whereas physical entanglement of the hyphae of microorganisms and roots which are labile or subject to rapid decomposition the major mechanisms in binding microaggregates into macroaggregates. Increases in macro- and mesopores volume which is a useful indicator of soil structure improvement caused by the incorporation of crop residues in both soils, i.e., Awagat and Shahpur, were comparable. Significantly lower bulk density and higher organic matter content because of crop-residue incorporation were in agreement with the higher macro- and mesoporosity. This effect of improved soil structure was more intense where the integrated nutrient management treatment was applied in combination with crop-residue incorporation (BHATTACHARYYA et al., 2007). Further, micropore (equivalent diameter >30 μm) volume was relatively uniform among the different treatments under crop residues incorporated and removed conditions in Awagat soil. While in Shahpur soil, INM treatment showed relatively lower microporosity than those of other nutrient treatments (Figure 4).

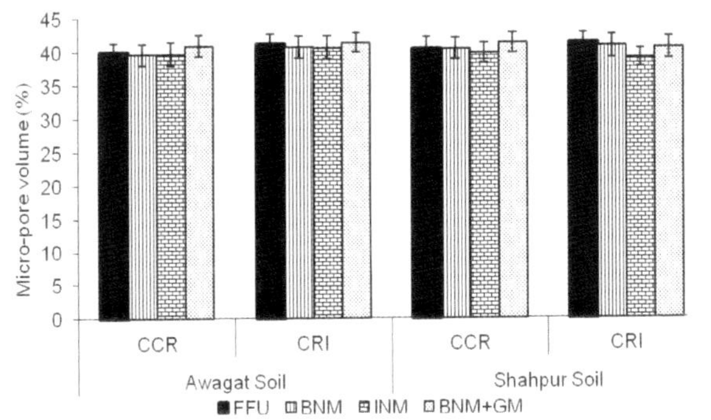

Figure 4. Effect of nutrient managements and crop residue incorporation on volume of micro-pores (equivalent diameter > 30 μm).

Infiltration

Cumulative infiltration as a function of different nutrient management treatments, with and without crop-residue incorporation, is illustrated in Figure 5. Empirical model; $I = at^b$, where a and b are empirical constants, was used to derive relationships between measured cumulative infiltration (I) and time (t) and values of constants as well as final infiltration rates (Table 4). The high r values showed strong relationship between measured and predicted infiltration values.

Beneficial effect of crop-residue incorporation on infiltration was consistent with the greater macro- and meso-porosities (Figure 3) and lesser bulk density (Table 2) in both the soils. However, overall impact of crop-residue incorporation was more pronounced (41% increase over crop residue removal) in the Awagat soil than in the Shahpur soil (38% increases over crop-residue removal).

The BNM, INM and BNM+GM treatments increased infiltration rates by 25, 50, 32%, respectively, over FFU treatment in Awagat soil. Relatively less increase, i.e., 16, 46, 23% in infiltration rates with BNM, INM and BNM+GM treatment over FFU was observed in Shahpur soil. Integrated nutrient management treatment increased infiltration the most: 56% in Awagat soil and 46% in Shahpur soil over FFU treatment when applied without crop-residue incorporation; with crop-residue incorporation, the increase was 50% in Awagat soil and 44% in Shahpur soil. The greater cumulative impact of addition of organic material (FYM, crop residues, green manure) in Awagat soil than in Shahpur soil was probably related to the differences in native organic matter contents and structure development of the both soils (LIU et al., 2005). The values of empirical constant *b* (slope of the cumulative infiltration line) were smaller where nutrient management treatments were applied without crop-residue incorporation than with crop-residue incorporation. Consequently, infiltration rates were lesser without crop-residue incorporation than with crop-residue incorporation.

Hydraulic conductivity

The percent increase in the saturated hydraulic conductivity was of the order of 17, 78, and 24% in the Awagat soil and 15, 69, and 25% in the Shahpur soil with the BNM, INM, and BNM+GM treatments, respectively, over FFU (Table 5). This increase could possibly be because of an increase in macro- and meso-porosity fraction (volume of larger water-conducting pores) of the 0-15 cm depth of the soil (Figure 2). The unsaturated hydraulic conductivity as a function of soil-water content is shown in Figure 6. Calculated results showed no difference between hydraulic conductivities of any of the nutrient-management treatments.

Table 4. Infiltration rate parameters as affected by crop residue incorporation after 5 years.

Nutrient Treatments	Awagat								Shahpur							
	CRR				CRI				CRR				CRI			
	a¶	b†	r!!	FIR‡	A	b	r	FIR	a	b	r	FIR	a	b	r	FIR
FFU	0.42	0.48	0.98	0.61	0.47	0.53	0.99	1.04	0.46	0.50	0.99	0.75	0.53	0.54	0.99	1.22
BNM	0.43	0.51	0.99	0.75	0.47	0.57	0.99	1.32	0.50	0.51	0.98	0.87	0.52	0.57	0.99	1.41
INM	0.44	0.52	0.99	0.95	0.48	0.59	0.99	1.52	0.52	0.55	0.99	1.12	0.50	0.59	0.99	1.76
BNM+GM	0.43	0.51	0.99	0.81	0.47	0.56	0.99	1.36	0.48	0.53	0.99	0.92	0.52	0.57	0.99	1.51

¶: imperial constant (intercept of the fitted equation); b†: imperial constant (slope of the fitted equation); ‡: final infiltration rate after 6 hours (cm h^{-1}); !!: correlation coefficient CRR: crop-residue removed; CRI: crop-residue incorporated.

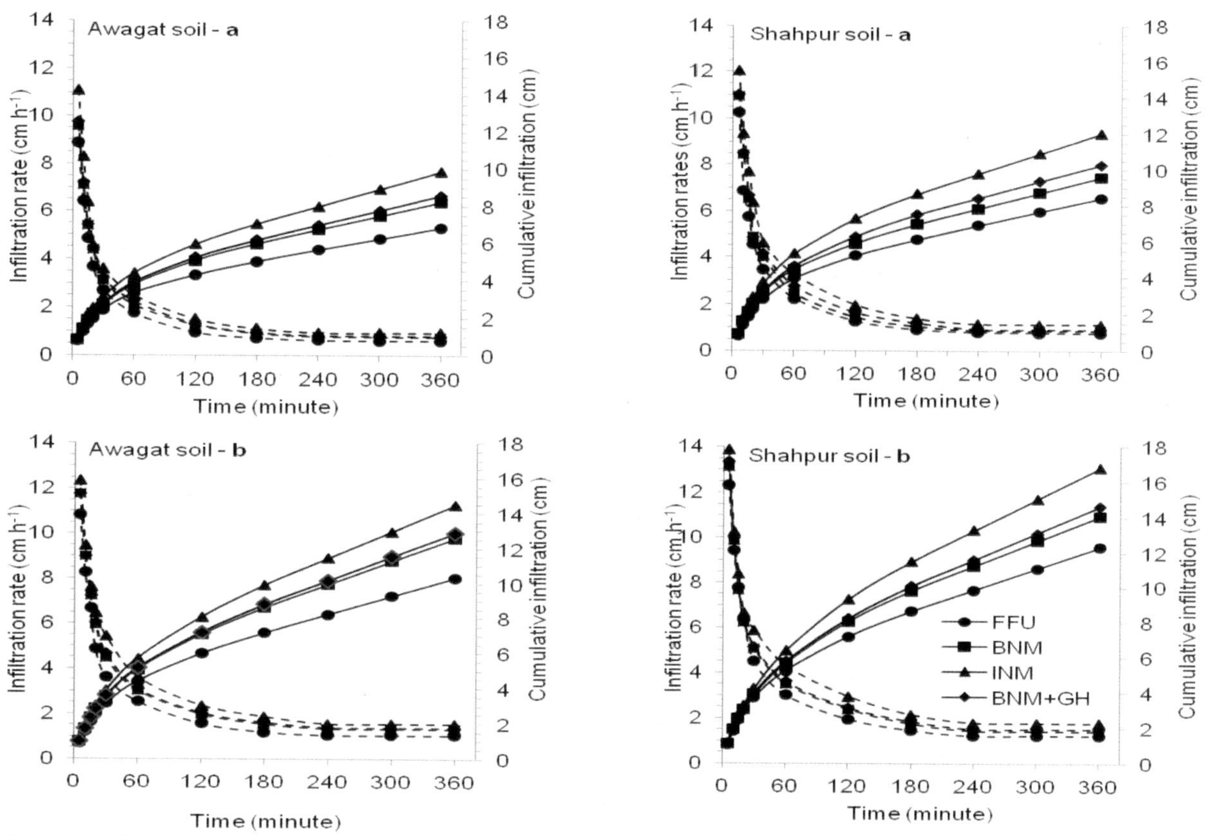

Figure 5. Infiltration and cumulative infiltration as affected by nutrient management treatments after 5 years: (a) crop residue removed and (b) crop residue incorporated.

Table 5. Effect of nutrient managements and crop residue incorporation on saturated hydraulic conductivity (mm day^{-1}) after 5 years.

Nutrient Treatments	Awagat						Shahpur					
	0-15 cm			15-30 cm			0-15 cm			15-30 cm		
	CRR	CRI	Mean	CRR	CRI	Mean	CRR	CRI	Mean	CRR	CRI	Mean
FFU	134	205	170	59	86	72	177	264	221	70	128	99
BNM	149	222	186	67	92	80	193	278	236	82	134	108
INM	174	258	247	75	109	92	220	359	290	103	150	127
BNM+GM	154	231	193	66	97	82	202	289	246	93	140	117
Mean	153	229		67	96		198	298		97	138	
LSD ($p \leq 0.05$)												
Treatments			9.5			6.3			14.0			6.9
Interaction			14.3			7.0			25.6			9.9

CRR: crop-residue removed; CRI: crop-residue incorporated.

However, there was a 7-fold decrease in conductivity, from 140 to 20 mm day^{-1} as volumetric water content decreased from 48 to 42%. This drastic decrease in unsaturated conductivity was presumably because of the emptying of inter-aggregate porosity (macro- and meso-pores) that cannot hold and conduct water at potentials ≥ -10 kPa (BOUMA, 1991; MAHMOOD-UL-HASSAN; GREGORY, 2002). The considerable difference between the unsaturated hydraulic conductivity of crop residues incorporated and removed plots near saturation were very much consistent with the increase in macro- and meso-porosity due to crop residues incorporation (Figure 3). Once the water content of a soil system is exceeded the intra-aggregate (microporosity or matrix porosity), water starts flowing through the inter-aggregate pore network. When the inter-aggregate pore network is started to conduct water, a very small increase in water content can then result in very large increase in unsaturated hydraulic conductivity. Then there was a relatively small change, from 20 to 2 mm d^{-1} over the range of 42 to 35% of water content. At water contents < 35%, the values of hydraulic conductivity were fairly constant. The finding of the experiment revealed that increase in macro- and meso-pores volumes (inter-aggregated pores) enhanced the profile water storage capacity; firstly by increasing infiltration, and secondly by increasing water retention due to increased SOM. The unsaturated hydraulic conductivity was higher where crop residues were incorporated than where crop residues were removed at given water content near saturation (means saturation values for crop residues incorporated plots was 51 and 49% for crop residues removed plots). However, at lower water content, the similar conductivity of crop residue-incorporated and removed plots was in line with the almost similar micro-porosity, i.e., 44.3 and 45.5%, respectively.

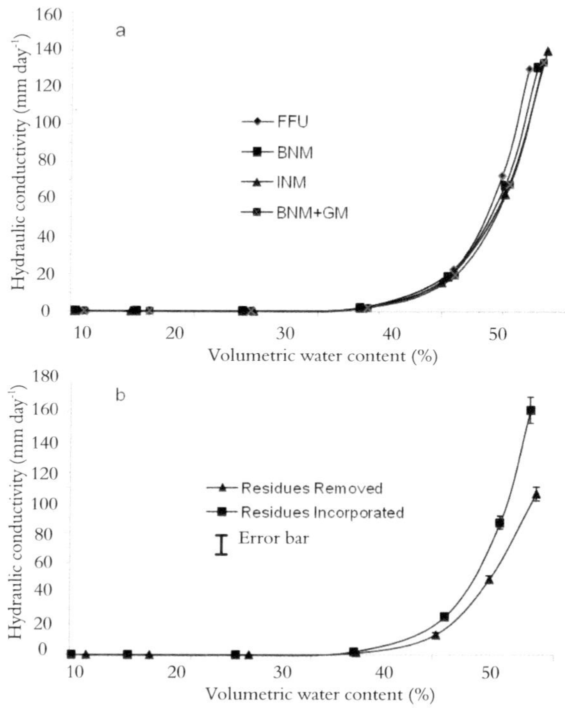

Figure 6. Unsaturated hydraulic conductivity as a function of water content as affected by (a) nutrient managementtreatments and (b) crop residue incorporation (means of two sites).

Conclusion

Results of this five-year field experiment on two predominant soils under cotton-wheat cropping system demonstrated that incorporation of cotton and wheat residue enhanced soil organic matter content and helped generate soil structure and improve soil physical and hydraulic properties in an arid subtropical continental climate. Based on these beneficial impacts, it is suggested that indiscriminate crop residue removal need to be discouraged. Further, its removal can be detrimental to future soil productivity and environmental quality. Alternatively, integrated nutrient management and crop-residue incorporation could help in averting organic matter decline, nutrient depletion, soil structure deterioration, and loss of biodiversity. As a consequence, it could help in improving and sustaining soil productivity. The beneficiaries would include: firstly farmers (by way of increased/sustained soil health and crop productivity); secondly the nation (by reducing environmental pollution caused by burning of crop residues); and thirdly the future generations (as a consequence of sustaining the precious non-renewable soil resource).

Acknowledgements

The selected data presented in this paper are part of the Pak-Kazakh Joint Research Project funded by Ministry of Science & Technology, Government of Pakistan, Islamabad. We are indebted to the Central Cotton Research Institute, Multan and the Pak-German Institute for Cooperative Agriculture, Multan for invaluable support in this field research endeavor. We thank Dr. M. Inayat Khan for guidance and support in experimental design, Mr. Muhammad Saleem (Late), Ghulam Haider and Shamas-ud-Din for assistance in laboratory and field work.

References

AKRAM, M.; AKHTAR, G. H.; ZAHEER, M. R.; AFZAL, M.; MIAN, M. A. **Reconnaissance soil survey of multan district**. Lahore: Soil Survey of Pakistan, 1969.

BELLAKKI, M. A.; BADANUR, V. P. Long term effect of integrated nutrient management on soil properties of vertisol and dry land agriculture. **Journal of the Indian Society of Soil Science**, v. 45, n. 3, p. 437-442, 1997.

BERTRAND, A. R. Rate of water intake in the field. In: BLACK, C. A. (Ed.). **Methods of soil analysis**. Part 1. Wisconsin: Soil Science Society of America, Inc. Madison, 1965. p. 197-209.

BHATTACHARYYA, R.; CHANDRA, S.; SINGH, R. D.; KUNDU, S.; SRIVASTVA, A. K.; GUPTA, H. S. Long-term farmyard manure application effects on properties of a silty clay loam soil under irrigated wheat–soybean rotation. **Soil Tillage Research**, v. 94, n. 2, p. 386-396, 2007.

BLANCO-CANQUI, H.; LAL, R. Crop residue removal impacts on soil productivity and environmental quality. **Plant Sciences**, v. 28, n. 3, p. 139-163, 2009.

BOUMA, J. Influence of soil macroporosity on environmental quality. **Advances in Agronomy**, v. 46, p. 1-37, 1991.

CAMPBELL, G. S. A simple method for determining unsaturated conductivity from moisture retention data. **Soil Science**, v. 117, n. 6, p. 311-331, 1974.

DANE, J. H.; TOPP, G. C. **Methods of soil analysis**. Part 4. Physical methods. Madison: SSSA Book Series No. 5, 2002.

DANIELSON, R. E.; SUTHERLAND, P. L. Porosity. In: KLUTE, A. (Ed.). **Methods of soil analysis**. Part 1. Physical and mineralogical methods. Madison: American Society of Agronomy-Soil Science Society of America, 1986. p. 443-461.

DE RIDDER, N.; VAN KEULEN, H. Some aspects of the role of organic matter in sustainable intensified arable farming systems in the West African semi-arid tropics. **Fertilizer Research**, v. 26, n. 1-3, p. 299-310, 1990.

DEGENS, B. P. Macro-aggregation of soils by microbial binding and binding mechanisms and factors affecting these: A Review. **Australian Journal of Soil Research**, v. 35, n. 2, p. 431-460, 1997.

GEE, G. W.; OR, D. Particle-size analysis. In: DANE, J. H.; TOPP, G. C. (Ed.). **Methods of soil analysis**. Part 4. Physical methods. Madison: SSSA Book Series No. 5, 2002. p. 255-293.

GHUMAN, B. S.; SUR, H. S. Tillage and residue management effects on soil properties and yields of rainfed maize and wheat in a subhumid subtropical climate. **Soil Tillage Research**, v. 58, n. 1-2, p. 1-10, 2001.

GROSSMAN, R. B.; REINSCH, T. G. Bulk density and Linear Extensibility. In: DANE, J. H.; TOPP, G. C. (Ed.). **Methods of soil analysis**. Madison: SSSA, 2002. p. 201-228. (Part 4. Physical Methods; Book Series No. 5).

GURSOY, S.; SESSIZ, A.; KILIC, H.; BAYRAM, N. Tillage system and cotton residue management effects on soil physical and chemical properties of an Anatolian clay loam sown within a wheat-cotton sequence. **Archives in Agronomy and Soil Science**, v. 57, n. 4, p. 391-400, 2011.

HAYNES, R. J.; SWIFT, R. J.; STEPHEN, R. C. Influence of mixed cropping rotations (pasture-arable) on organic matter content, water stable aggregation and clod porosity in a group of soils. **Soil Tillage Research**, v. 19, n. 1, p. 77-87, 1991.

HOOKER, M. L.; HERRON, G. M.; PENAS, P. Effect of residue burning, removal, and incorporation on irrigated cereal crop yields and soil chemical properties. **Soil Science Society of America Journal**, v. 46, n. 1, p. 122-126, 1982.

KARLEN, D. L.; WOLLENHAUPT, N. C.; ERBACH, D. C.; BERRY, E. C.; SWAN, J. B.; NASH, N. S.; JORDAHL, J. L. Long-term tillage effects on soil quality. **Soil Tillage Research**, v. 32, n. 4, p. 313-327, 1994.

LAL, R. Conservation tillage for sustainable agriculture. **Advances in Agronomy**, v. 42, p. 85-197, 1989.

LIU, A.; MA, B. L.; BOMKE, A. A. Effects of cover crops on soil aggregate stability, total organic carbon, and polysaccharides. **Soil Science Society of America Journal**, v. 69, n. 6, p. 2041-2048, 2005.

LUXMOORE, R. J. Micro-, meso- and macro-porosity of soil. **Soil Science Society of America Journal**, v. 45, n. 3, p. 671-672, 1981.

MAHMOOD-UL-HASSAN, M.; GREGORY, P. J. Dynamics of water movement on chalkland. **Journal of Hydrology**, v. 257, n. 1-4, p. 27-41, 2002.

MARSHALL, T. J.; HOLMES, J. W.; ROSE, C. W. **Soil physics**. 2nd ed. Cambridge: Cambridge University Press, 1996.

NELSON, D. W.; SOMMERS, L. E. Total carbon, organic carbon and organic matter. In: SPARKS, D. L. (Ed.). **Methods of soil analysis**. Wisconsin: Soil Science Society of America, Inc. Madison, 1996. p. 961-1010. (Part 3. Chemical methods).

NIMMO, J. R.; PERKINGS, K. S. Aggregate Stability and Size Distribution. In: DANE, J. H.; TOPP, G. C. (Ed.). **Methods of Soil Analysis**. Madison: SSSA, 2002. p. 802-817 (Part 4, Physical Methods; Book Series No. 5).

RAFIQUE, E.; MAHMOOD-UL-HASSAN, M.; RASHID, A.; CHAUDHARY, M. F. Nutrient balances as affected by integrated nutrient and crop residue management in cotton-wheat system in Aridisols. I. Nitrogen. **Journal of Plant Nutrition**, v. 35, n. 4, p. 591-616, 2012.

REYNOLDS, W. D.; ELRICK, D. E.; YOUNGS, E. G.; BOOLTINK, H. W. G.; BOUMA, J. Saturated and field-unsaturated water flow parameters: laboratory methods. In: DANE, J. H.; TOPP, G. C. (Ed.). **Methods of Soil Analysis**. Madison: SSSA, 2002. p. 802-817 (Part 4. Physical Methods; Book Series No. 5).

ROBERSON, E. B.; SARIG, S.; SHENNAN, C.; FIRESTONE, M. K. Nutritional management of microbial polysaccharide production and aggregation in an agricultural soil. **Soil Science Society of America Journal**, v. 59, n. 6, p. 1587-1594, 1995.

ROWELL, D. L. **Soil science**: methods and applications. U.K.: Longman, 1994.

SARWAD, I. M.; GULED, M. B.; GUNDLUR, S. S. Influence of integrated nutrient supply system for *rabi* sorghum-chickpea crop rotation on crop yields and soil properties. **Karnataka Journal of Agricultural Science**, v. 18, n. 3, p. 673-679. 2005.

SINGH, G.; JALOTA, S. K.; SIDHU, B. S. Soil physical and hydraulic properties in a rice-wheat cropping system in India: effects of rice-straw management. **Soil Use and Management**, v. 21, n. 1, p. 17-21, 2005a.

SINGH, Y.; SINGH, B.; TIMSINA, J. Crop residue management for nutrient cycling and improving soil productivity in rice-based cropping systems in the tropics. **Advances in Agronomy**, v. 85, p. 269-407, 2005b.

SINGH, G.; JALOTA, S. K.; SINGH, Y. Manuring and residue management effects on physical properties of a soil under the rice-wheat system in Punjab, India. **Soil Tillage Research**, v. 94, n. 1, p. 229-238, 2007.

SPARKS, D. L.; PAGE, A. L.; HELMKE, P. A.; LOEPPERT, R. H.; SOLTANPOUR, P. N.; TABATABAI. M. A.; JOHNSTON, C. T.; SUMNER, M. E. **Methods of soil analysis**. Part 3. Chemical methods. Wisconsin: Soil Science Society of America, Inc. Madison, 1996.

THORNLEY, J. H. M.; CANNELL, M. G. R. Soil carbon storage response to temperature: an hypothesis. **Annals of Botany**, v. 87, n. 5, p. 591-598, 2001.

TISDALL, J. M.; OADES, J. M. Organic matter and water stable aggregates in soils. **Journal of Soil Science**, v. 33, n. 2, p. 141-163, 1982.

WAGNER, S.; CATTLE, S. R.; SCHOLTEN, T. Soil-aggregate formation as influenced by clay content and organic-matter amendment. **Journal of Plant Nutrition and Soil Science**, v. 170, n. 1, p. 173-180, 2007.

ZELEKE, T. B.; GREVERS, M. C. J.; SI, B. C.; MERMUT, A. R.; BEYENE, S. Effect of residue incorporation on physical properties of the surface soil in the South Central Rift Valley of Ethiopia. **Soil Tillage Research**, v. 77, n. 1, p. 35-46, 2004.

15

Soil properties and cowpea yield after six years of consecutive amendment of composted tannery sludge

Ademir Sergio Ferreira de Araújo[1*], Luciano Moura Lima[1], Wanderley José de Melo[2], Vilma Maria dos Santos[1] and Fabio Fernando de Araujo[3]

[1]*Laboratorio de Qualidade do Solo, Centro de Ciências Agrárias, Universidade Federal do Piauí, Campus da Socopo, 64000-000, Teresina, Piauí, Brazil.* [2]*Faculdade de Ciências Agrárias e Veterinárias, Universidade Estadual de São Paulo, Jaboticabal, São Paulo, Brazil.* [3]*Universidade do Oeste Paulista, Presidente Prudente, São Paulo, Brazil. *Author for correspondence. E-mail: asfaruaj@yahoo.com.br*

ABSTRACT. This study evaluated soil properties and cowpea yield after six years of consecutive amendment of composted tannery sludge. The compost was applied annually at 0, 2.5, 5, 10, and 20 Mg ha^{-1} and at the end of the sixth year, the chemical and physical properties of the soil were evaluated using a randomized block design. The Cr, P, K, Ca, Na and organic C contents and the pH and cation exchange capacity increased linearly after six years of compost amendment. The soil bulk density decreased linearly while the aggregate stability index increased after compost amendment. As a consequence of the changes in the chemical and physical properties of the soil, cowpea yield showed a quadratic response to the tannery rates, with an estimated maximum cowpea yield at 8.3 Mg ha^{-1}. In conclusion, the soil chemical and physical properties improved after six years of composted tannery sludge amendment. However, the soil pH and the Cr and Na contents increased with composted tannery sludge amendment, which influenced the cowpea yield and resulted in a quadratic response to the compost.

Keywords: waste management, soil fertility, soil quality.

Introduction

The tannery industry generates and releases a high volume of solid wastes known as tannery sludge (Santos, Nunes, Melo, & Araújo, 2011). Due to the tanning process, tannery sludge can accumulate chromium (Cr), and the waste can contain high levels of organic matter and a significant amount of hydroxides, carbonates, and salts. Consequently, tannery sludge may cause environmental pollution if applied in its untreated form to the soil. However, if it is treated before application, it may instead be used to improve agricultural soils (Araújo, Monteiro, & Carvalho, 2007).

Composting has been used as an alternative and efficient process for recycling waste, including tannery sludge (Gonçalves, Araújo, Nunes, & Melo, 2014; Silva, Araújo, Nunes, Melo, & Singh, 2014; Araújo et al., 2015). Following the amendment of composted tannery sludge (CTS), its application to the soil can increase plant yield by supplying additional nutrients, increasing soil pH, and consequently reducing the exchangeable aluminum content (Santos et al., 2011). Because CTS contains high levels of organic matter, it may also improve the chemical and physical properties of the soil.

The improvement of chemical and physical properties is important for the plant-growing potential of the soil, particularly for sandy soils that generally lack the necessary fertility properties (Gonçalves et al., 2014). Previous studies showed that soil properties, soil organic matter, and crop yield could all be improved by the addition of different types of compost (Darby, Stone, & Dick, 2006; Mbarki et al., 2008; Arthur, Cornelis, & Razaghi, 2012). In particular, improving the organic matter in soil is known to promote many benefits, including increased soil fertility, improved soil structuration, and reduced soil bulk density (Lal, 2014).

Despite the known benefits to soil, studies focusing on the repeated application of composted wastes such as CTS under field conditions are scarce. Indeed, the limited information available from previous studies does not indicate a clear trend in the effects of this process on the chemical and physical properties of soil over time. Accordingly, we began a long-term study of CTS application in 2009 to evaluate its effects on soil properties after consecutive CTS amendment. In particular, we hypothesized that soil properties are affected by the consecutive and cumulative addition of CTS with respect to the high organic matter and inorganic compounds found in this waste. Therefore, in this study, we evaluated the effects of six years of repeated CTS amendment on the chemical and physical properties of soil and investigated cowpea yield in a treated sandy soil.

Material and methods

The experimental site is located at the Long-Term Experimental Field of the Agricultural Science Centre of the Federal University of Piauí, Brazil. The regional climate is dry tropical (Köppen) and is characterized by two distinct seasons: rainy summer and dry winter, with annual average temperatures of 30°C and rainfall of 1200 mm. The rainy season extends from January to April, when 90% of the total annual rainfall occurs. The soil is classified as a Neossolo Fluvico (Brazilian Soil Taxonomy) with the following composition at 0–20-cm depth: 10% clay, 28% silt, and 62% sand.

In each year of the study, the CTS was produced by mixing tannery sludge with sugarcane straw and cattle manure (ratio 1:3:1; v:v:v). The composting process was performed using the aerated-pile method for 85 days. The pile was 2 m long, 1 m wide, and 1.5 m high. The pile was turned twice a week during the first 30 days and twice a month during the remaining 55 days. At the end of the composting process, 20 subsamples were randomly collected from the CTS to produce a composite sample. The properties of the CTS, averaged across the six years, are shown in Table 1. The water content was determined after oven drying the samples at 105°C for 24h, the pH and electrical conductivity (EC) were measured directly, and total solids were measured by drying the samples at 65°C. The total organic C content was evaluated by dichromate oxidation of the samples under external heating (Nelson & Sommers 1996). The total N content was determined using the Kjeldahl method after sulfuric acid digestion of the samples. The total Ca, Mg, K, P, S, Na, Zn, Cu, Cd, Pb, Ni, and Cr concentrations were determined by atomic absorption spectrophotometry after nitric acid digestion of the samples in a microwave oven (United States Environmental Protection Agency [USEPA], 1996).

Table 1. Average of properties of CTS during the six years.

	CTS	Maximum limit permitted[a]
Water content (g kg^{-1})	680	
pH	7.5	
EC (dS m^{-1})	19.2	
Organic C (g kg^{-1})	201.2	
Total N (g kg^{-1})	15.1	
C/N ratio	13.3	
Total P (g kg^{-1})	4.91	
Total K (g kg^{-1})	2.90	
Total Ca (g kg^{-1})	121.18	
Total Mg (g kg^{-1})	7.21	
Total S (g kg^{-1})	10.20	
Total Na (g kg^{-1})	49.1	
Total Cu (mg kg^{-1})	16.38	1,500
Total Ni (mg kg^{-1})	23.26	420
Total Cd (mg kg^{-1})	1.93	39
Total Cr (mg kg^{-1})	1,943	1,000
Total Pb (mg kg^{-1})	40.31	300

[a] According with Conselho nacional do Meio ambiente (CONAMA, 2009).

CTS was applied annually from 2009 to 2013 at five rates: 0 (no CTS application), 2.5, 5, 10, and 20 t ha^{-1} of CTS (dry basis). The experimental site was arranged in a completely randomized design with four replicates. Each plot measured 20 m^2, with 12 m^2 of usable area for soil and plant sampling, and the rows were spaced 1.0 m apart. Each year, the CTS was applied 10 days before cowpea (*Vigna unguiculata*) sowing. The CTS was spread on the soil surface and incorporated into the 20-cm layer using a harrow. Cowpeas were grown at a density of 5 plants m^{-1} (approximately 62 000 plants ha^{-1}). Plant yield was evaluated at 60 days by sampling ten plants inside the plots and the grains were dried until a moisture content of 13%.

In this study (sixth year of CTS amendment), soil samples were collected 60 days after CTS amendment. In each plot, four samples were collected (0–20 cm), sieved, and stored prior to analysis. Soil pH was estimated in water (1:2.5 v:v) and measured using a pH meter (Tedesco, Gianello, & Bissani, 1995). The soil cation exchange capacity (CEC) and the P, K, Ca, Mg, and Na concentrations were evaluated according to Tedesco et al. (1995). Soil organic C was determined by wet combustion using a mixture of 5 mL of 0.167 mol L^{-1} potassium dichromate and 7.5 mL of concentrated sulfuric acid under heating (170°C for 30 min.). The total Cr concentration was analysed according to USEPA (1996) after the soil was digested with HNO_3, HCl, and H_2O_2. The concentration of Cr from the soil samples was analysed using atomic absorption spectrophotometry.

The soil bulk density (SBD) of the 0–20 cm surface layer was determined using the core method (Black, 1965). The soil total porosity (TP) was calculated using an equation based on the relationship between the bulk density and particle density (Black, 1965). The aggregate stability index (ASI) was measured according to the method of Black (1965) using an oscillation device. The aggregation stability index was calculated as the ratio of the mass of the total sample and the mass of the sample retained in the 0.212-mm sieve mesh, expressed in percent.

The data were submitted to analysis of variance (ANOVA) and the means were compared using Student's test (5% level) and regression analyses. All statistical analyses were performed using the SPSS (version 15.0) software package. The data variables were analysed using multivariate ordination non-metric multidimensional scaling (NMS) with Sorensen distances. Ordination was performed using the PC-ORD v. 6.0 program.

Results and discussion

After six years of CTS amendment, the chemical properties of the soil increased linearly (Figure 1). Considering the highest CTS application rate (20 Mg ha^{-1}), the Cr content increased 50-fold and the P content increased 6-fold compared with the respective contents in unamended soil. In addition, compared with the measurements in unamended soil, both the TOC and Na content increased approximately 2-fold, whereas the K content, soil pH, and CEC increased approximately 0.4-fold. The regression equations showed constant increases of 11.2, 0.77, 0.72, 0.2, 0.08, 0.04, 0.03, and 0.07 units per ton of CTS applied for Cr, P, K, TOC, pH, Ca, Na, and CEC, respectively.

The amendment of organic wastes usually contributes to an increase in the soil organic matter and consequently soil fertility over time (Arthur, Cornelis, & Razaghi, 2012). After six years of amendment, CTS increased the organic C content in the soil because the waste has a high content of organic matter (Table 1). Our results are in agreement with those of others studies, which reported increased organic C contents after compost amendment (Garba, Cornelis, & Steppe, 2011; Arthur, Cornelis, & Razaghi, 2012). The soil pH increased after six years of CTS amendment, which is contrary to the general belief that organic residues decrease soil pH via the production of acids formed during the decomposition of organic matter. The increased soil pH observed in our study may be explained by the high pH and Ca content found in the CTS (Table 1). The CTS has a high pH and Ca and Na contents because the industry uses large amounts of hydroxides, carbonates, and salts during the tanning process (Santos et al., 2011). However, the soil pH observed after six years of CTS amendment in our study was lower than that reported by Martines et al. (2010), who used natural tannery sludge with a high pH and Ca content. This suggests that composting stabilizes some of the chemical properties of tannery sludge.

The increased P and K contents observed in the soil are also related to the CTS composition (Table 1). Thus, the CTS can apparently supply P and K to the soil as the application rates increase; this was also reported by Reynolds, Drury,Tan, and Yang (2015), who evaluated municipal waste compost and observed an increase in the P and K concentration. Our results are in agreement with those of Ozores-Hampton, Stansly, and Salame (2011), who reported an increase in the soil P and K after repeated amendment of composted waste over 10 years. Similarly, Ribeirinho, Melo, Silva, Figueiredo, and Melo (2012) found an increase in the CEC after six years of sewage sludge amendment that was directly related to the increase in the waste rates and which occurred because of the increased negative charge produced by the organic matter in the waste. However, the increased soil Na content in our study due to the amendment of CTS is a major concern in relation to compost use (Arthur, Cornelis, & Razaghi, 2012) because increased Na may limit the growth of plants at the highest CTS application rates.

After six years of CTS amendment, the soil Cr content increased (Table 1), and the amendment of the largest amount of CTS produced Cr levels in the soil at the upper limit proposed by Brazilian regulations (CONAMA, 2009), i.e., 150 mg kg^{-1}. However, it should be noted that as the soil pH increases above 7.0, the Cr in the soil is likely to remain inert with low mobility (Gonçalves et al., 2014).

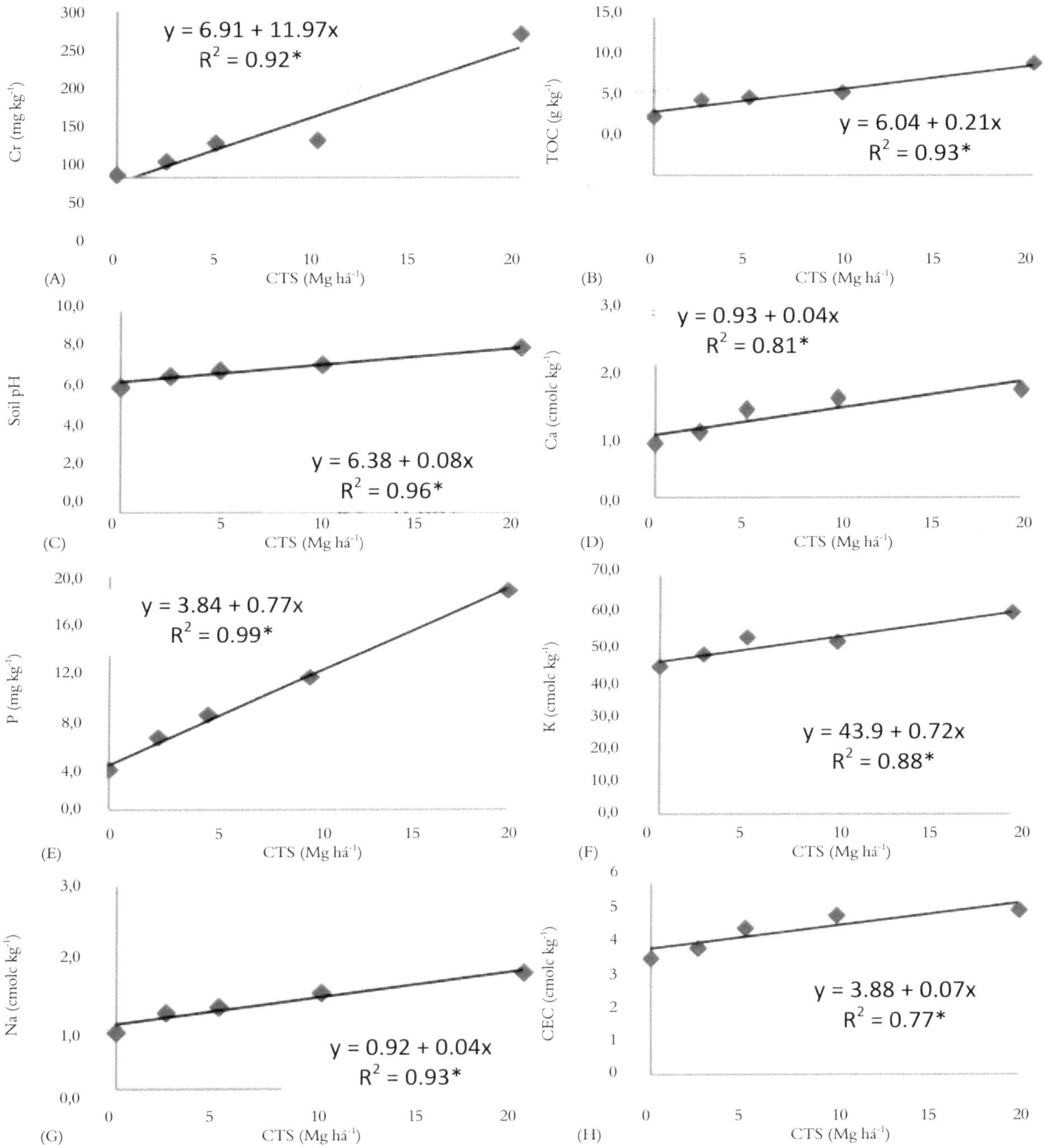

Figure 1. Soil chemical properties after 6 years of CTS amendment.

After six years of CTS amendment, the soil bulk density decreased linearly, whereas the aggregate stability index increased (Figure 2); however, there was no significant difference in the soil total porosity. Considering the highest CTS application rate, six years of CTS amendment resulted in a 15% decrease in soil bulk density and a 62% increase in aggregate stability index. The regression equations showed a constant decrease of 0.01 and an increase of 0.27 units per ton of CTS applied for the soil bulk density and aggregate stability index, respectively.

The decrease in the SBD after six years of CTS amendment may have occurred because of the increase in the organic C content of the soil. An increase in the soil organic matter after compost amendment is known to positively affect soil bulk density and may help to avoid soil compaction (Rivenshield & Bassuk 2007). Courtney and Mullen (2008) observed similar trends after the amendment of organic compost in sandy soils. In addition, the decreased SBD may also be related to the dilution of the denser mineral fraction by the CTS compost. According to Civeira and Lavado (2006), compost particles can penetrate into the soil matrix and decrease the soil density. Soil structuration improves with CTS amendment, and this change may be related to the permanent addition of organic compost. A previous study on composted municipal wastes showed a positive effect on aggregate stability (Reynolds, Drury, Tan, & Yang, 2015).

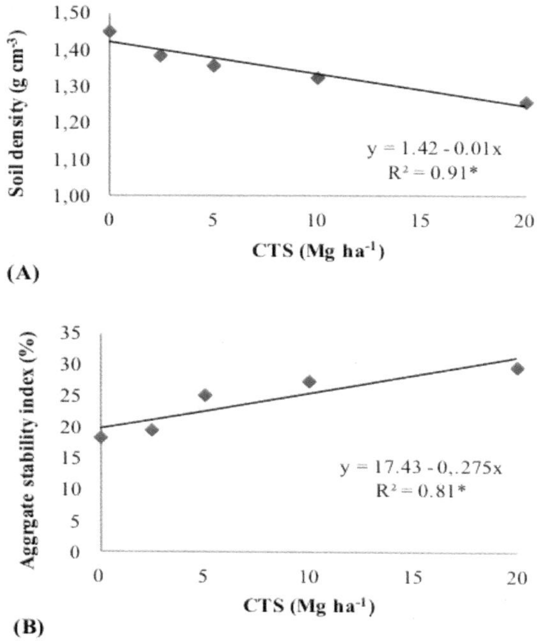

Figure 2. Soil physical properties after 6 years of CTS amendment.

Cowpea yield showed a quadratic response after six years of CTS amendment (Figure 3). According to the regression equation, the estimated maximum cowpea yield was 1.91 Mg ha^{-1} at an estimated CTS application rate of 8.3 Mg ha^{-1}.

Although the soil chemical and physical properties exhibited linear responses to the increase in the CTS rates, the cowpea yield showed a quadratic response and decreased when the CTS rate exceeded approximately 8.3 Mg ha^{-1}. This may have occurred due to the increase in the soil pH and Cr and Na contents, which might have limited the growth of the cowpea crop. High concentrations of Cr can affect the growth of plants and consequently reduce plant yield (Sundaramoorthy, Chidambaram, Unnikannan, & Baskaran, 2010). Previous studies have demonstrated the negative effects of Cr on cowpea (Gonçalves et al., 2014) and rice (Sundaramoorthy, Chidambaram, Unnikannan, & Baskaran, 2010). Similarly, Na can negatively affect plant yields via the reduction of plant growth because this element decreases water uptake, increases sodium toxicity in the shoot cells, and reduces photosynthesis (Tavakkoli, Rengasamy, & McDonald, 2010). With respect to the soil pH, cowpea grown at pH greater than 7.0 exhibits poor development caused mainly by deficiencies of Fe, Zn, and often Mn (Marschner, 2003). Goenaga, Gillaspie, and Quiles (2010) evaluated genotypes of cowpea in alkaline soils and observed that the plant height and pod length decreased by 50 and 37%, respectively, and they reported that high soil pH was a limiting factor for the adequate growth and development of cowpea.

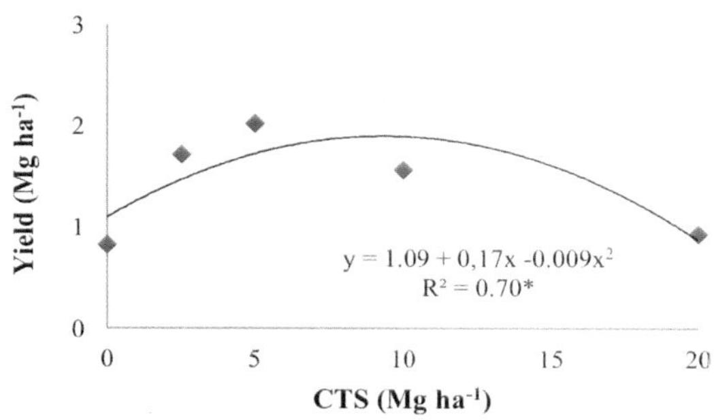

Figure 3. Cowpea yield after 6 years of CTS amendment.

NMS analysis explained approximately 91% of the variation and clustered three distinct groups according to the soil properties and cowpea yield (Figure 4). The first group comprised the unamended soil and was linked with soil density. The second group comprised the 2.5 and 5 Mg ha^{-1} CTS, which were grouped with cowpea yield. The third group comprised the 10 and 20 Mg ha^{-1} CTS, which were grouped with the soil chemical properties and ASI.

The NMS analysis showed different responses related to CTS amendment over the six years and suggested that there was an improvement in the chemical properties of the soil at the highest rate of CTS application. Previous studies with organic compost also found a strong improvement in soil fertility (Scherer, Metker, & Welp, 2011; Patel & Patra, 2014). In addition, the soil structure was better with the amendment of CTS at a high application rate, which indicates that high levels of CTS may improve the physical properties of soil in the long-term. However, CTS amendment at the lowest rate was clustered with plant yield, which indicates that the increase in the CTS rates did not promote high crop production, probably due to the accumulation of chromium and sodium, and the increase in soil pH.

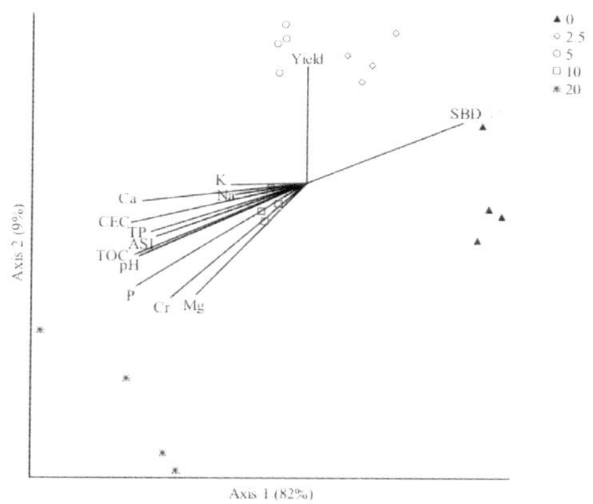

Figure 4. Non-metric multidimensional scaling of chemical and physical properties of soil after 6 years of CTS amendment.

Conclusion

In this study, the soil chemical and physical properties improved after six years of CTS amendment, and soil fertility was positively affected by CTS application. However, the soil pH and the Cr and Na contents increased with CTS amendment, which influenced the cowpea yield and resulted in a quadratic response to the CTS application. After six years of CTS amendment, the estimated maximum cowpea yield was 1.91 Mg ha^{-1} at an estimated CTS application rate of 8.3 Mg ha^{-1}.

Acknowledgements

This research was funded by "Conselho Nacional de Desenvolvimento Científico e Tecnológico" (CNPq 471347/2013-2) and "Coordenação de Aperfeiçoamento de Pessoal de Nível Superior" (PNPD-CAPES). Ademir S.F Araújo and Wanderley J. Melo are supported by a personal grant from CNPq-Brazil.

References

Araújo, A. S. F., Monteiro, R. T. R., & Carvalho, E. M. S. (2007). Effect of textile sludge composted on growth, nodulation and nitrogen fixation of soybean and cowpea. *Bioresource Technology, 98*(5), 1028-1032.

Araújo, A. S. F., Miranda, A. R. L., Oliveira, M. L. J., Santos, V. M., Nunes, L. A. P. L, & Melo, W. J. (2015). Soil microbial properties after 5 years of consecutive amendment with composted tannery sludge. *Environmental Monitoring and Assessment, 187*(1), 4153-4160.

Arthur, E., Cornelis, W., & Razaghi, F. (2012). Compost amendment of sandy soil affects soil properties and greenhouse tomatoes productivity. *Compost Science & Utilization, 2*(1), 215-221.

Black, C. A. (1965). *Methods of soil analysis. Part I. Physical and mineralogical properties*. Madison, WI: SSSA Inc.

Civeira, G., & Lavado, R. S. (2006). Organic matter addition effect on some hydrological properties in a degraded urban soil. *Ciencia del Suelo, 24*(1), 123-130.

Conselho Nacional do Meio ambiente [CONAMA]. (2009). *Define critérios e procedimentos para o uso de lodos de esgoto gerados em estações de tratamento de esgoto sanitário e seus produtos derivados*. Resolução n.º 375 Diário Oficial da União, Brasilia, DF, n.° 167. p. 141-146.

Courtney, R. G., & Mullen, G. J. (2008): Soil quality and barley growth as influenced by the land application of two compost types. *Bioresource Technology*, 99(8), 2913-2918

Darby, H. M., Stone, A. G., & Dick, R. P. (2006). Compost and manure mediated impacts on soil borne pathogens and soil quality. *Soil Science Society of American Journal*, 70(2), 347-358.

Garba, M., Cornelis, W. M., Steppe, K. (2011). Effect of termite mound material on the physical properties of sandy soil and on the growth characteristics of tomato (*Solanun Lycopersicum*) in semi-arid Niger. *Plant & Soil, 338*(1-2), 451-466.

Goenaga, R., Gillaspie Jr., A. G., & Quiles, A. (2010). Field screening of cowpea genotypes for alkaline soil tolerance. *HortScience, 45*(11), 1639-1642.

Gonçalves, I. C. R., Araújo, A. S. F., Nunes, L. A. P. L., & Melo, W. J. (2014). Soil microbial biomass after two years of consecutive application. *Acta Scientiarum. Agronomy, 36*(1), 35-41.

Lal, R. (2014). Organic matter, effects on soil physical properties and processes. *Encyclopedia of Earth Sciences Series, 17*(1), 528-534.

Marschner, H. (2003). *Mineral nutrition of higher plants*. London, UK: Academic Press.

Martines, A. M., Nogueira, M. A., Santos, C. A., Nakatani, A. S., Andrade, C. A., Coscione, A. R., ... Cardoso, E. J. B. N. (2010). Ammonia volatilization in soil treated with tannery sludge. *Bioresource Technology, 101*(12), 4690-4696.

Mbarki, S., Labidi, N., Mahmondi, H., Jdidi, N., & Abdelly, C. (2008) Contrasting effects of municipal compost on alfalfa growth in clay and in sandy soils: N, P, K content and heavy metal toxicity. *Bioresource Technology, 99*(15), 6745-6750

Nelson, D. W., & Sommers, L. E. (1996). Total carbon, organic carbon, and organic matter. In: C. A. Black (Ed.), *Methods of soil analysis. Part 3. Chemical methods* (961-1010). Madison, WI: Soil Science of America and American Society of Agronomy.

Ozores-Hampton, M., Stansly, P. A., & Salame, T. P. (2011). Soil chemical, physical, and biological properties of a sandy soil subjected to long-term organic amendments. *Journal of Sustainable Agriculture, 35*(1), 243-259.

Patel, A., & Patra, D. D. (2014). Influence of heavy metal rich tannery sludge on soil enzymes vis-à-vis growth of *Tagetes minuta*, an essential oil bearing crop. *Chemosphere, 112*(1), 323-332.

Reynolds, W. D., Drury, C. F., Tan, C. S., & Yang, X. M.

(2015). Temporal effects of food waste compost on soil physical quality and productivity. *Canadian Journal of Soil Science, 95*(1), 251-268.

Ribeirinho, V. S., Melo, W. J., Silva, D. H., Figueiredo, L. A., & Melo, G. M. P. (2012). Fertilidade do solo, estado nutricional e produtividade de girassol, em função da aplicação de lodo de esgoto. *Pesquisa Agropecuaria Tropical, 42*(1), 166-173.

Rivenshield, A., & Bassuk, N. L. (2007). Using Organic Amendments to Decrease Bulk Density and Increase Macroporosity in Compacted Soils. *Arborization Urban Forestry, 33*(1), 140-146.

Santos, J. A., Nunes, L. A. P. L., Melo, W. J., & Araujo, A. S. F. (2011). Tannery sludge compost amendment rates on soil microbial biomass of two different soils. *European Journal of Soil Biology, 47*(1), 146-151.

Scherer, H. W., Metker, D. J., & Welp, G. (2011). Effect of long-term organic amendments on chemi cal and microbial properties of a luvisol. *Plant, Soil & Environment, 57*(3), 513-518.

Silva, M. D. M., Araújo, A. S. F., Nunes, L. A. P. L., Melo, W. J., & Singh, R. P. (2014). Heavy metals in cowpea (*Vigna unguiculata* L.) after tannery sludge compost amendment. *Chilean Journal of Agricultural Research, 73*(2), 282-287.

Sundaramoorthy, P., Chidambaram, A., Ganesh, K. S., Unnikannan, P., & Baskaran, L. (2010). Chromium stress in paddy: (i) nutrient status of paddy under chromium stress; (ii) phytoremediation of chromium by aquatic and terrestrial weeds. *Compte Rendu Biologie, 333*(3), 597-607.

Tavakkoli, E., Rengasamy, P., & McDonald, G. K. (2010). High concentrations of Na^+ and Cl^- ions in soil solution have simultaneous detrimental effects on growth of faba bean under salinity stress. *Journal of Experimental Botany, 61*(15), 4449-4459.

Tedesco, M. J., Gianello, C., & Bissani, C. A. (1995). *Analises de solos, plantas e outros materiais*. Porto Alegre, RS: UFRGS.

United States Environmental Protection Agency [USEPA]. (1996). *Acid digestion of sediments, sludge's and soils*. Method 3050b. Washington, DC: EPA.

16

Prediction of soil chemical attributes using optical remote sensing

Aline Marques Genú[1*] and José Alexandre Melo Demattê[2]

*[1]Departamento de Agronomia, Universidade Estadual do Centro-Oeste, Rua Simeão Camargo Varela de Sá, 3, 85040-080, Guarapuava, Paraná, Brazil. [2]Departamento de Ciência do Solo, Escola Superior de Agricultura "Luiz de Queiroz", Universidade de São Paulo, Piracicaba, São Paulo, Brazil. *Author for correspondence. E-mail: agenu@unicentro.br*

ABSTRACT. Soil fertility variability management is one of the pioneering and important areas in which Precision Agriculture has been commercially applied. Consequently, the objective of this work was to predict soil chemical attributes through spectral responses. The 1,000 ha study area used for this report was located in Uberlândia, Minas Gerais State, Brazil. Thirty sampling points were established, at which the soil was collected at 3 different depths. The samples were chemically and physically analyzed and the radiometric data obtained in the 400 – 2500 nm range. Multiple regression equations were generated for sum of bases, cation exchange capacity, base saturation, aluminum saturation, pH, P, K, Ca, Mg, Al, and H, all using 60 soil samples. H, Al, m%, and pH were found to have R^2 values less than 0.50. Equations with an $R^2 > 0.50$ for the other attributes were tested for the 30 unknown soil samples, and the estimated values were obtained. These values were then compared with those determined by conventional analysis. The coefficients of correlation were higher than 50% for all attributes except P and V%. Results indicated that determining chemical attributes with models that are specific for the region is feasible.

Keywords: soil chemistry, reflectance, laboratory sensor.

Introduction

Agricultural producers consider the production area to be homogeneous; therefore, the amount of input for production is applied consistently over the entire area (CORÁ et al., 2004). However, the area typically has considerable differences in its characteristics, especially in regard to soil type, due to pedogenetic processes and factors of soil formation (SOUZA et al., 2008), which leads to variable productivity. For this reason, the identification of management zones that take into account the distribution of soil attributes is important to redefine management practices to minimize costs and maximize yield (BASNYAT et al., 2004).

The application of fertilizers is, without a doubt, one of the greatest costs involved in agricultural production, and therefore, determining the best possible cost-benefit ratio for this type of input is very important. In this respect, remote sensing has been shown to be a valuable source of information in soil fertility variability management, especially after the establishment of precision agriculture.

The traditional method used to determine soil fertility, and consequently recommend fertilizer applications, stems from the results of soil chemical analyses. The basic principle behind this method consists of sampling an area in such a way that analysis of those samples will reflect soil fertility in the area (THOMASSON et al., 2001).

Sample collection for fertility purposes consists of one compound sample comprising 20 subsamples for an area no larger than 20 ha, thus generating a considerable opportunity for error. In the case of precision agriculture, however, sampling consists of 1 to 2.5 points ha^{-1}, resulting in high cost. Demattê et al. (2001) demonstrated that for Oxisols in the State of São Paulo, the recommended sampling rate was 1 to 2 samples ha^{-1}. The authors also found that the analysis of soil samples had the greatest impact on production costs.

Therefore, new soil analysis methods that are less costly and faster have been investigated. The application of spectroscopy as a soil analysis method began after Bowers and Hanks (1965) demonstrated that it was possible to quantify soil attributes, such as moisture and organic matter content, by using reflected energy. In view of this, our work aimed to quantify soil chemical attributes using spectral data obtained in the laboratory and establishing their relationship to chemical management. It is expected that evaluation of the spectral data will show a correlation with the soil's chemical attributes, making quantification of these attributes possible.

Material and methods

Soil sampling and study area characterization

The 1,000 ha study area, located in Uberlândia, Minas Gerais State, Brazil, was defined by the geographical coordinates 19°27' - 19°23' latitude S and 47°57' - 47°51' longitude W. The climate was characterized by a rainy season from October to April and a dry period from May to September.

Thirty points were sampled at three depths, 0 - 20, 40 – 60, and 80 - 100 cm, referred to respectively as A, B, and C. The ninety collected samples were dried, ground and sieved. After that, chemical analyses were performed according to Brazilian guidelines reported in Raij et al. (1987) for pH, OM, Ca, Mg, K, P, Al and H + Al (potential acidity). The SB (sum of bases), CEC (cation exchange capacity), V% (base saturation) and m% (aluminum saturation) values were then calculated.

Obtaining the spectral data

The sample spectral data were obtained in the laboratory with an Infra Red Intelligent Spectroradiometer (IRIS) sensor ranging from 400 to 2,500 nm. The geometry of the system was based on positioning the sensor 27 cm above the sample, which was placed in a Petri dish measuring 9 cm in diameter. The lighting source, a 650 W halogen lamp, produced a non-collimated ray toward the target plane and was positioned 61 cm from the target at a zenithal angle of 20°. The standard reference utilized was a white spectralon plate with reflectance calibrated at 100 %. The spectral ratio between the radiation flow reflected by the reference surface and the sample generated the bidirectional reflectance factor (NICODEMUS et al., 1977).

Statistical analysis

The statistical analysis was performed between the reflectance data obtained in the laboratory for the different chemical attributes of the soils (SB, CEC, V%, m%, pH in $CaCl_2$, P, K, Ca and Mg, Al and H) studied. Twenty-two bands (*B*) and 13 heights (*H*) (in nm) were selected for the spectral data: *B1*: 420-481, *B2*: 481, *B3*: 481-596, *B4*: 596-710, *B5*: 710-814, *B6*: 814-975, *B7*: 975-1350, *B8*: 1350-1417, *B9*: 1417, *B10*: 1417-1449, *B11*: 1449-1793, *B12*: 1793-1831, *B13*: 1865-1927, *B14*: 1927, *B15*: 1927-2102, *B16*: 2101-2139, *B17*: 2139-2206, *B18*: 2206, *B19*: 2206-2258, *B20*: 2258, *B21*: 2258-2389, *B22*: 2389-2498; *H1*: 469-532, *H2*: 532-768, *H3*: 768-876, *H4*: 876-1353, *H5*: 1353-1411, *H6*: 1411-1439, *H7*: 1439-1783, *H8*: 1860-1923, *H9*: 1923-2120, *H10*: 2120-2206, *H11*: 2206-2258, *H12*: 2258-2389, and *H13*: 2389-2498, as suggested by Nanni and Demattê (2006).

Initially, variables (bands and heights) were selected for each soil attribute through stepwise analysis. Next, a multi-colinearity test was used to eliminate one or more of the variables that acted concurrently by estimating the same attribute that caused an overestimation of that attribute. Then, a statistical model was generated using 60 soil samples and a multiple linear regression analysis. These analyses were performed with the Statistical Analysis System (SAS, 1999). Estimated values were then generated for 30 samples that were meant for testing. These values were obtained only for attributes whose R^2 from the multiple regression equation showed values higher than 0.50. Next, a correlation analysis between values determined in the laboratory (DV) and those estimated (EV) by radiometry was conducted.

Results and discussion

The attributes m%, pH in $CaCl_2$, and Al did not have equations with satisfactory results, as their coefficients of determination were around 0.1 (Table 1). These results were lower than those reported by Dunn et al. (2002), Chang et al. (2001), Malley et al. (1999), or Nanni and Demattê (2006). Another attribute that showed an R^2 value smaller than 0.5 was hydrogen (0.42). It was interesting to note that the four attributes showing the smallest coefficients of determination were all related (pH and H, m% and Al). Therefore, it could be expected that as the base content increased, an increase in reflectivity also occurred, a result verified by Demattê et al. (2004). On the other hand, little information was available on the action of elements related to soil acidity, such as H, Al, and pH. The lack of precise results for those elements indicated that there was not a logical definition for their increases and decreases as a function of reflected energy.

For SB, V% and CEC, R^2 values of 0.83, 0.77, and 0.88 were obtained, respectively (Table 1). Other authors reported values that were similar to those observed (CHANG et al., 2001; GENÚ et al., 2010). However, Demattê and Garcia (1999) found coefficients that were smaller than 0.5 for the same three attributes. The potential CEC (pH 7.0) is an attribute determined by the sum of exchangeable bases, H, and Al. Therefore, the incident energy on the elements that make up CEC reflect the soil's retention capacity. In the area under consideration, CEC was low in most soil samples (approximately 42 $mmol_c\ dm^{-3}$). In other words, even when the data did not show great variation, the correlation was reasonable, indicating that CEC directly influenced reflected energy.

The exchangeable bases (K, Ca and Mg) and phosphorous showed R^2 values between 0.67 and 0.84 (Table 1), which were considered satisfactory and agreed with those obtained by Thomasson et al. (2001). However, Dematê and Garcia (1999) obtained values smaller than 0.4 for these four elements, whereas Malley et al. (1999) reported R^2 values higher than 0.9 for the same nutrients. The differences observed between results obtained in this work and other results cited in the literature could be related to the equipment utilized, which have different resolutions, or with selection of the best bands to carry out the tests, as observed by McBratney et al. (2006).

Additionally, the variation in results that were more or less reliable indicated that the deviations were related to the sampling locations and regions. While Demattê and Garcia (1999) correlated chemical elements with reflectance in soil samples from distinct regions in the State of Paraná, an area adding up to 25,620 ha with a total of 45 samples, Thomasson et al. (2001) in turn evaluated areas of approximately 273 ha with 724 samples. Nanni and Demattê (2006) followed the same line by studying an area of 280 ha with a high sampling density. In chemical terms, everything indicated that models should be evaluated for and constructed from smaller areas.

In order to verify if it was possible to utilize these equations to predict soil chemical attributes, those attributes with R^2 higher than 0.5 were tested using 30 samples that were not utilized to generate the model. These samples were used to make scatter plots comparing the values determined in the soil analysis laboratory (DV) with the values estimated by the spectral models.

It could be observed that the coefficients obtained in the correlations for the chemical attributes under study were all greater than 0.5 (Figure 1). The sum of bases (SB) and CEC also showed R^2 values greater than 0.5 (Figures 1a and b), which was in agreement with Nanni and Demattê (2006). Phosphorus (Figure 1d) had the smallest R^2 (0.40) when compared with potassium, magnesium, and calcium (Figures 1c, e, and f, respectively), but V% had the lowest coefficient at 0.35 (Figure 1g).

It was interesting to notice that the sum of bases (SB) and the exchangeable bases studied (K, Ca and Mg) were all related and had high correlation coefficients (Figures 1a, c, e, and f). However, both CEC and V% were also related, both having potential acidity (H + Al) involved in their determination. As observed in Table 1, the equations obtained had low R^2, indicating that there was interference with a spectral correlation of these attributes.

Table 1. Multiple linear regression for quantitative prediction (estimative) of soil chemical properties by reflected data (bands and heights).

Attribute	Regression Equation[(a)]	R^{2} [(b)]
SB[(c)]	14.81+(-353.73⋆B4)+(319.29⋆B6)+(2627.49⋆h13)+(-1224.25⋆h6)+(-201.45⋆h2)	0.83
CEC[(d)]	40.37+(-870.99⋆B4)+(710.46⋆B6)+(2187.78⋆h13)+(-698.67⋆h3)+(-733.42⋆h6)	0.88
V%[(e)]	34.89+(-227.99⋆B4)+(213.08⋆B7)+(-312.40⋆h2)+(2680.03⋆h13)+(-1324.68⋆h6)	0.77
m%[(f)]	-6.48+(330.77⋆h11)	0.10
pH[(g)]	4.77+(-16.59⋆h5)+(9.75⋆h10)	0.16
P[(h)]	4.48+(-284.54⋆B4)+(272.80⋆B6)+(1306.65⋆h13)+(-122.01⋆B7)+(-517.95⋆h6)+(123.52⋆B3)	0.79
K[(i)]	0.73+(-4.24⋆B9)+(-7.69⋆B4)+(9.05⋆B7)	0.67
Ca[(j)]	7.77+(-240.21⋆B4)+(218.68⋆B6)+(1825.66⋆h13)+(-859.10⋆h6)+ (-132.81⋆h2)	0.84
Mg[(k)]	5.77+(201.36⋆B6)+(711.92⋆h13)+(-291.91⋆h6)+(-229.18⋆B5)	0.75
Al[(l)]	0.01+(49.68⋆h8)+(-22.03⋆h10)	0.17
H[(m)]	27.41+(-336.52⋆B4)+(295.96⋆B6)	0.42

[a]B1...B22, h1...h13 bands and heights selected, respectively; [b]Significant at 5% of probability; [c]Sum of bases; [d]Cation exchange capacity; [e]Base saturation; [f]Aluminum saturation; [g]pH determined in $CaCl_2$; [h]Phosphorus; [i]Potassium; [j]Calcium; [k]Magnesium; [l]Aluminum; [m]Hydrogen.

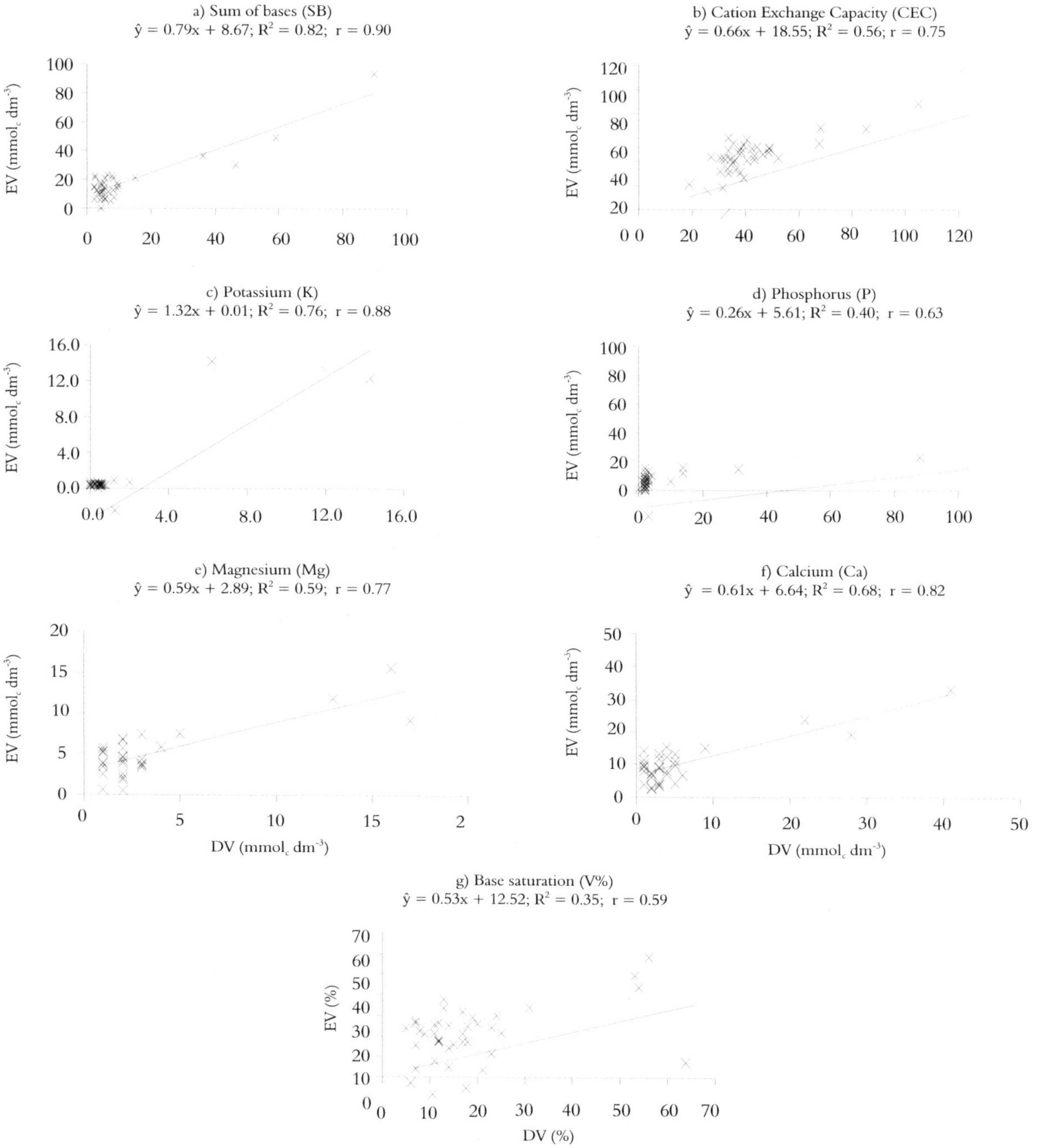

Figure 1. Correlations between values determined in the laboratory (DV) and estimated from spectral data (EV) for soil chemical attributes.

With the advent of Precision Agriculture and the need for optimizing and rationalizing soil sample costs, new research projects on remote sensing have been conducted, with a new emphasis placed on soil chemistry. Under this new perspective, Demattê et al. (2004) determined that the application of agro-industrial liquid residues from sugarcane promoted chemical alterations in the soil samples (increasing K and Ca, for example), which in turn were observed from reflectance. In this case, the authors verified a change in reflectance intensity and demonstrated the importance of the elements present in the soil solution on reflected energy.

Based on these questions, papers, such as those by Thomasson et al. (2001) and Dunn et al. (2002), indicated that electromagnetic energy, when studied in detail on a per band basis, was capable of detecting variations in exchangeable elements. This work, according to the results presented in Table 1 and Figure 1, verified the results obtained in these papers and indicated that the electromagnetic spectrum was influenced by these elements. Therefore, these elements are detectable depending on the evaluation methodology.

Conclusion

It was possible to estimate CEC, SB, K, Ca, and Mg content of soils for the region under study from the spectral responses and at levels comparable to those values found from specific models.

References

BASNYAT, P.; McCONKEY, B.; MEINERT, B.; GATKZE, C.; NOBLE, G. Agriculture field characterization using aerial photograph and satellite imagery. **IEEE Geoscience and Remote Sensing Letters**, v. 1. n. 1, p. 7-10, 2004.

BOWERS, S. A.; HANKS, R. J. Reflection of radiant energy from soils. **Soil Science**, v. 100, n. 2, p. 130-138, 1965.

CHANG, C. W.; LAIRD, D. A.; MAUSBACH, M. J.; HURBURGH JUNIOR, C. R. Near-infrared reflectance spectroscopy - principal components regression analyses of soil properties. **Soil Science Society of America Journal**, v. 65, n. 2, p. 480-490, 2001.

CORÁ, J. E.; ARAUJO, A. V.; PEREIRA, G. T.; BERALDO, J. M. G. Variabilidade especial de atributos do solo para adoção do sistema de agricultura de precisão na cultura de cana-de-açúcar. **Revista Brasileira de Ciência do Solo**, v. 28, n. 6, p. 1013-1021, 2004.

DEMATTÊ, J. A. M.; GARCIA, G. J. Avaliação de atributos de Latossolo Bruno e de Terra Bruna estruturada da região de Guarapuava, Paraná, por meio de sua energia refletida. **Revista Brasileira de Ciência do Solo**, v. 23, n. 2, p. 343-355, 1999.

DEMATTÊ, J. A. M.; DEMATTÊ, J. L. I.; CAMARGO, W. P.; FIORIO, P. R.; NANNI, M. R. Remote sensing in the recognition and mapping of tropical soils developed on topographic sequences. **Mapping Science and Remote Sensing**, v. 38, n. 2, p. 79-102, 2001.

DEMATTÊ, J. A. M.; GAMA, M. A. P.; COOPER, M.; ARAÚJO, J. C.; NANNI, M. R.; FIORIO, P. R. Effect of fermentation residue on the spectral reflectance properties of soils. **Geoderma**, v. 120, n. 3-4, p. 187-200, 2004.

DUNN, B. W.; BEECHER, H. G.; BATTEN, G. D.; CIAVARELLA, S. The potential of near-infrared reflectance spectroscopy for soil analysis – a case study from the Riverine Plain of south-eastern Australia. **Australian Journal of Experimental Agriculture**, v. 42, n. 5, p. 607-614, 2002.

GENÚ, A. M.; DEMATTÊ, J. A. M.; FIORIO, P. R. Análise espectral de solos da Região de Mogi-Guaçú (SP). **Semina: Ciências Agrárias**, v. 31, supl. 1, p. 1235-1244, 2010.

MALLEY, D. F.; YESMIN, L.; WRAY, D.; EDWARDS, S. Application of near-infrared spectroscopy in analysis of soil mineral nutrients. **Communications in Soil Science and Plant Analysis**, v. 30, n. 7-8, p. 999-1012, 1999.

McBRATNEY, A. B.; MINASNY, B.; VISCARRA ROSSEL, R. Spectral soil analysis and inference systems: a powerful combination for solving the soil data crisis. **Geoderma**, v. 136, n. 1-2, p. 272-278, 2006.

NANNI, M. R.; DEMATTÊ, J. A. M. Spectral reflectance methodology in comparison to traditional soil analysis. **Soil Science Society of America Journal**, v. 70, n. 2, p. 393-407, 2006.

NICODEMUS, F. E.; RICHMOND, J. C.; HSIA, J. J.; GINSBERG, I. W.; LIMPERIS, T. **Geometrical considerations and nomenclature for reflectance**. Washington, D.C.: U.S. Department of Commerce, 1977. (NBS Monograph 160).

RAIJ, B. Van; QUAGGIO, J. A.; CANTARELLA, H.; FERREIRA, M. E.; LOPES, A. S.; BATAGLIA, C. O. **Análise química do solo para fins de fertilidade**. Campinas: Fundação Cargill, 1987.

SAS-Statistical Analisys System. **Statistical analisys system**: user's guide version 8.2. Cary: Statistical Analysis System Institute, 1999.

SOUZA, G. S.; LIMA, J. S. S.; SILVA, S. A.; OLIVEIRA, R. B. Variabilidade espacial de atributos químicos em um Argissolo sob pastagem. **Acta Scientiarum. Agronomy**, v. 30, n. 4, p. 589-596, 2008.

THOMASSON, J. A.; SUI, R.; COX, M. S.; AL-RAJEHY, A. Soil reflectance sensing for determining soil properties in precision agriculture. **Transactions of the ASAE**, v. 44, n. 6, p. 1445-1453, 2001.

17

Temporal stability of soil moisture under effect of three spacings in a eucalyptus stand

José de Oliveira Melo Neto[1*], Carlos Rogério de Mello[2], Antônio Marciano da Silva[1], José Márcio de Mello[3], Marcelo Ribeiro Viola[2] and Sílvia de Nazaré Monteiro Yanagi[2]

[1]*Instituto de Ciência e Tecnologia, Universidade Federal de Alfenas, Cidade Universitária, Rodovia José Aurélio Vilela, BR-267, km 533, 11999, 37715-400, Poços de Caldas, Minas Gerais, Brazil.* [2]*Departamento de Engenharia, Universidade Federal de Lavras, Lavras, Minas Gerais, Brazil.* [3]*Departamento de Ciência Florestal, Universidade Federal de Lavras, Lavras, Minas Gerais, Brazil.* [*]*Author for correspondence. E-mail: jose.neto@unifal-mg.edu.br*

ABSTRACT. Soil moisture temporal variability is an important concept in understanding forest management. The objective of this study was to analyse the effect of planting density on soil profile moisture in a clonal *Eucalyptus* plantation. Soil moisture was observed every 15 days at 25 points distributed throughout the eucalyptus stand between November 2013 and October 2015 to characterize the moisture according to the following three types of spacings: 3 x 2, 3 x 3, and 3 x 5 m. At each point, the 0 - 10, 10 - 20, 20 - 30, 30 – 40, and 60 - 100 cm layers were evaluated using a Profile Probe PR2 model. Soil moisture showed high spatial variability in the 0 - 10 cm layer, whereas from 30 to 100 cm deep, the moisture tended to become more homogeneous throughout the area. The dry periods presented greater temporal soil moisture stability when compared to the rainy seasons. The 3 x 2 m spacing has greater temporal soil moisture stability when compared to the others, while the 3 x 3 m spacing presented greater temporal variability.

Keywords: planting density, oxisols, water storage in the soil profile, Minas Gerais.

Introduction

After interacting with the forest canopy and individual tree trunks, the throughfall and stemflows reach the forest floor. From this moment on, two important processes of the hydrological cycle occur, namely the direct runoff and infiltration of water into the soil profile. The infiltration and redistribution processes of water in the soil profile are affected by variability in the volume of water that interacts with the soil surface and the effects of vegetation cover. Thus, identifying the processes and mechanisms that influence the variability of soil moisture is an important tool in agricultural and forest crop management.

Understanding the variability of soil moisture, using both spatial and temporal scales, is essential in quantifying water storage variation in the soil profile; this information is fundamental in flood prediction and forecasting studies and agro-climatic modelling, especially in eucalyptus plantations (Albergel et al., 2010; Bolten, Crow, Jackson, Zhan, & Reynolds, 2010; Koster, Mahanama, Livneh, Lettenmaier, & Reichle,

2010; Famiglietti, Ryu, Berg, Rodell, & Jackson, 2008).

The spatial pattern of soil moisture can be represented by a small-scale component associated with soil-vegetation interactions with relief, and a large-scale component associated with precipitation and evapotranspiration demand. Guswa (2012) shows that the horizontal variability of soil moisture is mainly affected by the spatial variability in throughfall and stemflows, while the vertical variability is mainly associated with the physical soil characteristics and, especially, the distribution of the vegetation cover root system.

Temporal soil moisture variability analysis is a complex and dynamic process that generally requires an extensive sample set, resulting in a significant increase in monitoring costs over time. Vachaud, Silans, Balabanis, and Vauclin (1985) were the first to observe that, despite the spatiotemporal variability, some soil moisture spatial patterns are evident and that these patterns showed little variation over time. From these observations, the authors proposed the concept of temporal stability, which is the constant association between the spatial location and statistical measures that characterize a particular soil property over time (Van Wesenbeeck & Kachanoski, 1988).

The identification of patterns in time-stable soil moisture points means that these points have soil moisture equal to the average of the samples, i.e., points that are representative of the area, thereby reducing monitoring costs (Gao & Shao, 2012). However, Hu, Shao Han, Reicherdt, and Tan (2010) indicated that while most of the efforts applied to soil moisture temporal stability analysis have been focusing on the soil surface layer, few studies have evaluated the soil profile given the difficulty of monitoring soil moisture in depths and at various points in space.

In a study conducted by Western, Grayson, and Blöschl (2002), it was found that most of the theoretical soil moisture analysis, performed by applying dynamic statistical models, ignores the effects of soil moisture seasonality and have focused only on the temporal variability of precipitation and evapotranspiration.

The *Eucalyptus* genus is the most cultivated group of species in Brazil, especially in the state of Minas Gerais, which has the largest planted area in the country. The excellent performance of eucalyptus forests, mainly in terms of productivity, is associated with the species used for breeding and the application of forest management techniques that can provide alterations in wood quality, thus influencing final product quality and increasing its commercial value (Gonçalves, Oliveira, Silva, Nappo, & Tomazelo Filho, 2010).

The aim of this study was to analyse the effect by planting density in a clonal Eucalyptus stand in southern Minas Gerais State has on soil moisture and to analyse the effects of three spacings on the temporal stability of water content in the soil by layers.

Material and methods

The study was developed in a eucalyptus stand located in the experimental unit of the Forest Management Projects and Study Laboratory (LEMAF) on the Federal University of Lavras (UFLA) campus. The forest stand is formed by *Eucalyptus* hybrid clones, with species *urophylla* and *grandis* serving as the genetic basis. This hybrid is commonly identified as "Urograndis". Individuals were distributed in the study area using the following three spacings: 3 x 2 m (0.77 ha and 1025 individuals), 3 x 3 m (0.42 ha and 398 individuals) and 3 x 5 m (0.35 ha and 262 individuals).

Two soil classes are present in the area: Red-Yellow Latosol (LVA) covers 78.6% of the area (1.21 ha), and Dark-Red Latosol (LV) covers 21.4% of the area (0.33 ha). The LVA has a block structure with moderate drainage and Leucocratic granite-gneiss as the parent material. The LV is a granular structured soil, well drained, and has Mesocratic granite-gneiss as the parent material. Table 1 shows the averaged data for organic matter (OM), soil texture, and soil saturated hydraulic conductivity (Ksat) for both the A horizon and top of the B horizon for each studied soil class.

Table 1. OM, texture and Ksat mean data, by soil class, for the eucalyptus stand.

Soil Class	Horizon	(%)				Texture	Ksat (m day^{-1})
		OM	Sand	Silt	Clay		
LVA	A	3.63	45	10	45	Clayey	0.52
	Top of B	1.84	32	13	55	Clayey	
LV	A	3.99	27	9	64	Clayey	0.64
	Top of B	2.36	23	13	64	Clayey	

For soil moisture monitoring, we used a PR2 Profile Probe device from Decagon Devices® allowing the estimation of soil moisture based on volume (m^3 m^{-3}). Twenty-five probe access tubes were installed in order to represent the soil type diversity, slope and management methods used in the study area. The probe provides moisture data for the five following soil profile layers up to a metre deep: 0 - 10, 10 - 20, 20 - 30, 30 - 40, and 60 - 100 cm.

Figure 1 shows the spatial distribution of the soil moisture monitoring points in the eucalyptus stand.

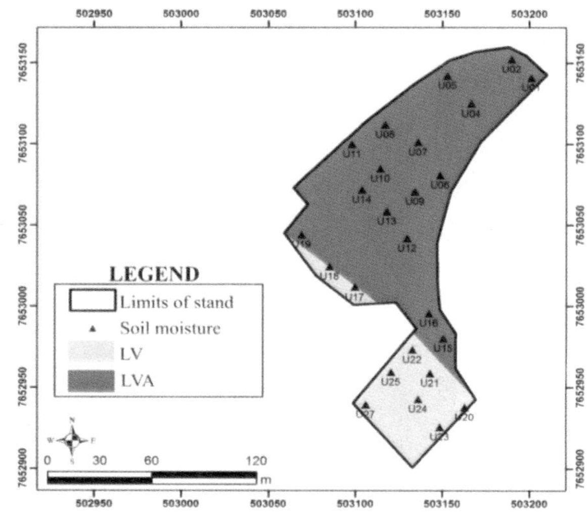

Figure 1. The spatial distribution of the soil moisture monitoring points in the eucalyptus stand.

From November 1, 2013 to October 31, 2015, 83 soil moisture readings were taken, with an average interval of seven days in the 2013/2014 hydrological year and 15 days for the 2014/2015 hydrological year. In the case of occurrences of rainfall events of relevant magnitude, additional readings were conducted to capture a more detailed variation of storage over time. Based on these readings, an average soil moisture behaviour analysis, recorded as a percentage, was conducted in each layer during the study period.

To analyse the temporal soil moisture stability, the relative difference technique proposed by Vachaud et al. (1985) was applied. This technique allows the characterization of the temporal persistence of the soil moisture pattern and identification of soil moisture monitoring points that are statistically representative of the study area. The relative difference (δ_{ij}) is expressed by Equation 1:

$$\delta_{ij} = \frac{\theta_{ij} - \overline{\theta_j}}{\overline{\theta_j}} \quad (1)$$

where: θ_{ij} corresponds to the soil moisture in position i in time j and θ_j represents the average moisture of all positions in time j.

For each position i, the mean temporal relative difference ($\bar{\delta}_i$) and its standard deviation ($\sigma(\bar{\delta}_i)$) were calculated according to Equations 2 and 3, respectively:

$$\overline{\delta_i} = \frac{1}{m}\sum_{j=1}^{m} \delta_{ij} \quad (2)$$

$$\sigma(\overline{\delta_i}) = \sqrt{\frac{1}{m-1}\sum_{j=1}^{m} \left(\delta_{ij} - \overline{\delta_i}\right)^2} \quad (3)$$

where: *m* is the number of readings (in this case, 83).

Subsequently, the data of the mean relative difference was plotted in ascending order, where the point closest to zero would be the representative point for the study area, presenting higher temporal stability. According to Brocca, Tullo, Melone, Moramarco, and Morbidelli (2012), the relative difference values both near zero and with low standard deviations indicate the possible representative soil moisture location.

The temporal stability study was applied to the five layers analysed, considered the entire study period, and was split into the dry and wet periods of the hydrologic years.

Results and discussion

On average, soil moisture increases as soil depth increases. The surface layer (0 - 10 cm) has the lowest average soil moisture among the layers, the greatest range observed in the time series, and high variability, with a 35.8% coefficient of variation. Table 2 presents the descriptive statistics for the average soil moisture for all evaluated layers in the eucalyptus plantation.

Table 2. Basic statistics for the average soil moisture for each studied layer in the eucalyptus stand.

Parameter	Layers (cm)				
	0 - 10	10 - 20	20 - 30	30 - 40	60 - 100
Average (%)	15.5	22.4	26.6	27.8	28.6
Maximum (%)	32.0	36.5	37.5	39.4	41.6
Minimum (%)	6.4	15.0	20.6	23.3	24.0
Standard Deviation (%)	5.6	5.0	4.1	3.5	3.3
CV (%)	35.8	22.2	15.3	12.7	17.5

The dispersion in the soil profile increases from the deepest layer (60 - 100 cm) to the uppermost layer (0 - 10 cm) as shown in Figure 2. The standard deviation confirms this statement, following the same behaviour and presenting greater deviation in the surface layer than in the deeper layer of the soil profile.

Figure 3 presents the average soil moisture time series for all layers monitored in the eucalyptus stand, and the series of rainfall events observed on monitoring days.

A wide fluctuation in soil moisture average data during the study period for the layers 0 - 10, and 10 - 20 cm can be observed. These layers feature a quick response to rainfall events, which are responsible for meeting the soil evaporative demand, and are the most explored by the root system of individuals, contributing to more accentuated moisture decay. Additionally, a greater accumulation of organic matter takes place in the surface layer, helping to maintain the soil structure preserved in these layers, thus contributing to a greater water flow, both at depth and laterally, and increasing the variability over time.

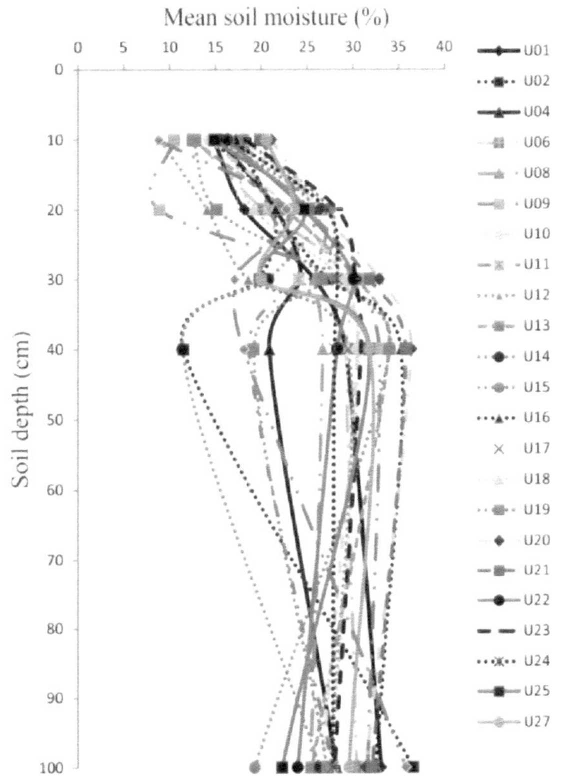

Figure 2. Average soil moisture, by layer, for the sampled points in the eucalyptus stand.

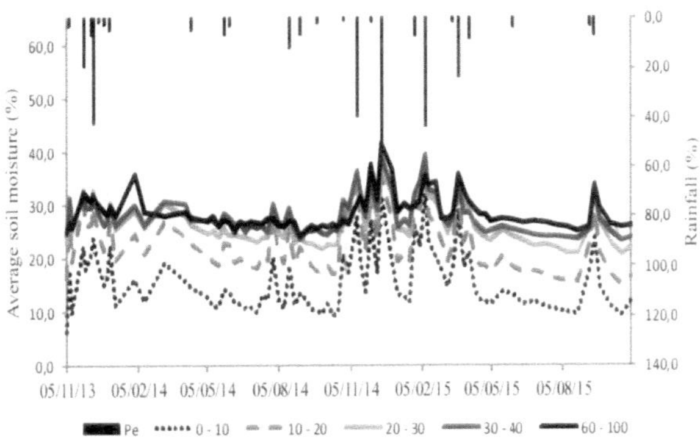

Figure 3. Time series of average soil moisture, by layer, in a eucalyptus stand.

Hu, Shao, Han, Reicherdt, and Tan (2010) and Trevisan, Salemi, Groppo, Silva, and Martinelli (2012) also observed similar behaviours by studying the soil moisture variability in a watershed in China and in a watershed with eucalyptus in Serra do Mar, in the state of São Paulo, respectively.

Figure 4(a, b, c, d, and e) show the mean relative differences and the respective standard deviations for soil moisture in a eucalyptus stand in the 0 - 10, 10 - 20, 20 - 30, 30 – 40, and 60 - 100 cm layers, respectively.

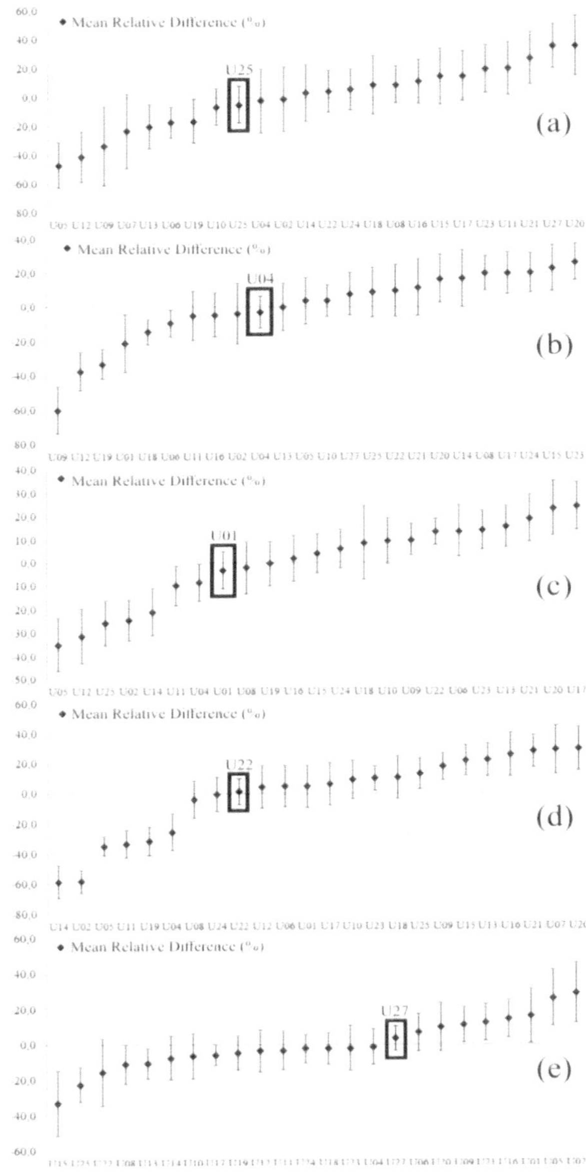

Figure 4. Mean relative difference and standard deviation of soil moisture for the 0 - 10 cm (a), 10 - 20 cm (b), 20 - 30 cm (c), 30 - 40 cm (d), and 60 - 100 cm (e) layers.

A wide range among the Mean Relative Difference (MRD) data was observed for all layers studied. These ranges were 83.1, 86.4, 59.4, 88.5, and 61.8% for the 0 - 10, 10 - 20, 20 - 30, 30 - 40, and 60 - 100 cm layers, respectively. Gao and Shao (2012) found ranges between 93.2 and 172.3% while studying the temporal patterns of soil moisture up to three metres deep in the soil profile predominantly used by grassland.

Brocca et al. (2012) suggested using the point which provides both a MRD module less than 5% and the lowest standard deviation as selection parameters for points representative of average soil moisture over an area. Following these criteria, points U25, U04, U01, U22, and U27 were indicated as representative of soil moisture patterns for the 0 - 10, 10 - 20, 20 - 30, 30 – 40, and 60 - 100 cm layers, respectively, in the studied eucalyptus stand. Points U05 and U20 presented the lowest and highest average soil moisture over time, respectively, among all studied layers. Point U05 is located in LVA with a higher soil density and lower saturated soil hydraulic conductivity and porosity when compared to U20 which is under LV. The results demonstrate that it was not possible to obtain a single stable point to represent the entire area and at all depths.

The average standard deviations, among all the points, were 17.5, 12.7, 9.7, 11.5, and 11.7% for 0 - 10, 10 - 20, 20 - 30, 30 - 40, and 60 - 100 cm layers, respectively. Gao and Shao (2012) and Hu, Shao, and Reichardt (2010) found similar behaviours for this statistical parameter when studying the temporal stability of soil moisture stratifying the profile in layers.

Figure 5 shows the correlations between the average soil moisture time series in the layers at the point indicated as their representative.

By analysing the results from Figure 5, there is, in fact, a strong relationship between the points identified as representative and average soil moisture in the analysed layers. A decrease in the average standard deviation of the MRD as the soil depth increases up to 30 cm followed by an increase up to the depth 100 cm is observed.

Of the five soil moisture pattern representative points for the layers studied, three are located in LV; of these points, two represent the deeper layers of the analysed soil profile. This soil class is better structured, when compared to LVA, having lower soil density and consequently, higher total porosity in the three layers studied.

Another crucial related physical-hydric characteristic is the hydraulic conductivity (Ksat), which was higher in LV. This characteristic favours drainage to deeper layers, allowing these layers to reach moisture equilibrium faster than in LVA.

The presence of an impervious layer, observed in the field at 20 - 30 cm depth, is also noteworthy, with lower Ksat values for the entire area. This condition favours soil moisture uniformity over the surface layer, resulting in a better condition for reaching a temporal stability pattern.

Table 3 shows the mean relative differences and standard deviations of the representative points of temporal soil moisture stability for the five layers studied and spacing in the eucalyptus stand.

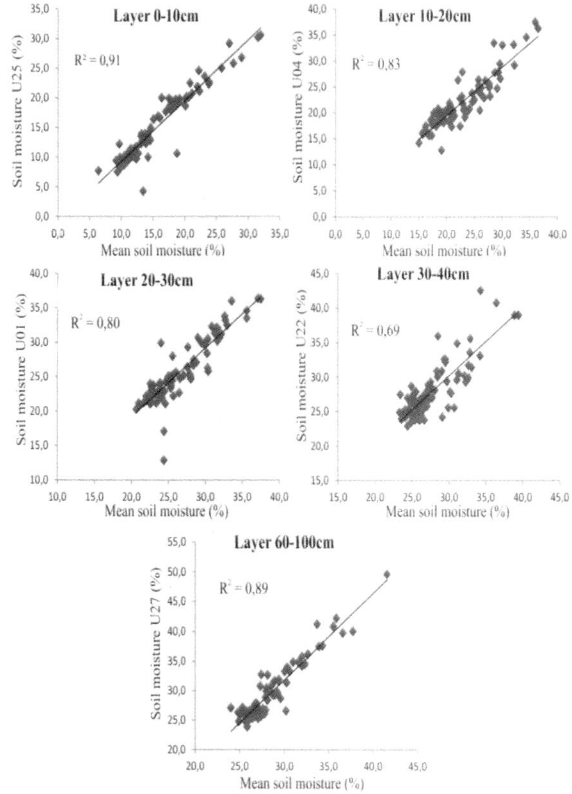

Figure 5. Correlation between the average soil moisture in the area studied and representative points found for each soil layer in the eucalyptus stand.

Table 3. Mean soil moisture relative difference in spacing in the eucalyptus stand.

Spacing (m)	3 x 2		3 x 3		3 x 5	
Layer (cm)	MRD (%)	SD (%)	MRD (%)	SD (%)	MRD (%)	SD (%)
0 – 10	3.2	19.7	-11.0*	17.7*	-4.9	12.6
10 – 20	-3.1	9.3	-4.4	13.0	15.5*	13.6*
20 – 30	-3.2	8.2	1.9	10.1	4.0	8.5
30 – 40	4.7	13.9	4.7	13.7	1.2	8.6
60 – 100	-4.8	9.7	-2.4	9.0	3.4	6.9

*This symbol indicates values with a MRD higher than the 5% criteria adopted.

According to the selection criteria adopted in this study, 3 x 3 m spacing at 0 - 10 cm depth and 3 x 5 m spacing in the 10 - 20 cm soil layer did not produce statistically representative points for average soil moisture. This condition indicates that the sampling points adopted were not enough to represent the soil moisture variability in these layers, requiring either a relocation or expansion of the monitoring network in these layers and spacing patterns.

The lowest average standard deviation observed in the soil profile was found in the 3 x 5 m spacing, followed by 3 x 3 and 3 x 2 m. For the 3 x 5 m spacing, other than a wider canopy opening, which facilitates the throughfall reaching the ground, there was a lower root system presence, which favours a greater persistence of soil moisture patterns and was probably related to less water consumption by the plants.

Such observations about the persistence of soil moisture patterns in a planted eucalyptus forest, associated with different spacings, may become another tool to assist in the genetic improvement and production of hybrids with high productivity characteristics, making them able to achieve the optimum production point associated with planting density and soil water availability. The lack of studies relating moisture patterns in soil profiles to the planting densities for eucalyptus stands is also a noteworthy point.

By evaluating the temporal stability of moisture in the soil profile for the rainy (Figure 6) and dry (Figure 7)

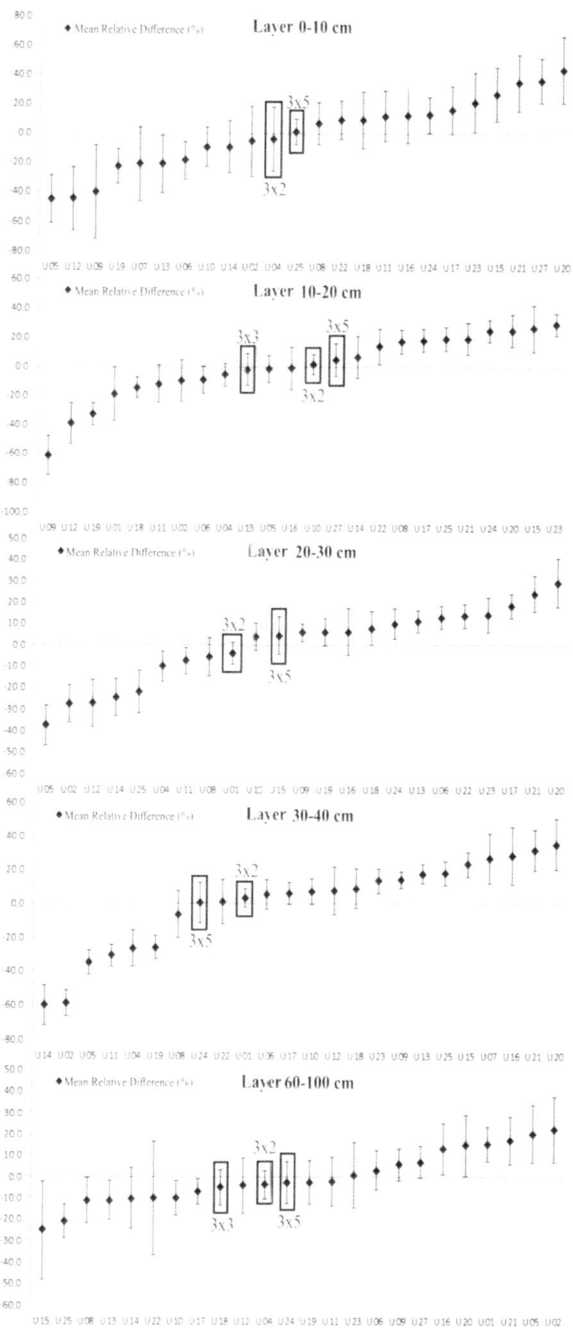

Figure 6. Soil moisture MRD for the rainy seasons in a eucalyptus stand.

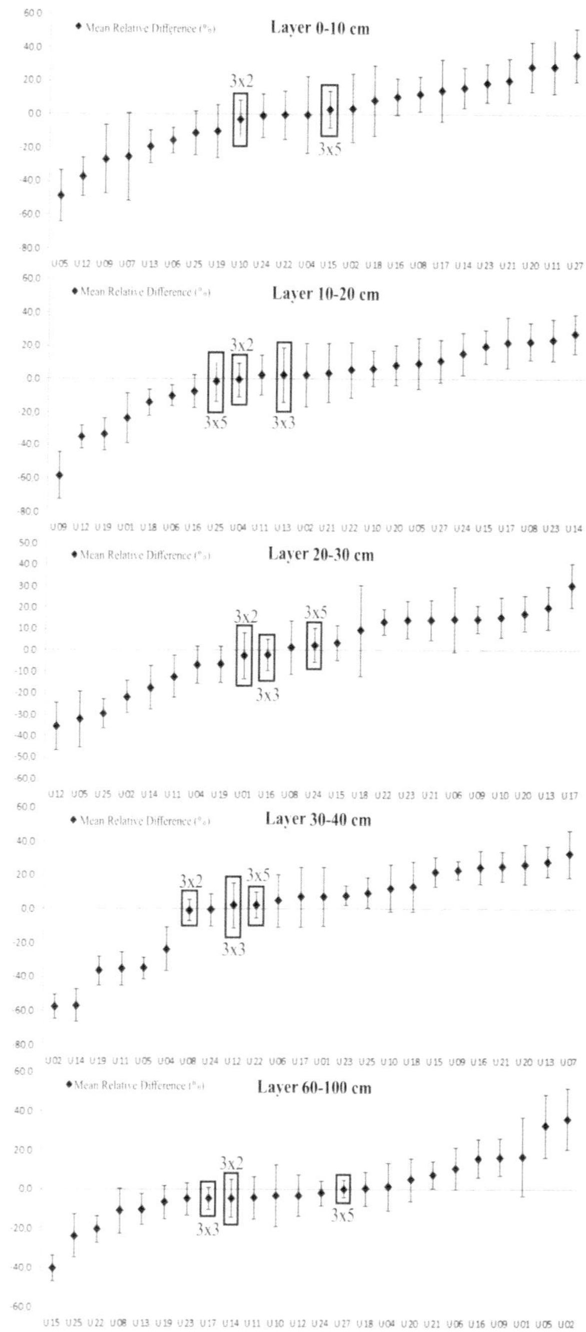

Figure 7. Soil moisture MRD for the dry seasons in a eucalyptus stand.

periods in the eucalyptus stand, we identified a behaviour similar to that observed for the entire study period. In general, the dry seasons had less MRD variation in relation to the rainy seasons, except for the 60 - 100 cm layer, which showed an inverse relationship with the rainy seasons with a lower range than the dry seasons.

For the rainy seasons, no average soil moisture representative points were identified for the 0 - 10, 20 – 30, and 30 - 40 cm layers in the 3 x 3 m spacing in the eucalyptus stand. As for the dry seasons, only the 0-10 cm layer did not present a soil moisture representative point in the 3 x 3 m spacing

Conclusion

The average soil moisture throughout the observation period showed high variability, mainly in the surface layer. Soil moisture tends to become more homogeneous from 30 to 100 cm in depth.

In the 0-10 and 10-20 cm layers, points representative of a soil moisture stability pattern were not identified for 3 x 3 and 3 x 5 m spacings, respectively, indicating the need for a refinement of the observation grid.

The dry periods had higher temporal soil moisture profile stability when compared to the rainy seasons, with less variation in the average relative differences and their standard deviations for all the studied spacings.

References

Albergel, C., Calvet, J. C., De Rosnay, P., Balsamo, G., Wagner, W., Hasenauer, … Mahfouf, J. F. (2010). Cross-evaluation of modeled and remotely sensed surface soil moisture with in situ data in Southwestern France. *Hydrology and Earth System Science*, *14*(11), 2177-2191.

Bolten, J. D., Crow, W. T., Jackson, T. J., Zhan, X., & Reynolds, C. A. (2010). Evaluating the utility of remotely-sensed soil moisture retrievals for operational agricultural drought monitoring. *Journal of Selected Topics in Applied Earth Observations and Remote Sensing*, *3*(1), 57-66.

Brocca, L., Tullo, T., Melone, F., Moramarco, T., & Morbidelli, R. (2012). Catchment scale soil moisture spatial-temporal variability. *Journal of Hydrology*, *422-423*(2), 63-75.

Famiglietti, J. S., Ryu, D., Berg, A. A., Rodell, M., & Jackson, T. J. (2008). Field observations of soil moisture variability across scales. *Water Resources Research*, *44*(1), 1-16.

Gao, L., & Shao, M. (2012). Temporal stability of soil water storage in diverse soil layers. *Catena*, *95*(8), 24-32.

Gonçalves, F. G., Oliveira, J. T. S., Silva, G. F., Nappo, M. E., & Tomazelo Filho, M. (2010). Parâmetros dendrométricos e correlações com propriedades tecnológicas em um híbrido clonal de *Eucalyptus urophylla* x *Eucalyptus grandis*. *Revista Árvore*, *34*(5), 947-959.

Guswa, A. J. (2012). Canopy vs. Root: Production and destruction of variability in soil moisture and hydrologic fluxes. *Vadose Zone Journal*, *11*(3), 1-13.

Hu, W., Shao, M., Han, F., Reicherdt, K., & Tan, J. (2010). Watershed scale temporal stability of soil water content. *Geoderma*, *158*(3-4), 181-198.

Hu, W., Shao, M., & Reicherdt, K. (2010). Using a new criterion to identify sites for mean soil water storage evaluation. *Soil Science Society of American Journal*, *74*(3), 762-773.

Koster, R. D., Mahanama, S. P. P., Livneh, B., Lettenmaier, D. P., & Reichle, R. H. (2010). Skill in streamflow forecasts derived from large-scale estimates of soil moisture and snow. *Nature Geoscience*, *3*(9), 613-616.

Trevisan, R., Salemi, L. F., Groppo, J. D., Silva, R. W. C., & Martinelli, L. A. (2012). Dinâmica da água em uma microbacia com cobertura florestal de eucalipto localizada na Serra do Mar no Vale do Paraíba do Sul. *Revista Brasileira de Recursos Hídricos*, *17*(4), 207-216.

Vachaud, G. A., Silans, A. P., Balabanis, P., & Vauclin, M. (1985). Temporal stability of spatially measured soil water probability density function. *Soil Science Society of American Journal*, *49*(4), 822-828.

Van Wesenbeeck, I. J., & Kachanoski, R. G. (1988). Spatial and temporal distribution of soil water in the tiled layer under a corn crop. *Soil Science Society American Journal*, *52*(2), 363-368.

Western, A. W., Grayson, R. B., & Bloschl, G. (2002). Scaling of soil moisture: A hydrologic perspective. *Annual Review of Earth and Planetary Sciences*, *30*(1), 149-180.

18

The use of multiple endmember spectral mixture analysis (MESMA) for the mapping of soil attributes using Aster imagery

Aline Marques Genú[1*], Dar Roberts[2] and José Alexandre Melo Demattê[3]

[1]*Departamento de Agronomia, Universidade Estadual do Centro-Oeste, Rua Simeão Camargo Varela de Sá, 3, 85040-080, Guarapuava, Paraná, Brazil.* [2]*Department of Geography, University of California, Santa Barbara, California, United State of America.* [3]*Departamento de Ciência do Solo, Escola Superior de Agricultura "Luiz de Queiroz", Universidade de São Paulo, Piracicaba, São Paulo, Brazil. *Author for correspondence. E-mail: agenu@unicentro.br*

ABSTRACT. Systematic, physically based acquisition of information regarding soils is required to meet increasing demand in agricultural and environmental systems. The objective of this work is to evaluate the use of multiple endmember spectral mixture analysis (MESMA) for mapping soil attributes within ASTER imagery. A total of 184 georeferenced soil samples were collected from Rafard, São Paulo State, Brazil. These points were overlain on the satellite image to collect spectral data. The laboratory and image information were then arranged and prepared by clustering samples into classes based on the following soil attributes: texture, organic matter, base saturation (V%), CEC and total iron. Following this classification, mean spectral curves were generated for each attribute class. Spectral curves were used as endmembers for the generation of maps using MESMA. Maps of the same attributes were also generated using geostatistical analyses. Based on the two generated maps, a cross-tabulation was used to evaluate the accuracy of MESMA for mapping soil attributes. Agreement was high for maps of the texture, organic matter, CEC and total iron. We conclude that the methodology used in this work was efficient for mapping soil attributes.

Keywords: remote sensing, spatial distribution, geostatistics.

Introduction

Mapping of the spatial distribution of soil properties is important for refining management practices and minimizing environmental problems. Knowledge of physical and chemical soil attributes is an important aspect of productivity analysis and the subsequent choice of management techniques for a given area (BASNYAT et al., 2004). In Brazil, for example, large agricultural areas occur on Oxisols, which are considered to be homogeneous from a pedological perspective but show differences in the spatial distribution of soil attributes, which may affect the management of such areas.

Conventional and non-conventional statistical methods for soil attribute mapping require soil samples that are collected in the field and analyzed in the laboratory, which require time and financial resources.

Data obtained through remote sensing can be used to characterize and map soil attributes in different ways, such as the implementation of direct measurements that are comparable to traditional forms, i.e., the generation of information regarding the spectral behavior of soils, as well as the use of spectral mixture analysis (SMA) to isolate and discriminate soil properties and generate images of its components (HUETE, 2004).

To achieve such results, the technique of mixture analysis is based on the fact that each pixel of the image is a single measure of an area that comprises multiple components, and the spectrum of this mixture is a linear combination of the reflectance of these components or endmembers. The endmembers are those elements whose presence and abundance in the image are being modeled (DENNISON; ROBERTS, 2003).

However, this method considers all pixels as a mixture of a single initial set of endmembers. Thus, it is possible to have a pixel this is modeled by endmembers that are absent. For example, when considering a scene described by the endmembers vegetation, soil and shade, the calculation of the mixture will be performed based on these three elements, although none of these components may be present for the pixel (CARVALHO JUNIOR et al., 2003).

To circumvent this problem, a new approach called multiple endmember spectral mixture analysis (MESMA) has been developed that identifies the best mixing model for each pixel. This technique has been used for mapping vegetation (DENNISON; ROBERTS, 2003), mineral deposits (CARVALHO JUNIOR et al., 2003) and wildfire fuels (ROBERTS et al., 2003); however, little is known regarding how MESMA might perform for soils.

The spectral response of soil in satellite images is affected by the soil chemistry and physical attributes, thereby making it possible to identify and map these properties. Therefore, the objective of this present work is to determine the potential use of MESMA in the spatialization of soil attributes in a satellite image.

Material and methods

Study area

The study area, covering approximately 184 ha, is located near the margins of the Capivari river in the Rafard municipality in the southwest region of São Paulo State, Brazil. The study area is bounded by the geographical coordinates 23°0'31.37" - 22° 58'53.97" south latitude and 53°39'47.81" - 53°3'25.65" west longitude (Figure 1).

The geology of the region belongs to Itararé formation, Tubarão group, and it is composed of sandstones with a heterogeneous granulation, argillites and shale of various colors from light gray to dark gray. The area includes basalts from the Serra Geral formation, the São Bento group and rubble near the river.

Soil sampling and analysis

The area was demarcated and staked as a regular 100 x 100 m grid to produce 184 points within a total of 184 ha. Each stake was georeferenced using GPS coordinates, and samples were collected from a depth of 0 to 20 cm. Soil samples were sieved through a 2 mm mesh and oven dried at 50°C for 48 hours. The physical and chemical analyses were performed on the soil sample fraction smaller than 2 mm.

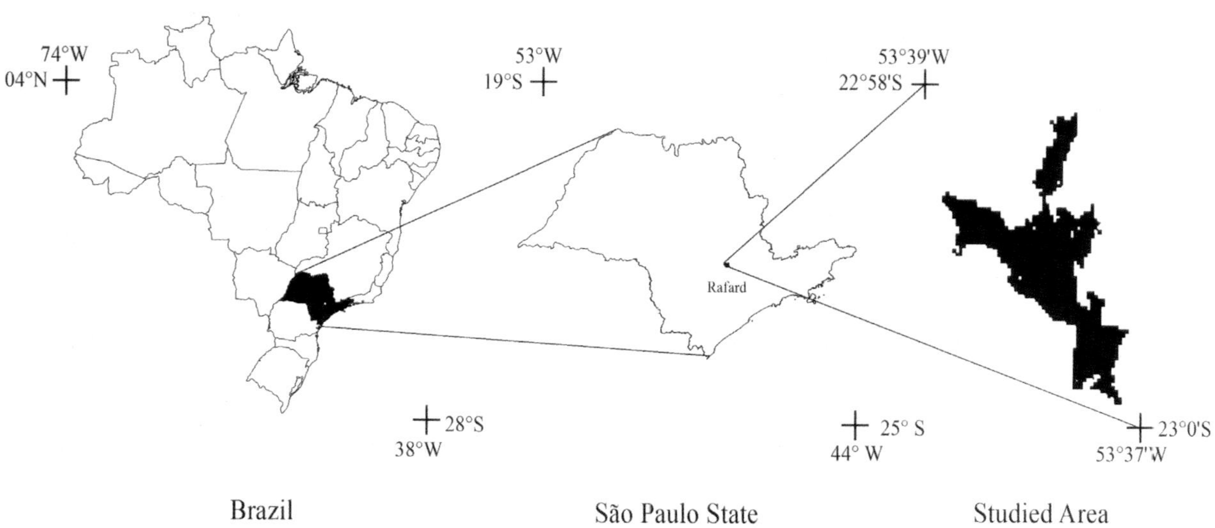

Figure 1. Location of the studied area.

For the physical analysis, the sand, silt and clay contents were determined using the densimeter method (CAMARGO et al., 1986). Chemical analyses included the pH, which was determined in water and KCl. The elements K^+, Ca^{2+}, Mg^{2+} and Al^{3+}, $H^+ + Al^{3+}$ and organic matter (OM) were determined according to Raij and Quaggio (1989).

The cation exchange capacity (CEC) and base saturation (V%) were calculated using the soil analysis results. Total iron concentrations (Fe_2O_3) were determined by sulfuric attack according to the methodology described by Camargo et al. (1986).

Subsequently samples were grouped according to a particular attribute. The evaluation of soil attributes using the spectral data was performed based on intervals for each attribute (GENÚ et al., 2011). The OM levels were designated as high (> 25 g dm^{-3}), medium (15 - 25 g dm^{-3}) or low (< 15 g dm^{-3}). Base saturation (V%) was classified as dystrophic (<50%) or eutrophic (> 50%). Total iron (Fe_2O_3) levels were categorized as hypoferric (< 80 g kg^{-1}), mesoferric (80 - 180 g kg^{-1}) or ferric (> 180 g kg^{-1}). The cation exchange capacity was considered to be very low (< 30 $mmol_c\,dm^{-3}$), low (30 - 60 $mmol_c\,dm^{-3}$), medium (60 - 100 $mmol_c\,dm^{-3}$), high (100 - 150 $mmol_c\,dm^{-3}$) or very high (> 150 $mmol_c\,dm^{-3}$). The following texture classes, defined based on the clay contents were used: sand (< 150 g kg^{-1}), sandy loam (150-250 g kg^{-1}), clay loam (250 - 350 g kg^{-1}), clay (350 – 600 g kg^{-1}) and very clayey (> 600 g kg^{-1}).

Spectral Information

The ASTER sensor of the TERRA satellite has 14 bands. Three of these bands are in the visible and near-infrared wavelengths, six are in the shortwave-infrared range and five are in the thermal infrared range. In this work, the analysis was restricted to bands situated in the visible, near infrared and shortwave-infrared ranges, including the following: ASTER 1 (green, 520-600 nm), ASTER 2 (red, 630-690 nm), ASTER 3 (near infrared, 760-860 nm), ASTER 4 (infrared, 1600-1700 nm), ASTER 5 (infrared, 2145-2185 nm), ASTER 6 (infrared, 2185-2225 nm), ASTER 7 (infrared, 2235-2285 nm), and ASTER 8 (infrared, 2295-2365 nm). The image was acquired on August 6, 2004.

Because bands 1, 2 and 3 have a spectral resolution of 15 m, and the others have a resolution of 30 m, bands 1-3 were resampled at 30 m using the ENVI (Environment for Visualizing Images) program (ESRI, 2005).

To convert the gray levels of the image to reflectance values, and to reduce the atmospheric effect in the ASTER satellite image, the ACORN (Atmospheric Correction Now) program was used. An additional geometric correction of the image was necessary to integrate the image with a cartographic map. Thus, the control points of the study area were collected using a Trimble PRO-XRS GPS receiver with submetric precision and post-processing. This correction was performed using ENVI (ESRI, 2005). The UTM projection system and the South American 1969 Datum (SAD-1969) were used.

The position information of the 184 sampling points was overlain on colored composites of the images for the collection of the orbital spectral data. To ensure that the points were located over areas of exposed soil, a Normalized Difference Vegetation Index (NDVI) image was generated, and the soil line method was used to verify that the soil was bare (NANNI; DEMATTÊ, 2006).

A number of different vegetation indices have been developed to quantitatively and qualitatively assess vegetation cover based on spectral data (THENKABAIL et al., 2004). The successful use of vegetation indices is attributed to the strong spectral contrast between chlorophyll absorption in the visible wavelength (400-740 nm) and strong scattering in the near-infrared (740-1400 nm). The NDVI is defined as the ratio between the red and near-infrared bands (THENKABAIL et al., 2004). For the ASTER image, bands ASTER 2 (red) and ASTER 3 (near infrared) were used to obtain the NDVI.

The soil line used to verify areas of exposed soil is a graphical relationship between red and near-infrared bands. The bands used for this analysis were the same those used for the vegetation index, i.e., ASTER 2 and 3.

Image analysis using MESMA

The first step in the analysis was the generation of an NDVI image, which was used to differentiate the vegetated areas from those with exposed soil. Next, the NDVI image was used as a mask (threshold of 0.30) such that the vegetation would be set to zero, thereby generating a new image.

Subsequently, the vector file of the sampling points was overlaid on the new image using ENVI 4.1 (ESRI, 2005). The reflectance of the pixels related to the sampling points was then obtained.

For each sampling point, the reflectance of the eight bands of the ASTER sensor were correlated with data from the laboratory analysis such that each of the 184 points were related to a value of the clay content (texture), CEC, OM, V% and total iron. Subsequently, the samples were grouped according to the attribute classes mentioned above. After grouping, a mean reflectance value was generated for each ASTER band in the attribute classes; i.e., the mean spectral curves were obtained for the classes of soil attributes.

Mean spectral curves were included in the ENVI program as spectral libraries (SLs). Five spectral libraries were generated, one for each attribute, and each library was composed of the mean spectral curves for the corresponding class (Figure 2).

MESMA analysis was performed using the image obtained earlier (i.e., after applying the mask) using the five generated spectral libraries as endmembers. The analysis was performed for each specific library for each attribute. This analysis generated an image in which each pixel is classified as belonging to an attribute class. If the spectrum in the SL did not match the spectrum in the image (i.e., the mixing constraints for the attribute class and RMS error were not attained), the pixel remained unclassified. These classified images were exported to ArcGIS 9.0 (ESRI, 2005) and named as MESMA maps.

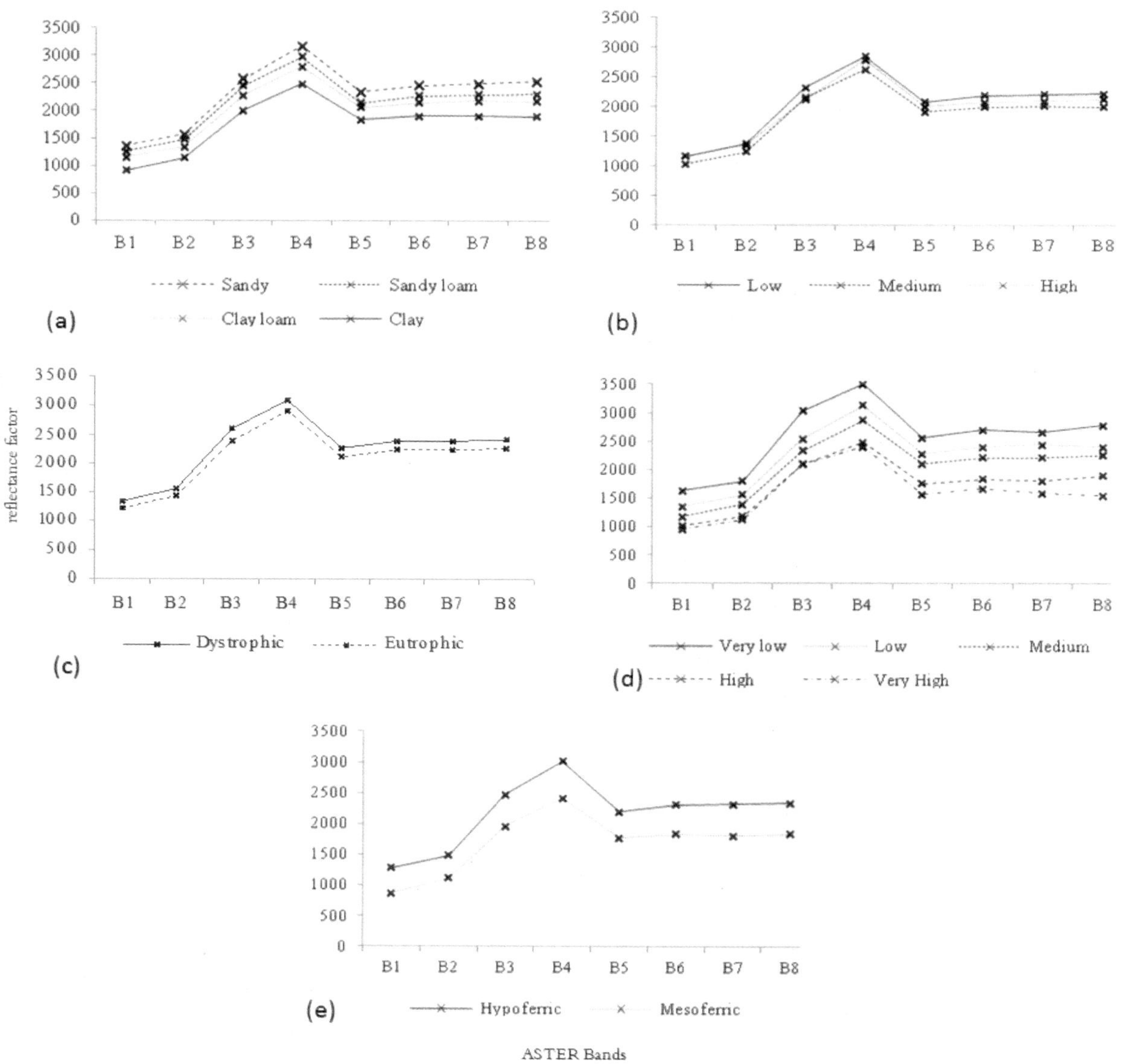

Figure 2. Spectral curves of soil attributes used as endmember for MESMA analysis. (a) Texture; (b) Organic Matter (OM); (c) Base saturation (V%); (d) Cation Exchange Capacity (CEC); (e) Total iron (Fe_2O_3).

Collection of attribute maps using geostatistics

The soil characteristics data and the geographical coordinates of each point sampling were imported to Matlab 7.0 (MATHWORKS, 2005) for the geostatistical analysis. Initially, a semivariogram of the soil attributes was generated to check for the existence of spatial continuity. Subsequently, the statistical model was adjusted to fit the data; based on the type of the adjusted model, nugget effect, range and sill (Table 1), an interpolation of the sampling points was performed using kriging, and a new map for each attribute was generated.

Table 1. Semivariogram parameters for kriging interpolation.

Soil Attribute	Model adjusted	Sill	Range
Organic Matter (OM)	Gaussian	50	700
Texture	Spherical	10000	700
Cation Exchange Capacity (CEC)	Exponential	2500	700
Base Saturation (V%)	Spherical	200	700
Total iron (Fe_2O_3)	Spherical	1000	700

Further maps were exported to ArcGIS 9.0, in which they were divided into the previously mentioned classes by assessing the layer properties and choosing the unique values category. For the purpose of discussion and to compare with the MESMA map, maps generated from the geostatistical analysis were named kriging interpolated maps.

Comparison of maps

The kriging-generated map included the entire study area (184 ha), whereas the MESMA map included only a part of the area (100 ha) due to the presence of vegetation. The geostatistical and MESMA maps were reconciled using the *clip* function available in the ArcToolbox of the ArcGIS software.

Both maps comprised the same area; therefore, the agreement between the maps was determined using ArcGIS 9.0 and the *raster calculator* function of the *spatial analyst* module, to determine the percentage of similarity between the two methods. This procedure was performed for each attribute to produce a confusion matrix.

Results and discussion

The generated texture maps (Figure 3) showed that the spatial distribution of the sand, sandy loam and clay classes is very similar.

The classes with the highest levels of accuracy were sand (56.5%) and clay (56.8%). Sandy loam was, in most cases, classified as sand (35.0%), and the values of clay loam, which did not show similarities between the maps, were confounded with the clay (42.7%) and sandy loam (31.7%) (Table 2).

Okin et al. (2001), working with AVIRIS images and the MESMA method, successfully discriminated between clay and sandy soils with approximately 90% accuracy. Similarly, Drake et al. (1999) successfully mapped the clay content of soils using Airborne Visible Infrared Imaging Spectrometer (AVIRIS) data.

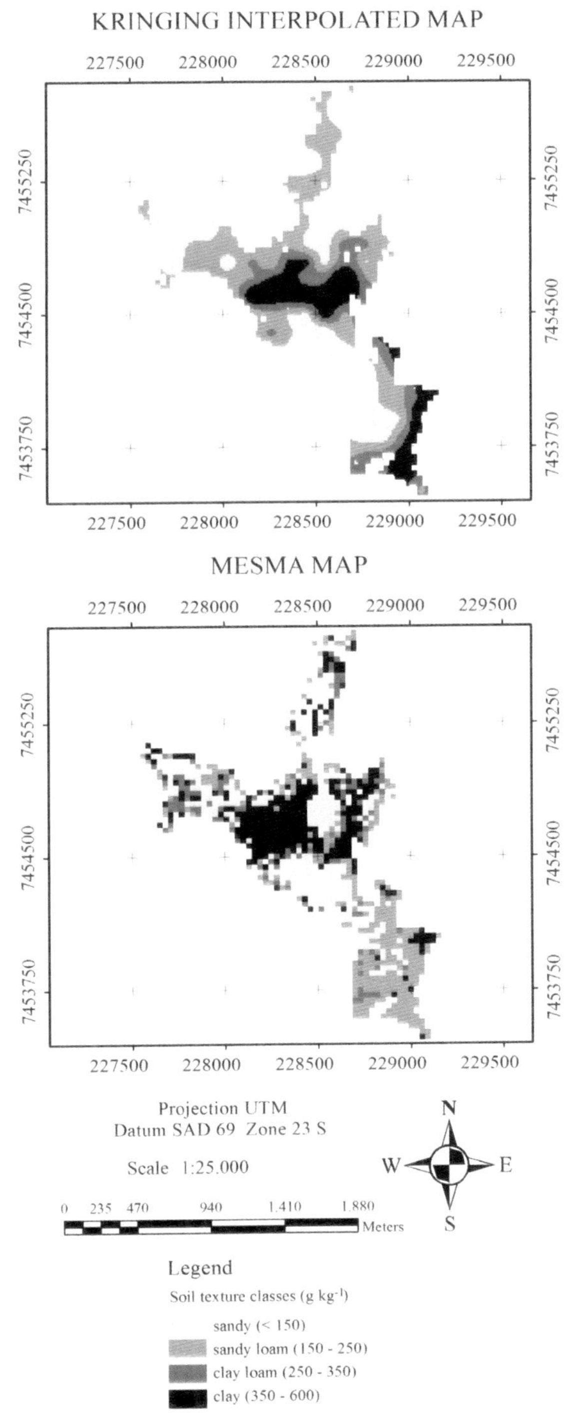

Figure 3. Kriging interpolated map and MESMA map of texture classes of the area.

Table 2. Percentage of similarity between the maps generated by MESMA and by kriging methods of the area for classes of soil attributes.

Attribute classes in the MESMA map	Attribute classes in the map generated by kriging				
	Texture				
	Sand	Sandy loam	Clay loam	Clay	
Sand	56.5	35.0	14.5	4.7	
Sandy loam	23.6	29.8	31.7	34.1	
Clay loam	8.8	14.7	11.1	4.4	
Clay	11.1	20.5	42.7	56.8	
Total (%)	100.0	100.0	100.0	100.0	
	Organic Matter (OM)				
	Low	Medium	High		
Low	58.7	32.2	4.7		
Medium	20.5	64.3	95.3		
High	20.8	3.5	0.0		
Total (%)	100.0	100.0	100.0		
	Base Saturation (V%)				
	Distrophic	Eutrophic			
Distrophic	96.1	57.2			
Eutrophic	3.9	42.8			
Total (%)	100.0	100.0			
	Cation Exchange Capacity (CEC)				
	Very low	Low	Medium	High	Very high
Very low	35.1	15.1	9.0	12.8	0.0
Low	16.7	21.2	13.4	2.7	0.0
Medium	25.9	55.0	57.0	20.1	1.4
High	22.3	5.3	8.3	25.4	32.6
Very high	0.0	3.4	12.3	39.0	66.0
Total (%)	100.0	100.0	100.0	100.0	100.0
	Total Iron (Fe_2O_3)				
	Hypoferric	Mesoferric			
Hypoferric	75.2	20.9			
Mesoferric	24.8	79.1			
Total (%)	100.0	100.0			

Texture is a factor that directly affects the soil spectral soil response. Higher soil clay content tends to exhibit lower soil reflectances (BEN-DOR et al., 2009). The spectral curve of the sandy soil showed a higher reflectance than the clay soil, which resulted in a high index of success for these two texture classes (Figure 2). However, for clay loam and sandy loam, the spectral curves used as endmembers (Figure 2) are very similar to those of clay and sandy soil, respectively, causing the highest levels of error (Table 2).

With respect to the OM, the maps show some similarity, especially for the medium and low classes. The majority of areas classified as belonging to the high organic matter class in the kriging map were confused with the medium organic matter class in the MESMA map (Figure 4). This result can be explained by the very similar spectral response of the high and medium OM classes (Figure 2).

This confusion of the OM classes is shown in Table 2, which indicates that 95.3% of the area with high OM in the kriging map belongs to the medium organic matter class in the MESMA map. The similarities between the maps for the high and medium classes were 58.7 and 64.3%, respectively.

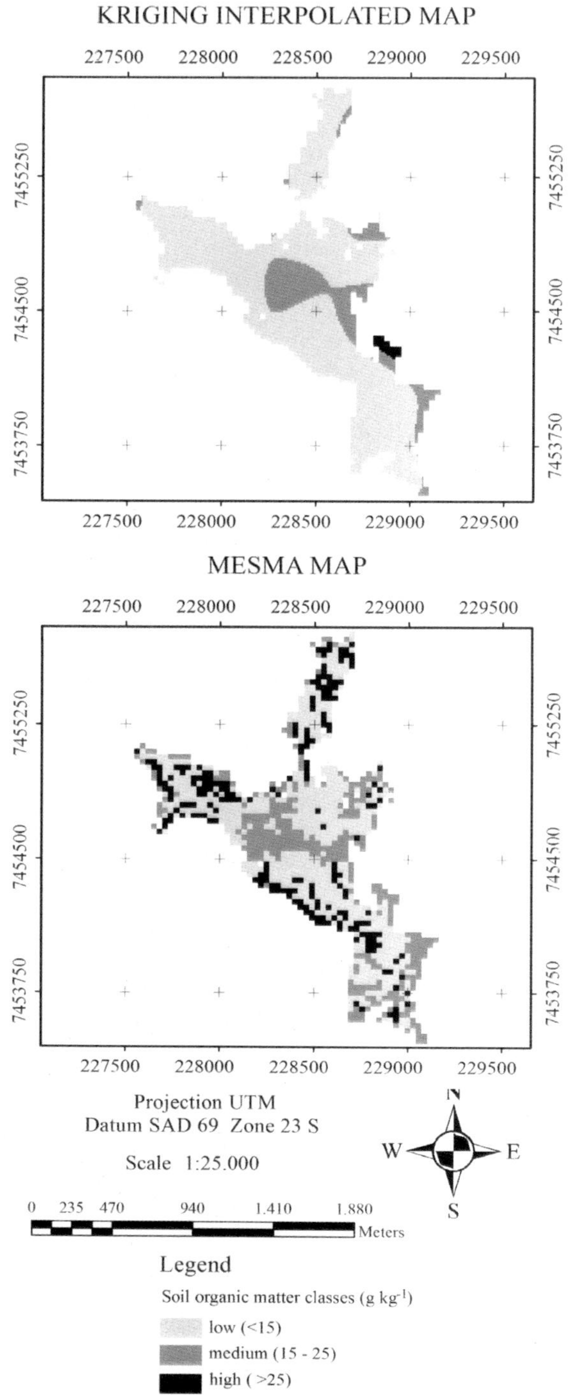

Figure 4. Kriging interpolated map and MESMA map of organic matter of the area.

Organic matter has a strong influence on soil reflectance. In general, an increase in OM content leads to reduced soil reflectance between 400 to 2500 nm (DEMATTÊ et al., 2003a), allowing for the differentiation of OM classes based on spectral orbital curves.

Palacios-Orueta et al. (1999), using a modified form of spectral mixture analysis applied to an AVIRIS image collected over two study areas located in California, was able to discriminate the OM contents of soils with a high degree of accuracy. According to these authors, despite the fact that the range in OM content was the same for both areas (0 to 6%), one of the study sites (La Jolla) had a higher number of pixels with values between 4 to 5%, whereas the other site (Serrano) had more pixels at the lower range between 2.5 to 3.5%.

The generated base saturation (V%) maps proved to be very distinct (Figure 5). In the krigring map, almost the entire area is eutrophic due to chemical soil management performed locally. In contrast, almost the entire area in the MESMA map was classified as dystrophic.

Due to the differences between maps, the dystrophic class that represents only 4.1% of the area in the kriging map when overlapped with the MESMA map was almost entirely classified in the same class (distrophic), resulting in the similarity shown in Table 2. According to Ben-Dor et al. (2009), there are soil properties that do not generate an absorption feature in the spectral curve, such as the pH and electrical conductivity. Perhaps this is the case for the base saturation, and therefore, the resulting maps generated using MESMA were distinct from the krigring map. Demattê et al. (2003b) indicated that higher base concentration and base saturation is associated with higher reflectance. However such behavior was not observed by Sousa Junior et al. (2008) using an ASTER image.

In contrast, the CEC maps were similar, especially for the medium class, which represents the majority of the area (Figure 6).

The very low and very high classes represented a smaller area of 1.4% and 3.1%, respectively. Nevertheless, there was a high rate of success for these classes with 35.1 and 66.0% agreement between the maps (Table 2).

The similarity between the maps shows that the low class, representing 38.8% of the area, was confused with the medium class with an index of similarity of 55%. The medium and high classes had higher rates of success. For the medium class, which occupies 45.1% of the study area, the classification accuracy was 57.0%, whereas the high class, which occupies 11.8% of the area, had a classification accuracy of 25.4% (Table 2).

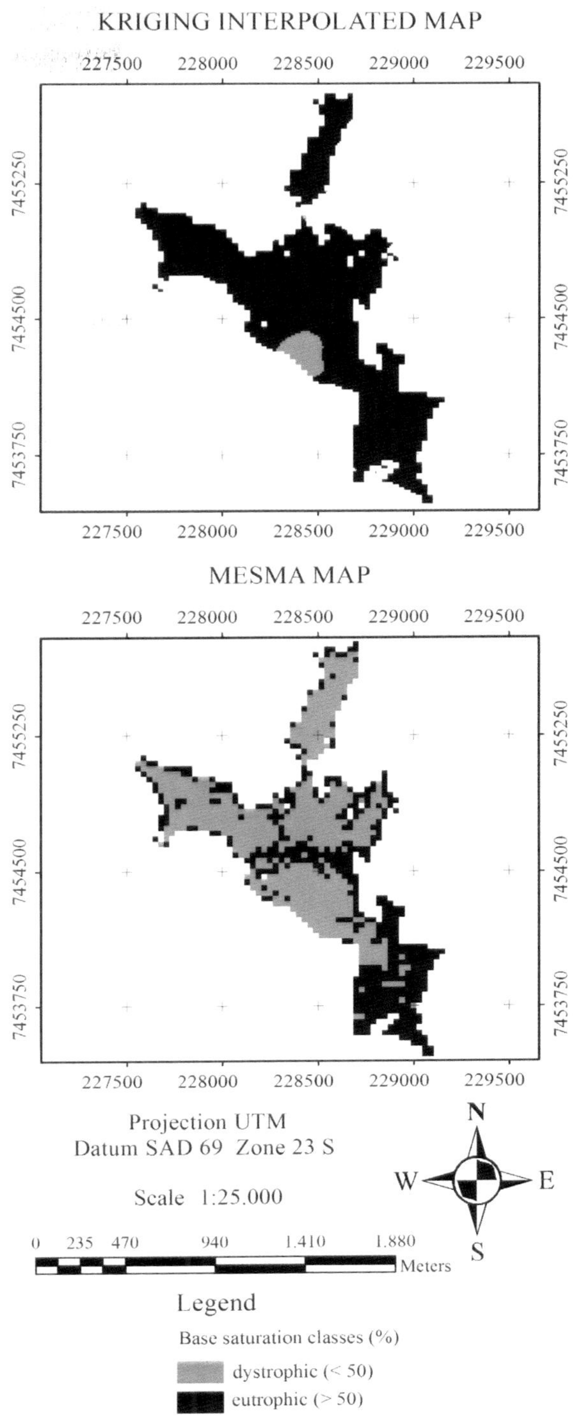

Figure 5. Kriging interpolated map and MESMA map of base saturation of the area.

CEC is a chemical attribute that is highly influenced by the type and quantity of clay minerals as well as the soil organic matter content. For example, clayey soils with high organic matter contents have higher CEC values than sandy soils with low organic matter levels.

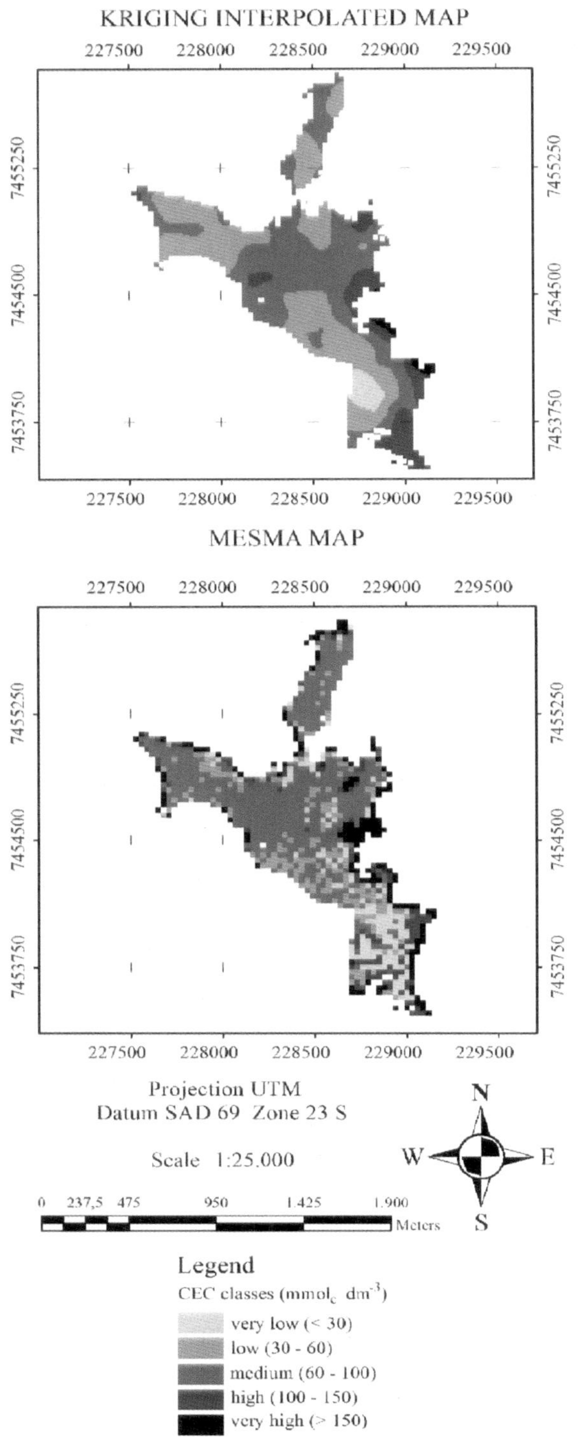

Figure 6. Kriging interpolated map and MESMA map of CEC of the area.

As observed by Ben-Dor et al. (2009), soil properties without features can be quantified and mapped based on the spectral curves due to their strong correlation with other properties that have absorption bands, including certain types of minerals and soil organic matter. This may explain the similarity between the CEC maps, as well as the dissimilarity between the V% maps.

The strong effect of iron on the spectral response of the soil resulted in a high extent similarity between the krigring and MESMA maps for this attribute (Figure 7).

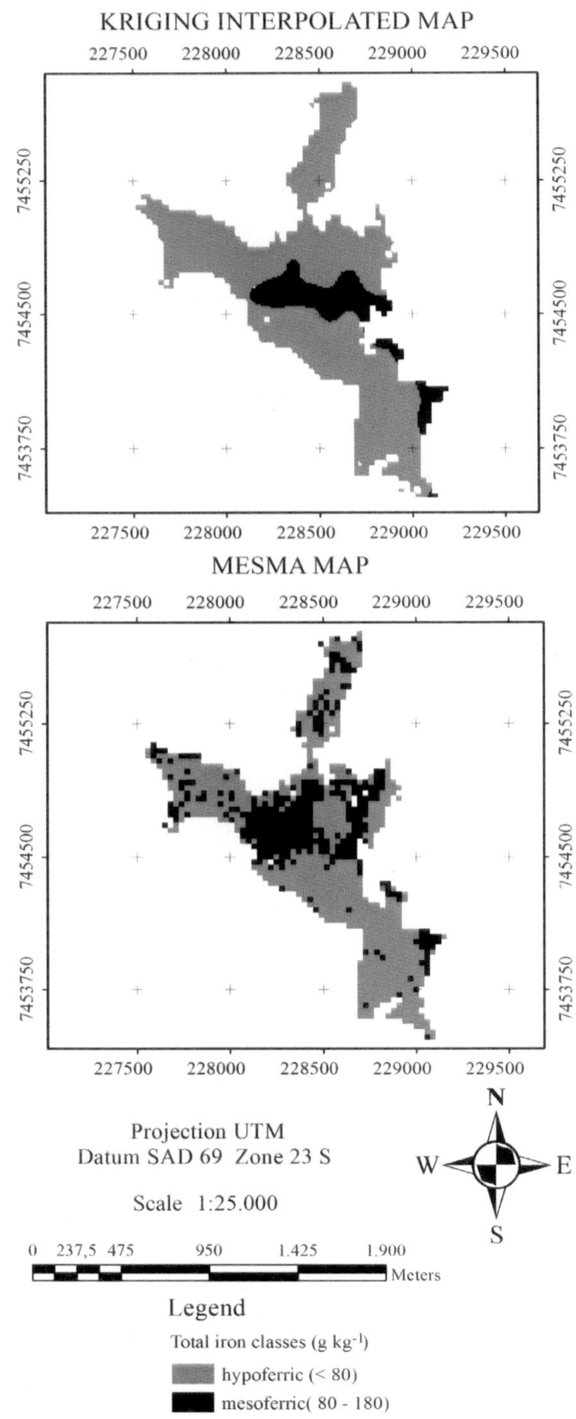

Figure 7. Kriging interpolated map and MESMA map of total iron of the area.

The total iron levels were divided into three classes, but only two of the classes were present in both of the maps because vegetation covered the areas with ferric samples.

The majority of the total area was classified as hypoferric (86.4%), and the index of similarity was 75.2%. For the mesoferric class, covering 13.67% of the area, the similarity index was 79.1% (Table 2).

It is interesting to note that the accuracy between the maps appeared to be higher when the number of classes is lower, as verified for the OM and total iron compared to the texture and CEC.

Total iron has a large effect on the spectral response of soils depending on its chemical form (DEMATTÊ et al., 2003a). Several features in the spectral curve can be attributed to the presence of iron because hematite and goethite show absorption bands at 530 nm and 480 nm, respectively. For the absorption band at 900 nm, the spectral curve becomes deeper as the iron content increases (DALMOLIN et al., 2005).

Palacios-Orueta et al. (1999) produced maps of iron levels that ranged from 2 to 6% in an AVIRIS image for two study areas in California using a modified spectral mixture analysis with accuracy determinations. It was also possible to assess the difference in the range of iron contents for the two areas of the Vale Serrano, which is mainly composed of basic igneous rocks. The higher iron contents ranged from 4 to 6%, whereas in the La Jolla Valley, which is composed of surficial sediments of sand and leaves, the iron contents were between 3 to 4%.

Conclusion

MESMA proved to be an efficient method for mapping texture, organic matter, total iron and CEC attributes and showed a high level of similarity when compared to the kriging method. However, these results are related to this specific area of study and must be tested in other areas with different soil types and attributes; this method can then be used as a tool in soil surveys and management practices.

References

BASNYAT, P.; McCONKEY, B.; MEINERT, B.; GATKZE, C.; NOBLE, G. Agriculture field characterization using aerial photograph and satellite imagery. **IEEE Geoscience and Remote Sensing Letters**, v. 1, n. 1, p. 7-10, 2004.

BEN-DOR, E.; CHABRILLAT, S.; DEMATTÊ, J. A. M.; TAYLOR, G. R.; HILL, J.; WHITING, M. L.; SOMMER, S. Using imaging spectroscopy to study soil properties. **Remote Sensing of Environment**, v. 113, n. 1, p. 38-55, 2009.

CAMARGO, A. O.; MONIZ, A. C.; JORGE, J. A.; VALADARES, J. M. **Métodos de análise química, mineralógica e física de solos do IAC**. Campinas: IAC, 1986. (Boletim Técnico, 106).

CARVALHO JUNIOR, O. A.; CARVALHO, A. P. F.; MENESES, P. R.; GUIMARÃES, R. F.; MARTINS, E. S. Análise de imagens hiperespectrais pelo método multiple endmember spectral mixture analysis (MESMA) em depósito supergênico de níquel. **Revista Brasileira de Geociências**, v. 33, n. 1, p. 63-74, 2003.

DALMOLIN, R. S. D.; GONÇALVES, C. N.; KLAMT, E.; DICK, D. P. Relação entre os constituintes do solo e seu comportamento espectral. **Ciência Rural**, v. 35, n. 2, p. 481-489, 2005.

DEMATTÊ, J. A. M.; EPIPHANIO, J. C. N.; FORMAGGIO, A. R. Influência da matéria orgânica e de formas de ferro na reflectância de solos tropicais. **Bragantia**, v. 62, n. 3, p. 451-464, 2003a.

DEMATTÊ, J. A. M.; PEREIRA, H. S.; NANNI, M. R.; COOPER, M.; FIORIO, P. R. Soil chemical alterations promoted by fertilizer application assessed by spectral reflectance. **Soil Science**, v. 168, n. 10, p. 730-747, 2003b.

DENNISON, P. E.; ROBERTS, D. A. Endmember selection for multiple endmember spectral mixture analysis using endmember average RMSE. **Remote Sensing of Environment**, v. 87, n. 1-2, p. 123-135, 2003.

DRAKE, N. A.; MACKIN, S.; SETTLE, J. J. Mapping vegetation, soil, and geology in semiarid shrublands using spectral matching and mixture modeling of SWIR AVIRIS imagery. **Remote Sensing of Environment**, v. 68, n. 1, p. 12-25, 1999.

ESRI-Environmental Systems Research Institute. **ArcGIS 9.0**. Redlands: Esri, 2005. (3 CD-ROM).

GENÚ, A. M.; DEMATTÊ, J. A. M.; NANNI, M. R. Geoprocessamento na avaliação espacial de variáveis do solo da região de Rafard-SP. **Revista Brasileira de Tecnologia Aplicada nas Ciências Agrárias**, v. 4, n. 1, p. 07-29, 2011.

HUETE, A. R. Remote sensing of soils and soil processes. In: USTIN, S. (Ed.). **Remote sensing for natural resources management and environmental monitoring**: Manual of remote sensing. 3rd ed. Hoboken: John Wiley, 2004. p. 3-52.

MATWORKS, INC. **Matlab 7.0**. Gouda: Matworks, 2005. (1 CD-ROM).

NANNI, M. R.; DEMATTÊ, J. A. M. Comportamento da linha do solo obtida por espectrorradiometria laboratorial para diferentes classes de solo. **Revista Brasileira de Ciência do Solo**, v. 30, n. 6, p. 1031-1038, 2006.

OKIN, G. S.; ROBERTS, D. A.; MURRAY, B.; OKIN, W. J. Practical limits on hyperspectral vegetation discrimination in arid and semiarid environments. **Remote Sensing of Environment**, v. 77, n. 2, p. 212-225, 2001.

PALACIOS-ORUETA, A.; PINZÓN, J. E.; USTIN, S. L.; ROBERTS, D. A. Remote sensing of soils in the Santa Monica Mountains: II. Hierarchical foreground and background analysis. **Remote Sensing of Environment**, v. 68, n. 2, p. 138-151, 1999.

RAIJ, B. Van.; QUAGGIO, J. A. **Métodos de análise de solo para fins de fertilidade**. Campinas: IAC, 1989. (Boletim Técnico 8).

RSI-Research Systems Inc. **ENVI 4.1**. Boulder: RSI, 2005. (2 CD-ROM).

ROBERTS, D. A.; DENNISON, P. E.; GARDNER, M. E.; HETZEL, Y.; USTIN, S. L.; LEE, C. T. Evaluation of the potential of hyperion for fire danger assessment by comparison to the airborne visible/infrared imaging spectrometer. **IEEE Transactions on Geoscience and Remote Sensing**, v. 41, n. 6, p. 1297-1310, 2003.

SOUSA JUNIOR, J. G.; DEMATTÊ, J. A. M.; GENÚ, A. M. Comportamento espectral dos solos na paisagem a partir de dados coletados por sensores terrestre e orbital. **Revista Brasileira de Ciência do Solo**, v. 32, n. 2, p. 727-738, 2008.

THENKABAIL, P. S.; ENCLONA, E, A.; ASHTON, M. S.; VAN DER MEER, B. Accuracy assessments of hyperspectral waveband performance for vegetation analysis applications. **Remote Sensing of Environment**, v. 91, n. 3-4, p. 354-376, 2004.

19

Comparison of methods to quantify soil microbial biomass carbon

Rafael Thomaz de Aquino Moura[1*], Marlon da Silva Garrido[2], Carla da Silva Sousa[3], Romulo Simões Cezar Menezes[4] and Everardo Valadares de Sá Barretto Sampaio[4]

*[1]Secretaria de Educação da Cidade do Recife, Av. Afonso Olindense, 1513, 50.810-000, Várzea, Recife, Pernambuco, Brazil. [2]Universidade Federal do Vale do São Francisco, Joazeiro, Bahia, Brazil. [3]Instituto Federal de Educação, Ciência e Tecnologia Baiano, Campus Uruçuca, Uruçuca, Bahia, Brazil. [4]Departamento de Energia Nuclear, Universidade Federal de Pernambuco, Recife, Pernambuco, Brazil. *Author for correspondence. E-mail: rafael_thomaz@hotmail.com*

ABSTRACT. Soil microorganism biomass is an important soil quality indicator. The microbial biomass of soil was determined by killing and lysing the soil microbes by fumigation with chloroform, irradiation with gamma rays, or irradiation with microwaves. Four soils with increasing carbon concentrations (5, 10, 15, and 30 g kg^{-1}) were analyzed using four methods: the direct application of chloroform, chloroform fumigation, microwave irradiation, and gamma ray irradiation with radiation doses of 15, 25, 35, 45, and 60 KGy. The fungi and bacteria in the soil were quantified by plate counting. Microwave irradiation and gamma irradiation with doses equal to or above 25 KGy killed all the soil microorganisms, but the chloroform methods did not. The carbon liberation increased with higher gamma doses, while the microbe mortality rates were the same, indicating that carbon was liberated from organic matter sources other than microorganisms. The biomasses determined by the microwave method correlated with those determined by the fumigation and 25 KGy gamma irradiation methods, but their values differed among all methods for at least one soil type. Despite this discrepancy, all methods were consistent in ranking microbial biomasses in increasing order of soil carbon concentrations, which corresponds with decreasing land use intensities.

Keywords: microwave; gamma irradiation; fumigation; incubation; microbial carbon.

Introduction

The two main environmental problems affecting agricultural activities in the semiarid region of the Brazilian Northeast are low water availability and low soil fertility, which each limit the productivity of the main crops. Therefore, agricultural incomes are low, and crops are cultivated with minimum external inputs, i.e., usually without fertilizer application (Menezes, Sampaio, Giongo, & Pérez-Marin, 2012). Consequently, soil organic matter is a major source of nutrients, and its decomposition is vital to supporting plant production (Moura et al., 2016).

Soil microorganisms are a key component of organic matter nutrient cycling (Brookes et al., 2008). Therefore, knowledge of the soil microbiota functions is important to properly manage the processes that control nutrient availabilities to the plants (Arcand, Helgason, & Lemke, 2016). Several microbiological parameters have been used as indicators of these processes, but no single parameter is appropriate for all situations due to the dynamic and complex nature of agroecosystems (Brookes et al., 2008; Leite et al., 2010). A good indicator needs to respond to soil perturbations, highlighting temporal changes in a rapid and low-cost manner.

Among soil quality indicators, microbial biomass is one of the best (Brookes et al., 2008; Jannoura, Bruns, & Joergensen, 2013; Makarov, Malyesheva, Maslov, Kuznetsovaa, & Menyailo, 2016). Soil microbial biomass (SMB), excluding plant roots and animals larger than 5 μm^3, corresponds with, on average, 2 to 5% of soil organic matter (Vance, Brooks, & Jenkinson, 1987). There is no consensus on the best method to quantify SMB under diverse edapho-ecological conditions, either because the results have not been very reliable, or the procedures are too labor intensive or too expensive (Hartmann, Frey, Mayer, Mäder, & Widmer, 2015; Makarov, 2016). Thus, it is advisable to use more than one method for SMB quantification.

Several methods are based on processes that cause the death and partial or total disintegration of the soil microorganisms (Powlson & Jenkinson, 1975; Araújo, 2010). After disintegration, some of the cell constituents can be extracted by even mild extractors. There are different methods to kill soil microorganisms, and the most common ones are chloroform fumigation (Brookes et al., 2008; King & Hofmockel, 2017) and microwave irradiation (Araújo, 2010; Souza et al., 2010); however, gamma irradiation has also been used (Powlson & Jenkinson, 1975). The most common method worldwide (Tang et al., 2016), which was developed by Jenkinson and co-workers (Powlson & Jenkinson, 1975) and later modified by Vance, Brookes, and Jenkinson (1987), consists of fumigation in a desiccator, followed by extraction with potassium sulfate and the determination of carbon by titration or colorimetry. Chloroform can also be applied directly to the soil (Witt, Gaunt, Galicia, Ottow, & Neue, 2000; Setia, Verma, & Marschner, 2012).

Comparisons of some of the different methods have been made, but they seldom verify the effectiveness of the methods in killing soil microbiota (Makarov et al., 2016). Chloroform fumigation does not kill all microorganisms, and the method is usually limited by the need for a large set of desiccators (Wolf, Dao, Scott, & Lavy, 1989). Microwave irradiation seems to be efficient in killing, and it is easily performed with commonly available equipment (Araújo, 2010). Gamma irradiation has seldom been used because the necessary equipment is only available at a few places; however, when it is used, the method seems to be efficient in killing and easily performed (Powlson & Jenkinson, 1975).

Considering the scarcity of data obtained on the gamma irradiation of tropical soils, the objectives of this study were: i) determine the proper gamma irradiation doses to kill soil microbiota and measure soil microbial biomass, and ii) compare the gamma irradiation method with the microwave radiation, chloroform fumigation, and direct chloroform application methods.

Material and method

Four soils samples with increasing carbon concentrations (4.8, 9.9, 15.8, and 29.2 g kg^{-1}) were collected in the Agroecological Center São Miguel (7°19'S and 33°51'W), the headquarters of the NGO ASPTA, in the municipality of Esperança, in the semiarid zone of Paraíba state, Brazil. All samples were collected within a radius of 500 m, and the soil was classified as Entisol, with a loamy sandy texture and 5% slope. The first three samples were collected in agricultural plots cultivated with increasing conservation practices: soil 1) continuous bean and potato cultivation without fertilization, soil 2) potato cultivation fertilized with manure, and soil 3) agroforestry system with maize and Gliricidia fertilized with manure. The last sample, soil 4, had the highest carbon concentration of the samples and was collected from Fluvic Entisol in the lower area of the landscape under apparently undisturbed dry forest vegetation. In each area, 20 individual samples were collected in the superficial soil layer (0 – 20 cm) and pooled into a single composite sample, which was analyzed for its chemical and physical characteristics (Table 1).

Table 1. Chemical and physical characteristics of soils with increasing carbon contents collected in the semiarid region of Paraíba, Brazil.

Soil	C	N	P	K	Ca	Mg	Na	pH (H_2O)	Density	Sand	Silt	Clay
	g kg^{-1}		mg kg^{-1}		cmol$_c$ kg^{-1}				g cm^{-3}		g kg^{-1}	
1	4.8	0.46	0.6	0.18	1.75	1.10	0.14	6.4	1.20	830	110	60
2	9.9	0.85	6.0	0.13	2.08	1.35	0.13	5.8	1.16	733	134	133
3	15.8	1.27	3.1	0.16	2.25	1.28	0.14	5.5	1.08	663	126	211
4	29.2	2.38	1.9	0.18	1.27	1.18	0.08	4.6	0.94	470	100	430

Twenty-gram subsamples were wetted to 40% of field capacity and incubated for 72 hours in polyethylene bags. These subsamples were then submitted to the four microbial killing procedures: fumigation with chloroform in desiccators, fumigation with chloroform applied directly to the soil, irradiation in a microwave oven and irradiation with gamma rays.

Part of the subsamples were initially submitted to gamma irradiation for 1.5, 2.5, 3.5, 4.5, and 6 hours in a ^{60}Co (GammaCell 220 Excel) irradiator to reach doses of 15, 25, 35, 45, and 60 KGy, respectively, to select the best dose for the second phase of the method comparison experiment. The soil microbial biomass was determined in these and in non-irradiated samples using the methodology recommended by Vance et al. (1987). The carbon concentrations were quantified by oxidation with potassium permanganate and colorimetry with a 495-nm absorbance wavelength (Bartlett & Ross, 1988).

To verify the efficiency of the different irradiation doses in killing the microbial population, the fungi and bacteria presences in the subsamples were determined. A 10 g portion of each irradiated subsample was placed in Erlenmeyer flasks with 90 mL of a 0.85% NaCl solution and shaken for 1 hour. Serial dilutions (10^{-1} to 10^{-5}) of the supernatant were made, and 100 μL aliquots were transferred, using a Drigalski spatula, to Petri dishes containing Sabouraud media with 100 mg mL^{-1} cycloheximide to isolate the fungi and TSA media with 100 mg mL^{-1} streptomycin to isolate the bacteria. All procedures were carried out in a laminar flow chamber. The Petri dishes were incubated at 30°C for 24 hours for the bacteria tests and for 48 to 72 hours for the fungi tests.

Next, subsamples were submitted to the four microbial killing methods. The subsamples that were gamma irradiated received a 25 KGy dose, as selected by the initial test. The subsamples submitted to the chloroform fumigation were placed in desiccators next to a beaker with 25 mL of chloroform and another beaker with water, which was introduced to maintain a high humidity level. Vacuum conditions were sustained for approximately 5 minutes, until the bubbling and volatilization of all the chloroform occurred, and the subsamples were maintained in the closed desiccator for 24 hours. The desiccator was then flushed, under vacuum, to remove excess fumigant (Vance et al., 1987). The direct application of chloroform was made by adding 1 mL chloroform to the soil subsamples and placing them under vacuum conditions for 24 hours, followed by the described flushing procedure. The microwave irradiation method was conducted with a commercial apparatus. The period of irradiation was calculated based on the tension, microwave frequency, and energy concentration values (Ferreira, Camargo, & Vidor, 1999; Araújo, 2010). In this method, microbes are killed by the increased temperature of their cytoplasm, which breaks their cell walls.

Data on the doses of gamma irradiation were submitted to an analysis of variance, according to a 5 x 4 factorial arrangement, which corresponded to the five doses and the four soils, with six replications in a completely randomized design. This was followed by a Tukey test at a 5% significance level. The interactions were split, and the doses underwent a regression analysis. Data on the microbial biomass determination methods were analyzed as a 4 x 4 factorial arrangement. The microbial counting data were transformed by the equation $\sqrt{x+1}$. A matrix of correlation was calculated for the microbial biomass data obtained with the two chloroform methods, the microwave method, and the 15 and 25 KGy irradiation methods. All analyses were performed using SISVAR version 4.8 (Ferreira, 2008).

Result and discussion

Irradiation with gamma rays at the lowest used dose (15 KGy) killed more than 99% of bacteria and fungi in all the soils (Table 2). This lethality may be adequate for the determination of soil microbial biomasses, but the radiation was not sufficient to sterilize the soils. The high dose (25 KGy), which was the dose level recommended by Powlson & Jenkinson (1975) to determine soil microbial carbon, sterilized all the soils, at least with respect to bacteria and fungi. Doses higher than 23 KGy had the same effect, as was found by Wolf et al. (1989), who irradiated soils with different total carbon contents using doses of 50 and 60 KGy. If sterilization is needed, 25 KGy is appropriate; 15 KGy killed only slightly less microorganisms than 25 KGy, so it should be sufficient for the determination of soil microbial biomasses if the sole requirement of the radiation is the death of the microorganisms.

If the effect of the radiation is restricted to the death and lyses of the soil microorganisms, the amounts of extracted carbon should be similar among all irradiation doses. However, the extracted carbon increased with the irradiation periods for all soils, although different patterns were observed for different soils (Figure 1). In the soil with the lowest organic carbon content, the increase was negligible, while in the other three soils, the extracted carbon increased more than four times from the shortest to the longest irradiation period. For two of the soils (2 and 4), the increases had linear tendencies, indicating that higher doses of radiation extracted more carbon. Since the microbial biomass was similar in subsamples from the same soil sample, and all dose levels killed practically all of the microorganisms, the gamma radiation of the higher doses liberated soil carbon fractions from sources other than the killed microorganisms. Therefore, using higher doses of gamma radiation could cause the overestimation of microbial biomass if the biomass calculations assumed that all extracted carbon belonged to the killed microorganisms. However, these results open the possibility of using gamma radiation to determine soil carbon fractions that are relatively labile and include more than just microbial biomass. It is interesting to note that the soil with the most intense agricultural use (soil 1), and the lowest carbon concentrations, had almost no increase in the extracted carbon with increases in the irradiation period, indicating that its carbon is retained in more recalcitrant fractions than the other soil types (Singh & Nandita, 2014) and, probably, it has little more labile carbon than the carbon from its microbial biomass.

Table 2. Colonies of bacteria and fungi in growth media inoculated with extracts from soils with increasing carbon concentrations, which were collected in the semiarid area of the Brazilian Northeast and submitted to increasing doses of gamma radiation.

Soil	Organic carbon	Doses (KGy)					
		0	15	25	35	45	60
	$g\ kg^{-1}$	Number of colony units x 10^3 gss^{-1} -					
		Bacteria					
1	4.8	243	0.063	0	0	0	0
2	9.9	650	0.050	0	0	0	0
3	15.8	850	0.078	0	0	0	0
4	29.2	300	0.037	0	0	0	0
		Fungi					
1	4.8	3	0.013	0	0	0	0
2	9.9	28	0.017	0	0	0	0
3	15.8	28	0.015	0	0	0	0
4	29.2	8	0.010	0	0	0	0

Another observation is that in the control, non-irradiated samples, the highest numbers of bacteria and fungi colonies formed in the inoculation from soil 3, which had the highest extracted carbon content among all the soils after an irradiation of 25 KGy, although it did not have the highest carbon concentration (Table 2). This confirms that other factors besides the organic matter content affect the reproduction capacity of soil microorganisms, and it reinforces the current opinion that microbial biomass is a useful indicator of soil quality under different management regimes (Jannoura et al., 2013; Makarov et al., 2016). This peculiar behavior of soil 3 may reflect better conditions for microbial growth in its collection area, which are produced by the combination of high soil mobilization (maize cultivation) associated with the application of manure and litter inputs and the favorable microclimate due to the presence of the trees in the agroforestry system.

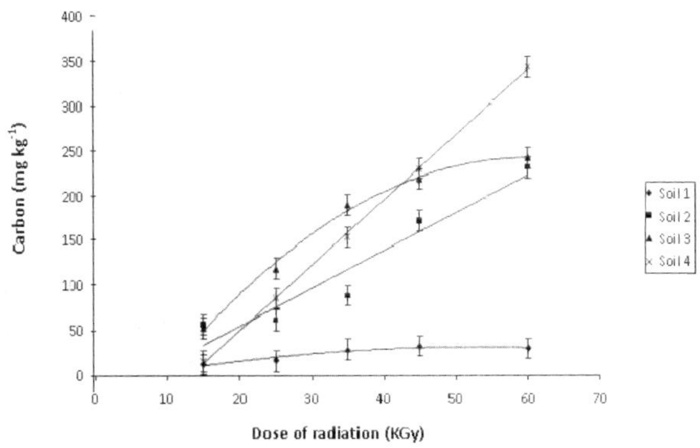

Figure 1. Carbon extracted from soils with increasing carbon concentrations collected in the semiarid area of the Brazilian Northeast and submitted to increasing doses of gamma radiation. Vertical bars represent minimum significant differences ($p < 0.05$).

The effect of the gamma irradiation can be further analyzed by comparing its results with those of other methods used to kill soil microorganisms and to determining their biomass data. Microwave radiation, like the 25 KGy gamma dose, killed all the bacteria and fungi, but the chloroform methods did not (Table 3). Since the original fumigation procedure was developed, it has been known that it does not kill all microorganisms. In fact, for the original method of fumigation, incubation, it was assumed that most of the microorganisms would be killed, and their biomass would be consumed by the new microorganism populations developing from the survivors of the fumigation, which would result in a flush of CO_2 evolution (Powlson & Jenkinson, 1975). Later, incubation was replaced by the immediate extraction with K_2SO_4 of the microbial biomass lysed by the fumigation (Brookes et al., 2008). By that period, Wolf et al. (1989) had already reported that chloroform fumigation within desiccators did not sterilize soil, while sterilization could be achieved with gamma radiation doses of 50 and 60 KGy.

Comparing fumigation with the direct application of chloroform, the consistent result for all soils was that higher numbers of bacteria colonies formed during the incubation after the fumigation, while a higher number of fungi colonies occurred during the incubation after the direct application. For both methods, the numbers of colonies that formed after the chloroform treatments were not related to the numbers that formed from unfumigated soils; nor were they related to the total carbon concentrations of the four soils. The causes of the differences in the bacteria and fungi survival rates after the two treatments and of the different survival proportions are not known and warrant further research. The direct chloroform application might result in higher sorption interactions with the soil clay and/or organic matter constituents than fumigation, possibly making its complete evacuation more difficult (Rotbart et al., 2017). This effect on the determination of microbial biomass may be only slightly greater than the small effect found by Rotbart et al. (2017) with fumigation but its effect on the microbial population recovery is not known.

Table 3. Colonies of bacteria and fungi formed in growth media inoculated with extracts from soils with increasing carbon concentrations collected in the semiarid area of the Brazilian Northeast and submitted to different microbial killing procedures.

Soil	Organic carbon	Control	Chloroform in the soil	Chloroform fumigation	Microwave radiation
	g kg^{-1}	Number of colony units x 10^3 gss^{-1}-			
			Bacteria		
1	4.8	243	23 Ba[1]	42 Aa	0 Ca
2	9.9	650	11 Bb	37 Aa	0 Ca
3	15.8	850	13 Bab	33 Aa	0 Ca
4	29.2	300	11 Bb	32 Aa	0 Ca
			Fungi		
1	4.8	3	5 Ac	0.5 Bb	0 Ca
2	9.9	28	10 Aa	3.0 Ba	0 Ca
3	15.8	28	8 Ab	0.5 Bb	0 Ca
4	29.2	8	8 Ab	0.5 Bb	0 Ca

Averages followed by the same capital letter in the row and small letter in the column are not significantly different by the Tukey test at the 5% probability level.

The microbial biomasses calculated for the soil that was cultivated for the longest period (soil 1), which also had the lowest soil carbon content, were all low and not significantly different using the two chloroform methods, the microwave method, and the two gamma irradiation methods with doses of 15 and 25 KGy (Table 4). For the undisturbed soil (soil 4), which had with the highest carbon content, the 25 KGy gamma dose, microwave and chloroform fumigation methods were not significantly different, and the direct application of chloroform to the soil was inferior to the irradiation methods but not different from the chloroform fumigation method, while the 15 KGy gamma radiation dose resulted in the lowest, and most statistically inferior, biomass value. For the two soils with intermediate carbon contents (soils 2 and 3), the 25 KGy radiation dose also resulted in the highest carbon levels during extractions, although the values for soil 2 were not significantly different from the methods that used microwave radiation and the direct application of chloroform to the soil. In both soils 2 and 3, the microbial biomasses determined by the gamma irradiation method with the 15 KGy dose were included in the group of methods that produced the lowest values, which also comprised the biomasses determined by chloroform fumigation.

Correlating the five measurement techniques (the four primary methods, including the two gamma dose levels of 15 and 25 KGy) using the biomass data of the four soils (Table 4) only resulted in significant correlation coefficients (> 0.95; $p < 0.05$) for the microwave and 25 KGy gamma irradiation pair and the microwave and chloroform fumigation pair. However, four is a low number of compared pairs, which may overshadow real correlations with lower coefficients, similar to that of the paired microwave and soil applied chloroform fumigation methods ($R = 0.93$; significant at 0.07). If this pair is included, the microwave method correlates with all others, except for the gamma irradiation method using a 15 KGy dose. In fact, the 15 KGy dose correlated poorly with all other measurement techniques.

Table 4. Microbial biomass carbon in soils with increasing carbon concentrations collected in the semiarid area of the Brazilian Northeast and submitted to different microbial killing procedures.

Soil	Organic carbon	Chloroform in the soil	Chloroform fumigation	Micro wave radiation	Gamma radiation 15 KGy	Gamma radiation 25 KGy
	g kg^{-1}	mg kg^{-1}				
1	4.8	14.8 Ab[1]	16.1 Ac	22.5 Ac	19.2 Ab	22.6 Ac
2	9.9	69.4 Aa	46.0 Bb	64.7 ABb	52.1 Ba	74.0Ab
3	15.8	63.6 BCa	56.5 Cb	81.6 Bab	49.4 Ca	107.3 Aa
4	29.2	71.8 Ba	85.3 ABa	92.8 Aa	19.7 Cb	97.1 Aa

1 - Averages followed by the same capital letter in the row and small letter in the column are not significantly different by the Tukey test at the 5% probability level

With a more detailed analysis of the biomasses determined by the five ways of measuring biomasses in the four soils with increasing carbon contents, more information can be extracted. The low biomasses determined by the 15 KGy gamma irradiation method, particularly for soil 4, indicates this dose seems to be inefficient in the lysis of all soil microorganisms, despite of killing most of them (Table 2). On the other hand, the 25 KGy gamma irradiation level always resulted in the highest calculated biomasses, and this may have resulted from the extraction of carbon from sources other than the microbial biomass, which is more evident with doses higher than 25 KGy; therefore, this method may overestimate the soil microbial biomass compared to the other methods. Consequently, this dose level cannot be unquestioningly recommended for the microbial biomass determination of all soils (Powlson & Jenkinson, 1975), and it is possible that an intermediate dose between 15 and 25 KGy works better, at least for tropical sandy soils relatively poor in total organic matter.

A similar doubt could be raised for the microwave method if it is compared with the fumigation treatment; not only were the absolute averages always higher for the former, they were significantly higher in the case of soil 3 (Table 4). However, both methods have been used extensively, and in some cases, the microwave method resulted in consistently lower C extraction levels (Araújo, 2010). The direct application of chloroform also resulted in higher carbon extraction than fumigation in the case of soil 2, and the average extraction value was very similar to those obtained with the microwave and 25 KGy irradiation methods. A higher carbon extraction level with direct application than with other methods was also reported by Setia et al. (2012). Either all three methods are not adequate for all soils, or the fumigation method, which is considered to be the standard, is also sometimes inadequate.

Despite the discrepancies among the average extraction levels of the methods, in general, they were all consistent in ranking the microbial biomasses in increasing order from the lowest to the highest total carbon content. This is consistent with the results of several authors who have found that intensively cultivated soils have lower microbial biomasses (Araújo, Santos, & Monteiro, 2008) and microorganism diversities (Kong, Scow, Córdova-Kreylos, Holmes, & Six, 2011) than soils under native vegetation or rotated crops (Souza et al., 2010; Leite et al., 2011; McDaniels, Tiemann, & Grandy, 2015; Hartmann et al., 2015; King & Hofmockel, 2017). From this perspective, all the methods were valid for detecting changes in the land use intensity, but their absolute microbial biomass results should be carefully interpreted. Independent of variations, the microbial biomass carbon of the soils determined by all four methods was always less than 1% of the total soil carbon concentration. These proportions, which are low compared to those reported for soils of the same type in other regions (Matias, Salviano, Leite, & Araújo, 2009; Souza et al., 2010) but similar to the values reported for the same Brazilian region (Araújo et al., 2008), can be attributed to the semiarid climate conditions prevailing in the area where the soils were sampled. These low proportions also minimize any possible need to stress the importance of the absolute biomass values.

Conclusion

Microwave and gamma irradiation equal to or above 25 KGy killed all microorganisms, but the chloroform methods did not. Higher numbers of bacteria colonies formed during incubation after chloroform fumigation than after the direct application, while the opposite occurred with fungi colonies.

Biomasses determined by the microwave method correlated with those determined by the fumigation and 25 KGy irradiation methods, but the values differed among all methods for at least one soil type. Despite this discrepancy, all methods were all consistent in ranking the microbial biomasses in increasing order from the lowest to the highest total carbon content.

References

Araújo, A. S. F. (2010). Is the microwave irradiation a suitable method for measuring soil microbial biomass? *Reviews in Environmental Science and Bio/Technology*, *9*(4), 317-321. doi: 10.1007/s11157-010-9210-y

Araújo, A. S. F., Santos, V. B., & Monteiro, R. T. R. (2008). Responses of soil microbial biomass and activity for practices of organic and conventional farming systems in Piauí state, Brazil. *European Journal of Soil Biology*, *44*(2), 225–230. doi: 10.1016/j.ejsobi.2007.06.001

Arcand, M. M., Helgason, B. L., & Lemke, R. L. (2016). Microbial crop residue decomposition dynamics in organic and conventionally managed soils. *Applied Soil Ecology*, *107*(1), 347-359. doi: 10.1016/j.apsoil.2016.07.001

Bartlett, R. J., & Ross, D. S. (1988). Colorimetric determination of oxidizable carbon in acid soil solutions. *Soil Science Society of America Journal*, *52*(4), 1191-1192. doi: 10.2136/sssaj1988.03615995005200040055x

Brookes, P. C., Cayuela, M. L., Contin. M., De Nobili, M., Kemmitt, S. J., & Mondini, C. (2008). The mineralization of fresh and humified soil organic matter by the soil microbial biomass. *Waste Management*, *28*(4), 716–722. doi: 10.1016/j.wasman.2007.09.015

Ferreira, A. S., Camargo, F. A. O., & Vidor, C. (1999). Utilização de microondas na avaliação da biomassa microbiana do solo. *Revista Brasileira de Ciência do Solo*, *23*(4), 991-996. doi: 10.1590/S0100-06831999000400026

Ferreira, D. F. (2008). SISVAR: um programa para análises e ensino de estatística. *Revista Symposium*, *6*(2), 36-41.

Hartmann, M., Frey, B., Mayer, J., Mäder, P., & Widmer, F. (2015). Distinct soil microbial diversity under long-term organic and conventional farming. *ISME Journal*, *9*(5), 1177-1194. doi: 10.1038/ismej.2014.210

Jannoura, R., Bruns, C., & Joergensen, R. G. (2013). Organic fertilizer on pea yield, nutrient uptake, microbial root colonization and soil microbial indices in organic farmer systems. *European Journal of Agronomy*, *49*(1), 32-41. doi: 10.1016/j.eja.2013.03.002

Kong, A. Y. Y., Scow, K. M., Córdova-Kreylos, A. L., Holmes, W. E., & Six, J. (2011). Microbial community composition and carbon cycling within soil microenvironments of conventional, low-input, and organic cropping systems. *Soil Biology and Biochemistry*, *43*(1), 20–30. doi: 10.1016/j.soilbio.2010.09.005

King, A. E., & Hofmockel, K. S. (2017). Diversified cropping systems support greater microbial cycling and retention of carbon and nitrogen. *Agriculture, Ecosystems and Environment*, *240*(1), 66-76. doi: 10.1016/j.agee.2017.01.040

Leite, L. F. C., Oliveira, F. C., Araujo, A. S. F., Galvão, S. R. S., Lemos, J. O., & Silva, E. F. L. (2010). Soil organic carbon and biological indicators in a Acrisol under tillage systems and organic management in Northeastern Brazil. *Australian Journal of Soil Research*, *48*(3), 258–265. doi: 10.1071/SR09122

Makarov, M. I., Malyesheva, T. I., Maslov, M. N., Kuznetsovaa, E. Y, & Menyailo, O. V. (2016). Determination of carbon and nitrogen in microbial biomass of Southern-Taiga soils by different methods. *Eurasian Soil Science*, *49*(6), 685-695. doi: 10.1134/S1064229316060053

Matias, M. C. B. S., Salviano, A. A. C., Leite, L. F. C., & Araujo, A. S. F. (2009). Biomassa microbiana e estoques de C e N do solo em diferentes sistemas de manejo, no Cerrado do Estado do Piauí. *Acta Scientiarum. Agronomy*, *31*(3), 517–521. doi: 10.4025/actasciagron.v31i3.687

Mcdaniel, M. D., Tiemann, L. K., & Grandy, S. (2014). Does agricultural crop diversity enhance soil microbial biomass and organic matter dynamics? A meta-analysis. *Ecological Applications*, *24*(3), 560–570. doi: 10.1890/13-0616.1

Menezes, R. S. C., Sampaio, E. V. S. B., Giongo, V., & Perez-Marin, A. M. (2012). Biogeochemical cycling in terrestrial ecosystems of the Caatinga Biome. *Brazilian Journal of Biology*, *72*(3), 643-653. doi: 10.1590/S1519-69842012000400004

Moura, P. M., Althoff, T. D., Oliveira, R. A., Souto, J. S., Souto, P. C., Menezes, R. S. C., & Sampaio, E. V. S. B., (2016). Carbon and nutrient fluxes through litterfall and nutrient fluxes at four succession stages in a very dry tropical forest. *Nutrient Cycling in Agroecosystems*, *105*(1), 25-38. doi: 10.1007/s10705-016-9771-4

Powlson, D. S., & Jenkinson, D. S. (1975). The effects of biocidal treatments on metabolism in soil. II. Gamma irradiation, autoclaving, air-drying and fumigation. *Soil Biology and Biochemistry*, *8*(3), 179-188. doi: 10.1016/0038-0717(76)90002-X

Rotbart, N., Borisover, M., Bukhanovsky, N., Nasonova, A., Bar-Tal, A., & Oren, A. (2017). Examination of residual chloroform interference in the measurement of microbial biomass C by fumigation-extraction. *Soil Biology & Biochemistry*, *111*(1), 60-65. doi: 10.1016/j.soilbio.2017.03.018

Setia, R., Verma, S.L., & Marschner, P. (2012). Measuring microbial biomass carbon by direct extraction - comparison with chloroform fumigation-extraction. *European Journal of Soil Biology*, *53*(1), 103-106 doi: 10.1016/j.ejsobi.2012.09.005

Singh, M. K., & Nandita, G. (2014). Variations in soil microbial biomass in the dry tropics: impact of land-use change. *Soil Research*, *52*(3), 299-306. doi: 10.1071/SR13265

Souza, E. D., Costa, S. E. V. G. A., Anghinoni, I., Lima, C. V. S., Carvalho, P. C. F., & Martins, A. P. (2010). Biomassa microbiana do solo em sistema de integração lavoura-pecuária em plantio direto, submetido a intensidades de pastejo. *Revista Brasileira de Ciência do Solo*, *34*(1), 79-88. doi: 10.1590/S0100-06832010000100008

Tang, X., Placella, S. A., Daydé, F., Bernard, L., Robin, A., Journet, E. P., Justes, E., & Hinsinger, P. (2016). Phosphorus availability and microbial community in the rhizosphere of intercropped cereal and legume along a P-fertilizer gradient. *Plant and Soil*, *407*(1), 119-134. doi: 10.1007/s11104-016-2949-3

Vance, E. D., Brookes, P. C., & Jenkinson, D. S. (1987). An extraction method for measuring soil microbial biomass C. *Soil Biology and Biochemistry*, *19*(6), 703-707. doi: 10.1016/0038-0717(87)90052-6

Witt, C., Gaunt, J. L., Galicia, C. C., Ottow, J. C. G., & Neue, H. U. (2000). A rapid chloroform-fumigation extraction method for measuring soil microbial biomass carbon and nitrogen in flooded rice soils. *Biology and Fertility of Soils*, *30*(5-6), 510–519. doi: 10.1007/s003740050030

Wolf, D. C., Dao, T. H., Scott, H. D., & Lavy, T. L. (1989). Influence of sterilization methods on selected soil microbiological, physical, and chemical-properties. *Journal of Environmental Quality*, *18*(1), 39-44. doi: 10.2134/jeq1989.00472425001800010007x

Incidence of epigeal nest-building termites in *Brachiaria* pastures in the Cerrado

Maria Inês Lopes de Oliveira[1], Didier Brunet[2], Danielle Mitja[2], Wellington dos Santos Cardoso[3], Norton Pólo Benito[4], Maria de Fátima Guimarães[5] and Michel Brossard[2*]

*[1]Programa de Pós-graduação em Agronomia, Universidade Estadual de Londrina, Londrina, Paraná, Brazil. [2]Institut de Recherche pour le Développement, Unité Eco&Sols, BP 64501, 34394, Montpellier cedex 5, France. [3]Faculdade da Terra de Brasília, Brasília, Distrito Federal, Brazil. [4]Núcleo Temático de Segurança Biológica, Embrapa Recursos Genéticos e Biotecnologia, Brasília, Distrito Federal, Brazil. [5]Departamento de Agronomia, Centro de Ciências Agrárias, Universidade Estadual de Londrina, Londrina, Paraná, Brazil. *Author for correspondence. E-mail: michel.brossard@ird.fr*

ABSTRACT. This study aimed to determine the number of termite epigeal nests and estimate the soil turning capacity of termites in cultivated pasture environments and in a native vegetation area. Surveys were conducted in four areas: three pasture areas and a Cerrado area, measuring 5 ha each. For each nest, the height, the perimeter and diameter at the base were measured. The volume of each mound was calculated. Turned soil was determined by randomly sampling 30 termite mounds in a pasture area for bulk density determinations. The mean number of termite mounds per ha was 408; the highest number of termite nests (672 per ha) was found in the 10-year-old pasture. In terms of termite mound density, there was no difference between the Cerrado, 3-year-old pasture, and 3-year-old intercropped pasture treatments. The total area occupied by termite nests ranged from 0.4 to 1.0%. The mean bulk density of the nest materials was 1.05 g cm^{-3}. The total nests volume ranged from 4 to 11 m^3 ha^{-1}. The total mass of the soil turned by the termites ranged from 3.4 to 13.4 ton. ha^{-1}.

Keywords: turned soil volume, termite, soil macrofauna.

Introduction

The Brazilian Cerrado covers an area of 204.7 million hectares (ADÁMOLI et al., 1986). In this area, cultivated pastures represent 49.5 million ha (SANO et al., 2000), of which over 34 million show a loss of productivity (KLUTHCOUSKI et al., 1999). Studies have demonstrated that modifications in the soil's productive functions directly affect the soil macrofauna and that these organisms respond to many anthropic interventions made to the environment (LAVELLE; SPAIN, 2001). Termites play an important role in the soil macrofauna.

Termite ecology, particularly spatial distribution aspects and their relationship with environmental factors, have been the object of many investigations, most of them in Africa, Australia, and United States (ABE et al., 2000; BACHELIER, 1978). The soil macrofauna inventories of the Brazilian Cerrado indicate dominance by termites, both in density and biomass (BENITO et al., 2004; CONSTANTINO, 2005; DIAS et al., 1997). Termites build biogenic structures (galleries, nests and chambers), thus changing the physical properties of the soils they inhabit (BACHELIER, 1978). By means of their mechanical actions into the soil, they contribute, by the formation of stable aggregates, to the protection of the organic matter against a fast mineralization (DECÄENS et al., 2003; LAVELLE; SPAIN, 2001).

There has been great concern in Brazil on the part of farmers with regard to the high incidence of "mound termites" (*cupins de montículo*) in pastures, that is, species whose colonies build epigeal nests in the soil. A substantial part of geophagous termites build their colonies inside the soil. Nevertheless, although a pasture dotted with mounds causes great visual impact, the area occupied by nests, considering a mean area of 0.5 m^2 per mound, with a density of 200 termite mounds per ha, would decrease the pasture area just by 1% (EMBRAPA, 1996).

Although the edaphic macrofauna is relevant for ecosystems to function properly, few studies have been conducted to evaluate the effects of management practices on the soil macrofauna, especially in Cerrado soils. This study aimed to determine the number of epigeal nests and assess the soil turning capacity of termites in cultivated pasture environments and in a native vegetation area.

Material and methods

The study was carried out in a farm at 1000 m a.s.l. on the Brazilian central Plateau (15° 14' S, 47° 42' W) in Goiás State. The climate is sub-humid tropical (mean annual temperature of 23°C) with a rainy season from September to April (mean annual precipitation of 1170 mm). Dry and hot spells, lasting from 5 to 15 days, may occur during this season. The drought period lasts around 4 to 6 months. The soil is a homogeneous dark red, clayey Ferralsol, Latossolo Vermelho according the Brazilian classification (EMBRAPA 2006) containing 650 to 750 g kg^{-1} clay, with an acid pH (between 4 and 5).

Four areas measuring 5 ha each were selected, three under pasture and one under native Cerradão: i) Cerradão, ii) "past 3" (*Brachiaria brizantha* pasture, renovated three years earlier), iii) "paststy 3" (*Brachiaria brizantha* pasture intercropped with *Stylosanthes guianensis* var. *vulgaris*, cv Mineirão, renovated three years earlier), and iv) "past 10" (10-year-old *Brachiaria brizantha* pasture, established after deforesting a native forest savanna area). The "past 3" and "paststy 3" areas are parcels resulting from the renovation of a 7-year-old low productivity pasture. They were renovated using a heavy harrow at a depth of 15 cm; the *Brachiaria* grass was not reseeded and developed from the soil seed bank and the preexisting roots. A fertilization of 90 kg ha^{-1} P_2O_5 as Gafsa phosphate and 74 kg ha^{-1} sulfur as flower of sulfur was applied. The pastures were managed by adjusting the grazing pressure according to forage availability. The soil macrofauna was initially characterized in these areas (BENITO et al., 2004).

Within each area, four plots of 1000 m^2 were delimited to evaluate the incidence of the termite mounds on the pastures. The height, perimeter, largest diameter and smallest diameter at the base of each mound were measured with a measuring tape. The basal area was obtained by the mean between the largest and the smallest diameter, via the formula used to calculate the area of the ellipse, $A = \pi * a * b$, where "a" is half of the largest diameter and "b" is half of the smallest diameter. In order to estimate turned soil density, cube-shaped samples were taken at random with a hoe and a handsaw, from 28 termite mounds in the "past 10" area, and their volumes were then calculated. The samples were weighed and oven-dried at 105°C during 72h, and then weighed again after drying. Turned soil volume was calculated using the equation $V = 1/3 * \pi R^2 * h$, where R (m) is mound radius and h (m) is mound height.

Soldiers from 30 epigeal nests were sampled in the older pasture to determine genera found, since a nest may contain more than one species, in addition to the species that built the nest.

The data were compared inside each class by Kruskal-Wallis test, Dunn test was processed using Bioestat (AYRES et al. 2007)

Results and discussion

Termites genera

Twelve Isoptera morphospecies were found within 11 genera, most of them of the Subfamily Nasutitermitinae: *Cornitermes* sp.1, *Labiotermes* sp.1, *Procornitermes* sp.1, *Procornitermes* sp.2, *Armitermes* sp.1, *Cyranotermes* sp.1, *Embiratermes* sp.1, *Nasutitermes* sp.1, *Syntermes* sp.1, and *Curvitermes* sp.1 and subfamily *Termitinae*: *Orthognatotermes* sp.1, *Spinitermes* sp.1.

Epigeal nest density

In this study, the number of termite nests ranged from 195 to 672 ha^{-1} (Figure 1). This result is similar to those found by Kaschuk et al. (2006), who obtained 270 to 518 termite nests in five areas, of Cambisol and Nitisol. The "past 10" had the highest number of termite nests, 672 ha^{-1}, differing from the other areas. The number of termite nests did not differ between the renovated pastures and the Cerradão. According to Constantino (2005), this is related to pasture age, pasture management and the fact that termites are able to digest cellulose, the food being more abundantly supplied in older pastures. Another important aspect to be taken into consideration is pasture age at evaluation time. It can be observed that the older the pasture, the greater the number of termite nests per ha.

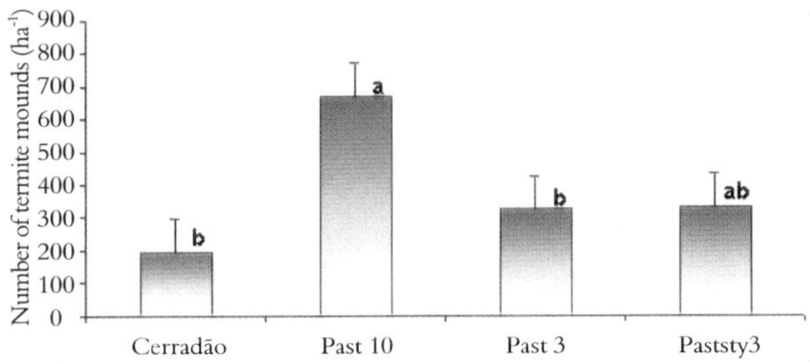

Figure 1. Total number of termite mounds per ha in soils under *Brachiaria* pastures and in the Cerradão area (values followed by the same letter are not significantly different at $p < 0.05$, Dunn test).

Area occupied by epigeal nests

The "Past 10" had the smallest mean basal area occupied by each mound (Table 1), although this area showed a high number of mounds. There were no differences between the Cerradão, "Past 3", and "Paststy 3" areas.

One of the concerns regarding the number of mounds is a potential reduction in the usable area of pastures Fernandes et al. (1998) pointed out that pastures could be considered to be degraded due to the presence of termite mounds in the area. However, according to Czepak et al. (2003), there is no proof of damage caused by termite mounds in pastures, and a high number of termite nests may not be associated with significant reductions in grazing areas.

Cosenza and Carvalho (1974) stated that in some pastures with infestations of about 160 termite nests per ha, there were no reductions in pasture yield, quality or plant cover when compared with non-infested areas.

In this study, it was observed that the total area occupied by termite nests ranged from 0.4 to 1% (Table 1), which does not correspond to a considerable area loss to the point of hampering pasture development. The need must be pointed out for adequate management that contemplates a balance between the quantity of termite mounds and the pasture.

The epigeal nest density per sampled area and the mean area of each nest define the total area occupied per ha. It is known that the number of epigeal nests tends to increase in areas that are not adequately managed. Therefore, older pastures will tend to present higher infestation levels, as observed in Figure 1. The main limiting factor for the development of those species will be the available forage supplied by the pasture.

Table 1. Number of termite mounds per area according their height, area, volume and mass (in each row, means followed by the same letter are not significantly different at $p < 0.05$, Dunn test).

	Cerradão	Past10	Past3	Paststy3
Total number of mounds per area (n/4000 m^2)	78	270	130	134
Height (m)				
≤ 0.15	30 a	212 a	52 bc	62 b
0.15 - 0.83	48 a	58 a	78 a	72 a
Basal Area (m^2)				
≤ 0.25	55 c	240 a	81 bc	103 b
0.25 - 0.5	11 b	15 b	39 a	22 ab
> 0.5	12 a	15 a	10 a	9 a
Area (%) per ha	0.4	0.7	0.8	1.0
Volume (m^3)				
≤ 0.015	53 c	238 a	70 bc	95 b
0.015 - 0.125	22 b	27 b	55 a	36 ab
> 0.125	3 a	5 a	5 a	3a
Mass (kg)				
≤ 15	53 c	238 a	69 bc	95 b
15 - 125	21 b	26 b	55 a	36 ab
> 125	4 a	6 a	6 a	3 a

Height of mounds

Several nests up to one meter in height were found in the studied areas. The tallest termite nests were found in the Cerrado area (0.5 - 1 m tall), without difference between "past 3" and "paststy 3"; however, a higher number of termite nests was observed in the "past 10" treatment, with heights between 0 and 0.1 m (Table 1). The smaller height could be related to pasture age (10 years); during this time, the species evolve and the colonies abandon their nests, which gradually break up by the effect of erosion, with an important role played by cattle trampling.

Turned soil mass and volume

The termites transport soil particles from different depths to the surface and deposit them in mounds, forming the epigeal nests. With a mean bulk density of 1.05 g cm^{-3}, the mass of the soil turned by the termites ranged from 3.4 to 13.4 ton. ha^{-1} which corresponds to a volume from 4 to 11 m^3 ha^{-1} (Figure 2). We can consider that a part of these masses are over estimated, because many species constructed with organic matter the central part of the nests. These data do not corroborate the results of Kaschuk et al. (2006), who found higher turned soil volume values, ranging from 8.11 to 63.3 m^3 ha^{-1}; others results showed 7.5 m^3 ha^{-1} year^{-1} (SARCINELLI et al., 2009). So, the variability of soil masses transported to the epigeal nests has to be related to the diversity of species, soil type and especially to different weather conditions observed in the survey locations. The highest turned soil mass, 13.4 ton. ha^{-1} was obtained on "past 10" plot; although this plot showed a higher number of termite nests, it had the mounds with one of the largest occupied areas (0.7 %), which denotes different species from those observed in the other plots. The smallest soil mass was found in the Cerradão plot, which had the lowest incidence of epigeal nests. We suggest, as previously observed (BENITO et al., 2004), that under the cultivated pastures with a long time, here 10 years of pastured system, even when they are in decline from a cattle production point of view, the soil termite community is reestablished.

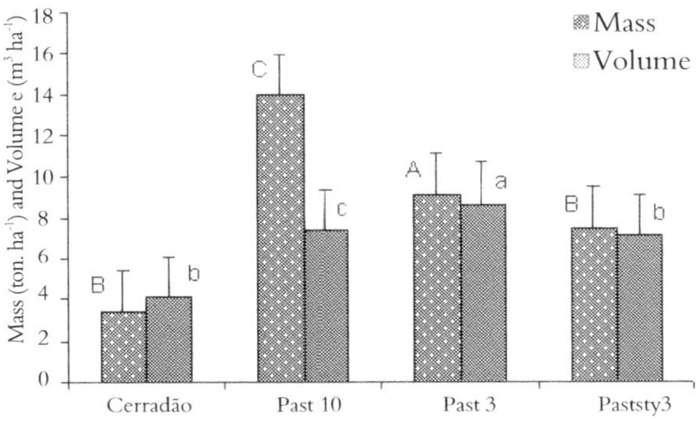

Figure 2. Turned soil mass and volume per ha in soils under *Brachiaria* pastures and Cerradão area (means followed by the same letter are not significantly different at $p < 0.05$, Dunn test).

Conclusion

Termites do not induce loss of grazing areas, termite's mounds represented from 0.4 to 1% of the surfaces. The mean soil volumes in epigeal mounds ranged from 4 to 11 m^3 ha^{-1}, highest value was observed for a period of 3 years. In the old pasture the number of termite mounds was higher with a slightly lower occupied area than in younger pastures that tend to be more homogeneous in terms of mounds density and size.

According to results of the renewed pastures, the structure and the size of termite mounds populations were influenced by trophic inputs.

Acknowledgements

The authors thank Prof. Inês Cristina de Batista Fonseca for helpful to the statistics.

References

ABE, T.; BIGNELL, D. E.; HIGASHI, M. **Termites:** evolution, sociality, symbioses, ecology. Dordrecht: Kluwer Academic Publishers, 2000.

ADÁMOLI, J.; MACEDO, J.; AZEVEDO, L. G.; MADEIRA NETTO, J. Caracterização da região dos cerrados. In: GOEDERT, W. J. (Ed.). **Solos dos Cerrados**: tecnologias e estratégias de manejo. Planaltina: Embrapa-CPAC; São Paulo: Nobel, 1986. p. 33-74.

AYRES, M.; AYRES JÚNIOR, M.; AYRES, D. L.; SANTOS, A. A. S.; AYRES L. L. **Bioestat Versão 5.0.** Aplicações estatísticas nas áreas de ciências e bio-médicas. Belém do Pará,: Sociedade Civil de Mamirauá, 2007.

BACHELIER, G. **La faune des sols, son écologie et son action**. Paris: Orstom, 1978.

BENITO, N. P.; BROSSARD, M.; PASINI, A.; GUIMARÃES, M. F.; BOBILLIER, B. Transformations of soil macroinvertebrate populations after native vegetation conversion to pasture cultivation (Brazilian Cerrado). **European Journal of Soil Biology**, v. 40, p. 147-154, 2004.

CONSTANTINO, D. Padrões de diversidade e endemismo de térmitas no bioma cerrado. In: SCARIOT, A.; SILVA, J. C. S.; FELFILI, J. M. (Org.). **Cerrado**: ecologia, biodiversidade e conservação. Brasília: Ministério do Meio Ambiente, 2005. p. 319-333.

COSENZA, G. W.; CARVALHO, M. M. Controle e nível de dano do cupim de montículo em pastagens. **Revista da Sociedade Brasileira de Zootecnia**, v. 3, n. 1 p. 1-12, 1974.

CZEPAK, C.; ARAÚJO, E. A.; FERNANDES, P. M. Ocorrência de espécies de cupins de montículo em pastagens no estado de Goiás. **Pesquisa Agropecuária Tropical**, v. 33, n. 1, p. 35-38, 2003.

DECÄENS, T.; LAVELLE, P.; JIMÉNEZ, J. J.; ESCOBAR, G.; RIPPSTEIN, G.; SCHNEIDMADL, J.; SANZ, J. I.; HOYOS, P.; THOMAS, R. J. Impacto del uso de la tierra en la macrofauna del suelo de los Llanos Orientales de Colombia. In: JIMÉNEZ, J. J.; THOMAS, R. J. (Ed.). **El arado natural**: las comunidades de macroinvertebrados del suelo en las savanas neotropicales de Colombia. Cali: Centro Internacional de Agricultura Tropical, 2003. p. 21-45. (Publicación CIAT, 336).

DIAS, V. S.; BROSSARD, M.; ASSAD, M. L. L. Macrofauna edáfica invertebrada em áreas de vegetação nativa da região de Cerrados. In: LEITE, L. L.; SAITO, C. H. (Ed.). **Contribuição ao conhecimento ecológico do cerrado**. Brasília: UnB, 1997. p. 168-173.

EMBRAPA-Empresa Brasileira de Pesquisa Agropecuária. Centro Nacional de Pesquisa de Gado de Corte. **Cupim de montículo em pastagens.** Campo Grande, 1996. (Embrapa-CNPGC. Divulga, 18).

EMBRAPA-Empresa Brasileira de Pesquisa Agropecuária. Centro Nacional e Pesquisa em Solos. **Sistema Brasileiro de Classificação de Solos**. Brasilia: Embrapa-SPI; Rio de Janeiro: Embrapa Solos, 2006.

FERNANDES, P. M. C.; CZEPAK, C.; VELOSO, V. R. S. Cupins de montículos em pastagens: prejuízo real ou praga estética? In: FONTES, L. R.; BERTI FILHO, E. (Ed.). **Cupins**: o desafio do conhecimento. Piracicaba: Fealq, 1998. p. 187-210.

KASCHUK, G.; SANTOS, J. C. P.; ALMEIDA, J. A.; SINHORATI, D. C.; BERTON-JUNIOR, J. F. Termite activity in relation to natural Grassland soil attributes. **Scientia Agricola**, v. 63, n. 6, p. 583-588, 2006.

KLUTHCOUSKI, J.; OLIVEIRA, I. P.; YOKOYAMA, L. P.; DUTRA, L. G.; PORTES, T. A.; SILVA, A. E.; PINHEIRO, B. S.; FERREIRA, E.; CASTRO, A. M.; GUIMARÃES, C. M.; GOMIDE, J. C.; BALBINO, L. C.

Sistema barreirão: recuperación renovación de pasturas degradadas utilizando cultivos anuales. In: GUIMARÃES, E. P.; SANZ, J. I.; RAO, I. M.; AMÉZQUITA, M. C.; AMÉZQUITA, E. (Ed.). **Sistemas agropastoriles en sabanas tropicales de América Latina.** Cali: Centro Internacional de Agricultura Tropical; Brasília: Empresa Brasileira de Pesquisa Agropecuária, 1999. p. 195-231. (Publicación CIAT, 313).

LAVELLE, P.; SPAIN, A. V. **Soil ecology**. Dordrecht: Kluwer Academic, 2001.

SANO, E. E.; BARCELLOS, A. O.; BEZERRA, H. S. Assessing the spatial distribution of cultivated pastures in the Brazilian savanna. **Pasturas Tropicales**, v. 22 n. 3, p. 2-15, 2000.

SARCINELLI, T. S.; SCHAEFER, C. E. G. R ; LYNCH, L. S.; ARATO, H. D.; VIANA, J. H. M.; ALBUQUERQUE FILHO, M. R.; GONÇALVES, T. T. Chemical, physical and micromorphological roperties of termite mounds and adjacent soils along a toposequence in Zona da Mata, Minas Gerais State, Brazil. **Catena**, v. 76, p. 107-113, 2009.

21

Net mineralization nitrogen and soil chemical changes with application of organic wastes with 'Fermented Bokashi Compost'

Cácio Luiz Boechat[1*], Jorge Antonio Gonzaga Santos[2] and Adriana Maria de Aguiar Accioly[3]

*[1]Departamento de Ciência do Solo, Universidade Federal do Rio Grande do Sul, Av. Bento Gonçalves, 7712, 91540-000, Porto Alegre, Rio Grande do Sul, Brazil. [2]Centro de Ciências Agrárias Biológicas e Ambientais, Universidade Federal do Recôncavo da Bahia, Cruz das Almas, Bahia, Brazil. [3]Empresa Brasileira de Pesquisa Agropecuária, Centro Nacional de Pesquisa de Mandioca e Fruticultura, Cruz das Almas, Bahia, Brazil. *Author for correspondence. E-mail: clboechat@hotmail.com*

ABSTRACT. The use of organic wastes in agricultural soils is one of the possible ways to employ these materials. The aims of this study were to evaluate the effectiveness of organic wastes and Fermented Bokashi Compost (FBC), to establish the most efficient use of organic wastes for a soil, changing the net nitrogen mineralization and soil chemical properties. The experimental design was completely randomized in a 6 x 2 x 5 factorial, being five organic wastes plus an control (soil without waste), with or without FBC, evaluated at 0, 7, 42, 70 and 91 days of incubation, with three replicates, under laboratory conditions. The organic wastes enhanced the soil chemical properties and increased nitrogen concentration in soil. However, the net nitrogen mineralization was affected by C/N ratio of wastes and incubation time. The FBC mixed with the wastes accelerated and enhanced organic matter degradation, resulting in quickly available quantity of net nitrogen. The wastes can be considered potentially useful as organic fertilizer but their usefulness appears to depend on knowing the C/N ratio of each one. The FBC can be used when one wants a more accelerated degradation, resulting in a quicker quantity of available nutrients to the plants.

Keywords: efficacy of microorganisms, sewage sludge, degradation, fertility.

Introduction

The use of organic wastes from production processes or in biofertilizer form as a source of plant nutrients in agricultural soils is one of the ways to employ these materials, as they increase fertility, and improve soil characteristics and physic-chemical properties, and also microbial activity (BRADY; WEIL, 2002).

One of the world's most commonly used biofertilizers in this context is the so-called 'Effective microorganisms' (EM), referring to an undisclosed mixture of naturally occurring microorganisms that supposedly has beneficial properties in a wide range of applications (HIGA, 2002).

Bokashi is the Japanese term for 'fermented' organic matter and is equivalent to compost used in traditional organic farming which is mostly prepared with the addition of EM. The Fermented Bokashi Compost is a mixture of several types of organic matter subjected to predominantly lactic fermentation. It can be prepared under complete anaerobic or aerobic conditions. The latter means that partial anaerobic conditions occur in the middle of the compost pile while the outer layers remain aerobic. The preparation of Anaerobic Bokashi is prepared in closed vessels, while aerobic Bokashi is prepared similarly to traditional composting with additional usage of a cover such as a jute bag, straw mat, or similar material (KYAN et al., 1999).

However, Tiruo Higa suggested a very broad application range of EM preparations and has reported beneficial effects in different environments such as soils, plants and water. It has been proposed that EM preparations help to produce substances acting as antioxidants, such as inositol, ubiquinone, saponin, low-molecular polysaccharides, polyphenols and chelating agents. These substances may inhibit harmful microbial species enhance the proliferation of beneficial microorganisms and detoxify harmful substances simultaneously (HIGA, 2001 apud MAYER et al., 2010).

Only small and often contradictory effects of the application of EM preparations on crop yields and plant development have been reported (KHALIQ et al., 2006; OKORSKI et al., 2008; VAN VLIET et al., 2006). EM-applications have also been found to have no effect (KHALIQ et al., 2006; VAN VLIET et al., 2006). Similar results were found by Schenck zu Schweinsberg-Mickan and Müller (2009) to N mineralization and microbial biomass C and N in soil without amendments and with application of wheat straw and coarse meal of yellow lupins. Few effects have been observed related to the EM carrier substrate molasses. Okorski and Majchrzak (2007) observed the highest soil fungal diversity when EM was applied together with pesticides in a field experiment planted with peas.

Many studies have been performed to investigate the effect of organic wastes on the physicochemical characteristics of soil (ARAUJO, 2011; DE MARIA et al., 2010), their biological properties (BUENO et al., 2011) and crop yields (ARAUJO, 2011; PIGOZZO et al., 2008).

There are only a few studies on the effects of EM on decomposition of organic wastes in soil and on soil quality in general. Our aim was to verify whether the application of Fermented Bokashi Compost mixed with five different types of organic wastes, promotes soil chemical improvement and higher net nitrogen mineralization under laboratory conditions.

Material and methods

Soil and wastes characterization

Standard quality assurance chemical analyses data for the soil and wastes was obtained from Embrapa Laboratory, in Brazil. Some chemical analysis according to Embrapa (1999) and total C, ammonium nitrogen, nitrate nitrogen and total N were analyzed using a methodology cited by Tedesco et al. (1995).

The soil was an Oxisol Ustox (USDA, 1999) or according to the Brazilian System of Soil Classification 'Latossolo Amarelo distrófico' (EMBRAPA, 2006) with 172, 754 and 74 g kg^{-1} clay, sand and silt respectively in the 0–20 cm layer. The chemical analysis results are shown in Table 1. Before soil incubation with treatments, about 100 kg of soil were sieved (Ø ≤ 4.75 mm), homogenized, moistened (70% of field capacity) and stored in dark plastic bags at room temperature for 21 days to restore the microbial communities.

The organic wastes used in the experiment were: Treated pulp mill sludge (PMS) - The raw material for this industry is newspaper, books, bank and office composts and other recyclable paper. The organic wastes were treated using physical, chemical and biological activated sludge processes to remove biochemical oxygen demand and concentration of resin with the addition of lime (CaO) in order to eliminate pathogens and promote organic waste stabilization.

Petrochemical complex sludge (PS) - Sludge derived from wastes generated by industries located in a petrochemical complex. The industrial sludge is derived from aerated biological treatment and treatment to reduce heavy metal concentrations.

Treated urban sewage sludge (USS) – This organic waste was the result of sewage treatment that used physical processes to remove coarse solids, biological activated sludge to reduce biochemical oxygen demand, plus the addition of lime (CaO), aiming at eliminating pathogens and stabilizing wastes.

Treated dairy industry sewage sludge (DSS) - Organic waste generated by the dairy industry to manufacture cheese, and to process butter and milk. The organic wastes are aerated in a pond to reduce the organic load and eliminate pathogens, and chemical stabilization is performed applying lime (CaO).

Organic compost from the fruit pulp industry (FPW) - Organic compost derived from the remains of fruit peel and seeds for frozen fruit pulp production. The organic composts piled for primary fermentation processes and partial decomposition of the material and stabilization of organic load for three years.

The chemical properties of the organic composts can be seen in Table 2.

Table 1. Chemical characteristics of soil used in the experiment.

pH^a	pH^b	P	Ca	Mg	CEC^c	$N-NH_4^+$	$N-NO_3^-$	SOC^d	SOM^e	N-Kjeldahl	C/N ratio
.....	.....	$g\ dm^{-3}$		$cmol_c\ dm^{-3}$			$mg\ kg^{-1}$		$g\ kg^{-1}$		.....
5.20	4.18	0.002	1.12	0.23	2.97	57.12	100.80	3.53	6.08	1.05	3.36

[a]In H2O (ratio 1:2.5); [b]In CaCl2; [c]Cation exchange capacity; [d]Soil organic carbon and [e]Soil organic matter.

Table 2. Chemical characteristics of organic wastes used in the experiment.

Parameter	Unit			Organic wastes[a]		
(Total value)		PMS	PS	USS	DSS	FPW
pH^b	.....	8.30	7.40	5.67	6.90	5.40
P	$g\ dm^{-3}$	0.28	4.04	9.49	15.00	0.51
Ca	$cmol_c\ dm^{-3}$	190.23	116.80	59.86	198.20	28.20
Mg	$cmol_c\ dm^{-3}$	16.61	5.48	27.18	13.05	18.50
Organic carbon	$g\ kg^{-1}$	236.40	34.40	235.00	161.60	232.40
Organic matter	$g\ kg^{-1}$	407.55	59.31	405.14	278.60	400.60
N-Ammonium	$mg\ kg^{-1}$	263.20	750.12	8619.80	6182.40	460.60
N-Nitrate	$mg\ kg^{-1}$	171.08	855.40	421.12	36.96	881.72
N-Kjeldahl	$g\ kg^{-1}$	3.72	4.49	32.63	19.20	19.50
Cu	$mg\ kg^{-1}$	162.20	373.20	334.02	141.79	101.72
Fe	$mg\ kg^{-1}$	3,240.86	7,640.86	7,364.59	11,752.69	701.08
Mn	$mg\ kg^{-1}$	50.34	83.58	113.39	292.28	68.43
Ni	$mg\ kg^{-1}$	0.70	1.10	0.77	1.13	0.27
Cd	$mg\ kg^{-1}$	0.11	0.12	0.12	0.12	0.10
Pb	$mg\ kg^{-1}$	37.61	32.48	8.55	44.44	3.42
Cr	$mg\ kg^{-1}$	6.22	4.22	3.82	6.83	0.60
C/N ratio	.....	63.55	7.66	7.20	8.42	11.92

[a]Based on dry mass; [b]In H2O (ratio 1:2.5); DSS - Treated dairy industry sewage sludge; USS - Treated urban sewage sludge; PS - Petrochemical complex sludge; PMS - Treated pulp mill sludge and FPW - Organic compost from the fruit pulp industry.

Incubation experiment

A 91 days incubation experiment was conducted aiming to evaluate soil chemical changes and measure N mineralization of five wastes at application rates of 27.0 (PMS), 22.2 (PS), 3.0 (USS), 5.2 (DSS) and 5.2 (FPW) Mg ha^{-1} dry weight (defined based on the supply of 100 kg N ha^{-1} calculated based in the N- Kjeldahl), associated or not with organic compost enriched with effective microorganisms (EM-4) called Fermented Bokashi Compost (FBC) at an application rate of 1.5 Mg ha^{-1} (MOKITI OKADA FUNDATION, 1998). The FBC chemical analysis results were pH $_{(H2O)}$ 7.6, total N, P, K, Ca and Mg; 12.0, 4.5, 5.8, 20.5, 2.9 mg kg^{-1}, respectively.

Treatments consisted of 150 g dry weight of sieved soil samples placed in small plastic cups. Organic wastes were mixed added to soils at desired application rates, mixed vigorously for 2 min. to ensure uniformity within and between samples, and deionized water was added to moisten soils to 70% of field capacity.

Cups were then placed in a BOD incubator in the absence of light, at controlled temperature (25 ± 0.20°C) and humidity maintained close to 70% of field capacity. The moisture content was checked every two days by weighing and adjusted with deionized water. The experimental units were evaluated at 0, 7, 42, 70 and 91 days after hatching, with three replications. The chemical soil analysis was performed after the last day of incubation.

Measurements of net nitrogen mineralization

The mineral N (NH_4^+ and NO_2^- + NO_3^-) and some chemical attributes were determined immediately after assembly treatment (time 0).

The following equations were used to estimate (a) Net Nitrogen mineralization (NNm) and (b) Net Nitrogen mineralization accumulated (Nma), which represents the amount of organic N of the wastes that were mineralized and the sum for all dates:

$$NNm = ((Nres2 - Nres1) - (Nt2 - Nt1) / Nad) \star 100 \quad (a)$$

$$Nma = \text{date } 0\,(NNm) + \text{date } 1\,(NNm) + \\ + \text{date } n\,(NNm) \quad (b)$$

where: NNm (%) corresponds to N mineralization; Nres1 and Nres2 (mg 100 g^{-1}) quantities of soil mineral N in the treatments with organic waste at the beginning and end of each evaluation interval, respectively; Nt1 and Nt2 (mg 100 g^{-1}) quantities of soil mineral N in control (Blank soil) at the beginning and end of each evaluation interval, respectively, Nad (mg 100 g^{-1}) organic N added by waste; Nma (%) corresponds to accumulated N mineralization and date is the time interval between incubation followed analysis.

Analysis of data

Treatments and time effects data were analyzed as a completely randomized design in a factorial 6 x 2 x 5, being five organic wastes incorporated into the soil plus control, with or without FBC, evaluated at 0, 7, 42, 70 and 91 days of incubation and soil data were evaluated at initial and final time of incubation, with three replicates, under laboratory conditions. All data were analyzed using analysis of variance (ANOVA) in the SAS program (SAS, 2004). When differences between treatments were significant, means were compared by Tukey test at the 0.01 probability level.

Results and discussion

Chemical properties of soil

Comparison of means showed that chemical attributes of the soil were affected by organic waste treatments compared to control (Table 3). The soil chemical characterization initially showed that soil presented a higher acid pH and lower P, K, Ca, Mg, Al and organic matter concentrations (TOMÉ JÚNIOR, 1997). The FBC added to the wastes affected soil chemical attributes. After 91 days of incubation, the PMS, PS and DSS treatments in the presence or not of FCB, increased the soil pH. The initial treatment of these wastes basically consists of a biological process with limestone or lime ($CaCO_3$ or CaO) addition to eliminate pathogens and promote organic waste stabilization, so that carbonate reacts with hydrogen ions present in soil solution (liquid phase of the soil) to form water and CO_2 molecules, in addition to increasing Ca and Mg concentrations and P availability (Table 3).

The USS and FPW treatments did not present significant pH changes, mainly due to their chemical characteristics (Table 2). The pH in the treatments did not change statistically in the presence of FBC or not. Moreira and Fageria (2010) studying alfalfa (*Medicago sativa L.*) under tropical conditions, affirm that optimum acidity indices for maximum dry matter yield were pH 5.4, base saturation 57%, Ca saturation 40%, and Mg saturation 24%. Therefore, to waste application rates, an initial study of soil pH neutralization power and Ca and Mg saturation index should be performed to soil system analysis. Soil pH ranged between 4.10 and 7.6, where 4.10 and 4.27 (USS and FPW) didn't differ of pH of control (blank soil) with FBC and without FBC ranged between 4.30 and 6.44, where 4.30 (FPW) didn't differ of pH of control (Table 3). However, when the results of both blank soil pH were compared, no significant difference was observed with FCB, as reported by Zydlik and Zydlik (2008). Moreover, concentrations of Ca and Mg are very important in the analysis to ensure complete neutralization of exchangeable Al in the soil. How was observed to PMS, PS and DSS.

The average concentrations of available P in PMS, PS, USS, DSS and FPW treatments with FCB were 2.25, 6.6, 4.55, 4.25 and 2.75 times higher than the control, respectively, and without FCB they were 4.28, 12.85, 4.35, 6.9 and 3.03 times higher, respectively (Table 3). However, in the presence of FCB the concentrations of P available follow the decreasing order; PS > DSS = USS > PMS = FPW > control and quantities of P available in the absence of FCB were observed in the following decreasing order: PS > DSS > PMS = USS = FPW > control (Table 3).

The differences between initial concentration of P in USS and the difference between P available found in USS treatment with FBC and without FBC are associated with the pH in the solution. They occurred because some microorganisms present in Fermented Bokashi Compost are more adaptable to acid conditions than native microbes resulting in higher degradation and P available in the soil (SHEN et al., 2011).

Table 3. Chemical characteristics changes of soil treated with organic wastes from different sources (based on dry mass) mixed or not with Fermented Bokashi Compost after incubation for 91 days.

Attribute	pH[b]	P	K^+	Ca^{2+}	Mg^{2+}	Al^{3+}	Na^+	H+Al[c]	SOM[d]
Unit	.....	mg dm^{-3}			cmol$_c$ dm^{-3}				g kg^{-1}
				With Fermented Bokashi Compost					
Blank soil	4.23[a] cA	6.67 dA	0.063 bA	0.30 cB	0.30 abA	0.53 aA	0.020 bA	2.75 aA	6.55 abA
PMS	7.60 aA	15.00 cA	0.057 bA	3.00 aB	0.20 bA	0.0 cA	0.043 aB	0.0 cA	8.97 aA
PS	6.00 bA	44.00 aB	0.057 bA	2.00 bA	0.30 abA	0.0 cA	0.030 bA	0.88 bA	7.07 abB
USS	4.10 cA	30.33 bA	0.060 bA	0.47 cA	0.33 aA	0.50 abA	0.023 bA	2.31 aA	7.04 abB
DSS	6.10 abA	28.33 bA	0.060 bA	2.23 bA	0.30 abA	0.0 cA	0.030 bA	0.81 bA	6.35 cB
FPW	4.27 cA	18.33 cA	0.090 aA	0.33 cA	0.30 abA	0.40 bA	0.023 bA	2.27 aA	7.42 bA
				Without Fermented Bokashi Compost					
Blank soil	4.28 bA	4.30 cA	0.042 bB	0.33 cA	0.22 bcA	0.43 aB	0.014 bA	2.60 aA	5.45 cB
PMS	6.40 aA	18.41 bcA	0.041 bA	3.45 aA	0.17 cA	0.0 bA	0.061 aA	0.08 cA	7.77 bB
PS	6.44 aA	55.24 aA	0.042 bA	2.14 bA	0.30 bA	0.0 bA	0.022 bB	1.11 bA	7.95 abA
USS	4.69 abA	18.70 bcB	0.051 abA	0.54 cA	0.41 aA	0.44 aB	0.021 bA	2.85 aA	8.71 aA
DSS	5.55 abA	29.67 abA	0.041 bB	2.25 bA	0.21 bcA	0.0 bA	0.027 bA	0.92 bcA	7.68 bA
FPW	4.30 bA	13.03 bcB	0.066 aB	0.38 cA	0.33 abA	0.41 aA	0.018 bA	2.65 aA	7.80 bA
C.V.[e] (%)	10.98	36.63	22.45	34.53	15.39	31.28	18.66	27.80	11.52

[a]Mean values followed by the same lower letter do not differ between wastes and by the same capital letter do not differ between with or without Fermented Bokashi Compost within the same waste (Tukey test, $p < 0.01$); [b]In H2O (1:2.5); [c]Exchangeable acidity; [d]Soil organic matter; [e]Coefficient of variation; DSS - Treated dairy industry sewage sludge; USS - Treated urban sewage sludge; PS - Petrochemical complex sludge; PMS - Treated pulp mill sludge and FPW - Organic compost from the fruit pulp industry.

The soil K contents in soil treated with wastes increased in lowest proportion on both dates, FPW treatment being the highest concentration found, however K concentrations still remained low (TOMÉ JÚNIOR, 1997). In order to use organic wastes with successfully in agriculture will be necessary to apply K fertilizer or another element source (MELO; MARQUES, 2000).

The Na concentration in the FCB presence or not, increased only in PMS treatment (Table 3). The exchangeable acidity was relatively higher in the control (blank soil), USS and FPW treatments. As discussed previously, it was related to reduced pH capacity. Most Oxisols in Brazil are acidic and present low fertility (FAGERIA; BALIGAR, 2008). Theoretically, soil acidity is quantified on the basis of pH and Al^{3+} concentrations in soils (MOREIRA; FAGERIA, 2010).

The SOM degradation in the treatments was closely related to the C/N ratio, organic carbon quality (labile fractions) and quantity of each organic waste added to soil. Schenck zu Schweinsberg-Mickan and Müller (2009) found same behavior working with different EM and organic composts. Reductions in soil organic matter were observed using treatments PS, USS and DSS in the presence of FCB, compared to without FCB.

The C/N ratio of these treatments is lower and their organic fractions may be more labile and thus easier for the microorganisms to take up. Further, according to Zydlik and Zydlik (2008) the EM accelerates organic matter degradation.

The control and PMS treatments were better without FCB; this may occur because there are less or not specialized microorganisms, resulting in a higher quantity of available nitrogen and lower nitrogen competition by native microorganisms to degradation of the high quantity of carbon. In the FPW treatment no statistical differences with or without FCB were observed. Possibly, this occurred because FPW is an organic waste composted in piles for three years and it may have higher concentrations of stable or humified organic matter.

Net nitrogen mineralization

As expected, the organic wastes added to soil led to differences in net mineral nitrogen contents and reflected differences in the quality of the materials. The application of organic wastes increase net mineral nitrogen significantly ($p < 0.01$) over control and Fermented Bokashi Compost (FBC) accelerated the mineralization of nitrogen making available in the mineral form an higher amount of N in the first 7 days of incubation, coming to be higher without FBC, only 70 days after start incubation (Table 4). However, observed that the mineralization of nitrogen in the soil was highly enhanced when the organic waste was amended with FCB. Aryal et al. (2003) reported that performance of organic materials and effective microorganisms present in FCB was not as expected when they were applied alone.

When FCB was applied with PMS, a decrease of 5.71 times in net nitrogen quantity was recorded as compared to nitrogen quantity found in the PMS without FCB after 7 days of incubation. This probably occurred because the high C/N ratio (Table 2) of this waste stimulated the microorganisms present in FCB together with native microorganisms to use the mineral nitrogen in waste degradation occurring a net immobilization during decomposition of the added organic waste in the treatment PMS with FCB. However, observed a better nitrogen mineralized distribution over time with FBC application (Table 4).

Table 4. Net nitrogen mineralized (NNm) of treatments with organic wastes from different sources (based on dry mass) with or without Fermented Bokashi Compost during incubation period.

	Period in days			
Treatment	7	42	70	91
(%)	With Fermented Bokashi Compost			
Blank soil[a]	0.96[b] cA	-0.30 cA	1.62 cA	-1.21 eB
PMS	-24.73 dB	2.34 bA	9.37 bB	21.95 aA
PS	24.30 bA	18.94 aA	7.99 bB	6.28 dA
USS	50.53 aA	-6.10 dB	-8.36 dB	9.24 cA
DSS	54.12 aA	-16.94 eB	21.13 aB	5.27 dA
FPW	24.52 bA	-14.55 eB	3.44 cB	13.45 bA
	Without Fermented Bokashi Compost			
Blank soil	-0.21 cA	0.38 cA	-0.72 eB	1.49 aA
PMS	-4.33 dA	-4.92 dB	31.08 cA	-27.09 dcB
PS	9.16 bB	-2.34 dB	56.24 aA	-29.32 dB
USS	-14.93 eB	41.87 aA	24.00 dA	-25.94 cB
DSS	12.43 aB	10.21 bA	34.89 bA	-11.34 bB
FPW	11.38 abB	2.70 cA	33.21 bcA	-27.83 cdB
C.V.[c]	17.11			

[a]Soil without wastes; [b]Mean values followed by the same lower letter do not differ significantly on the same date and followed by the same capital letter do not differ significantly with or without Fermented Bokashi Compost (Tukey test, $p < 0.01$); [c]Coefficient of variation; DSS - Treated dairy industry sewage sludge; USS - Treated urban sewage sludge; PS -Petrochemical complex sludge; PMS - Treated pulp mill sludge and FPW - Organic compost from the fruit pulp industry.

On the other hand, application of FCB in PS treatment resulted in a better distribution of mineral nitrogen on all dates, showing the degradation capacity of the microorganism present in FCB working together with native microorganisms (Table 4). It was expected because FCB is a mixture of different microorganisms. However, in the PS treatment without FBC, the mineral nitrogen mineralization occurred only at 7 and 70 days of incubation. In the DSS and USS treatments an increase was observed on all dates in the presence of FCB, except at 42 and 70 days of incubation. The increase in FPW treatments with FCB was observed at 7 and 91 days of incubation. The relatively low response of FPW to FCB, only responds well in the presence of sufficient labile organic matter. This compost was stabilized by three years in typical aerated fermentation piles and probably has a most recalcitrant carbon fraction or the lack of labile carbon components (SONG et al., 2011).

A preliminary field study at the University of Fort Hare did not detect a substantial contribution to crop yield by the recommended application of EM in combination with commercial compost (FATUNBI; NCUBE, 2009). This was attributed to the low quality C constituent in the compost used, which is typical of most matured compost. Composted organic wastes are low in soluble C and may not be able to effectively support proliferation of a decomposer community (FATUNBI; NCUBE, 2009).

Therefore, the net N mineralization and immobilization observed in these organic waste soils are in accordance with the common understanding of turnover of organic materials characterized by high and low C/N ratios (PALM et al., 2001; SENEVIRATNE, 2000).

The results do not agree with those reported by Schenck zu Schweinsberg-Mickan and Müller (2009). These authors detected no effect related to the addition of added living microorganisms (EM) in soil fertility and increased plant growth under a specific experimental condition, where authors related characteristics of the organic waste with time of incubation.

In Figure 1, observed that in the presence of FBC the net nitrogen mineralized accumulated (Nma) was higher after 7 days of incubation for all wastes, except PMS. However, were obtained without FBC high amounts of Nma only after 70 days of incubation (Figure 2).

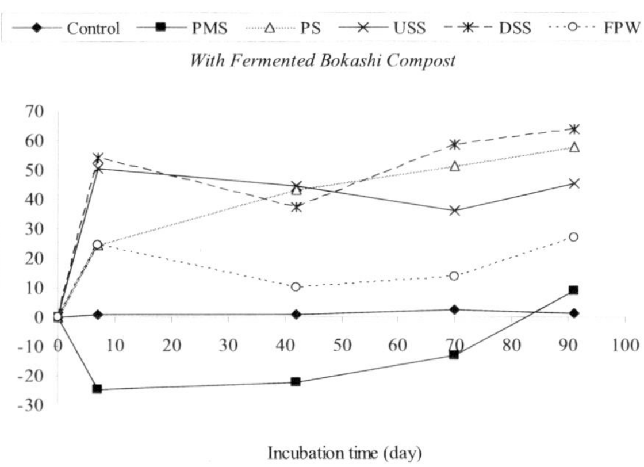

Figure 1. Net nitrogen mineralized accumulated (Nma) of treatments with organic wastes with Fermented Bokashi Compost during incubation period; DSS - Treated dairy industry sewage sludge; USS - Treated urban sewage sludge; PS - Petrochemical complex sludge; PMS - Treated pulp mill sludge and FPW - Organic compost from the fruit pulp industry.

This corroborates the hypothesis that FBC accelerates the nitrogen availability. Sahain et al. (2007) observed that Fermented Bokashi Compost called of biostimulant increased the values of soil elements as compared with the untreated soil. This treatment produced the highest significantly values of N, P, K, Fe, and Zn and while the control treatment was significantly the lowest values.

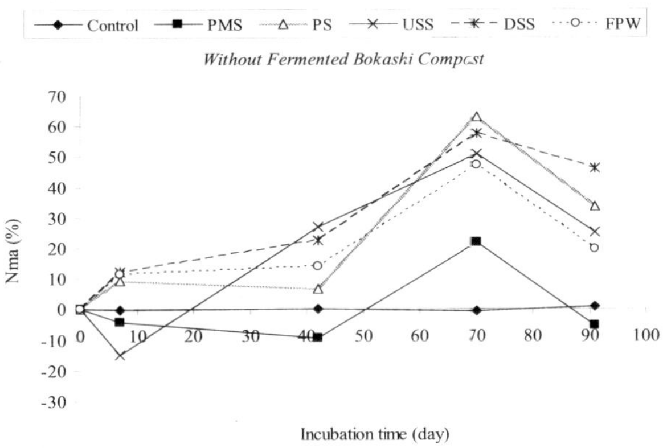

Figure 2. Net nitrogen mineralized accumulated (Nma) of treatments with organic wastes without Fermented Bokashi Compost during incubation period; DSS - Treated dairy industry sewage sludge; USS - Treated urban sewage sludge; PS - Petrochemical complex sludge; PMS - Treated pulp mill sludge and FPW - Organic compost from the fruit pulp industry.

Conclusion

The expectation that Fermented Bokashi Compost with 'effective microorganism' (EM-4) would increase and accelerate organic waste degradation was confirmed here. Positive effects of Bokashi on net N mineralization and soil fertility evolution could be identified as a combination with the organic waste quality effects. Possible suppressive effects of a higher C/N ratio in soil microbial biomass and activity are observed.

Acknowledgements

The authors would like to thank CAPES for providing scholarship for masters of the first author and UFRB for collaborating in this study. The authors also thank the support of the EMBRAPA – CNPMF.

References

ARAUJO, F. F. Disponibilização de fósforo, correção do solo, teores foliares e rendimento de milho após a incorporação de fosfatos e lodo de curtume natural e compostado. **Acta Scientiarum. Agronomy**, v. 33, n. 2, p. 355-360, 2011.

ARYAL, U. K.; XU, H. L.; FUJITA, M. Rhizobia and AM fungal inoculation improve growth and nutrient uptake of bean plants under organic fertilization. **Journal of Sustainable Agriculture**, v. 21, n. 3, p. 29-41, 2003.

BRADY, N. C.; WEIL, R. R. **The nature and properties of soils**. 13th ed. New Jersey: Prentice Hall, 2002.

BUENO, J. R. P.; BERTON, R. S.; SILVEIRA, A. P. D.; CHIBA, M. K.; ANDRADE, C. A.; DE MARIA, I. C. Chemical and microbiological attributes of an Oxisol treated with successive applications of sewage sludge. **Revista Brasileira de Ciência do Solo**, v. 35, n. 4, p. 1461-1470, 2011.

DE MARIA, I. C.; CHIBA, M. K.; COSTA, A.; BERTON, R. S. Sewage sludge application to agricultural land as soil physical conditioner. **Revista Brasileira de Ciência do Solo**, v. 34, n. 3, p. 967-974, 2010.

EMBRAPA-Empresa Brasileira de Pesquisa Agropecuária. **Manual de análises químicas de solos, plantas e fertilizantes**. Brasília: Embrapa, 1999.

EMBRAPA-Empresa Brasileira de Pesquisa Agropecuária. **Sistema brasileiro de classificação de solos**. Brasília: Embrapa Produção de informação; Rio de Janeiro: Embrapa Solos, 2006.

FAGERIA, N. K.; BALIGAR, V. C. Ameliorating soil acidity of tropical Oxisols by liming for sustainable crop production. **Advances in Agronomy**, v. 99, p. 345-431, 2008.

FATUNBI, A. O.; NCUBE, L. Activities of effective microorganism (EM) on the nutrient dynamics of different organic materials applied to soil. **American-Eurasian Journal of Agronomy**, v. 2, n. 1, p. 26-35, 2009.

HIGA, T. **Revolution to save the earth**: with Effective Microorganisms (EM) to solve the problems of our world. Kevelaer: OLV (Organic Farming Verlag), 2002.

KHALIQ, A.; ABBASI, M. K.; HUSSAIN, T. Effects of integrated use of organic and inorganic nutrient sources with effective microorganisms (EM) on seed cotton yield in Pakistan. **Bioresource Technology**, v. 97, n. 8, p. 967-972, 2006.

KYAN, T.; SHINTANI, M.; KANDA, S.; SAKURAI, M.; OHASHI, H.; FUJISAWA, A.; PONGDIT, S. **Kyusei nature farming and the technology of effective microorganisms, guidelines for practical use**. Bangkok: APNAN; Atami; INFRC, 1999.

MAYER, J.; SCHEID, S.; WIDMER, F.; FLIEẞBACH, A.; OBERHOLZER, H. How effective are 'Effective microorganisms® (EM)'? Results from a field study in temperate climate. **Applied Soil Ecology**, v. 46, n. 2, p. 230-239, 2010.

MELO, W. J.; MARQUES, M. O. Potencial do lodo de esgoto como fonte de nutrientes para as plantas. In: BETTIOL, W.; CAMARGO, O. A. (Ed.). **Impacto ambiental do uso agrícola do lodo de esgoto**. Jaguariúna: Embrapa Meio Ambiente, 2000. p. 45-67.

MOKITI OKADA FOUNDATION. **Microorganismos efetivos (EM) na agricultura**. São Paulo: Mokiti Okada Foundation, 1998.

MOREIRA, A.; FAGERIA, N. K. Liming influence on soil chemical properties, nutritional status and yield of alfalfa grown in acid soil. **Revista Brasileira de Ciência do Solo**, v. 34, n. 4, p. 1231-1239, 2010.

OKORSKI, A.; MAJCHRZAK, B. Fungi isolated from soil before the seeding and after harvest of pea (*Pisum sativum* L.) after application of bio-control product EM. **Acta Agrobotanica**, v. 60, n. 1, p. 113-121, 2007.

OKORSKI, A.; OLSZEWSKI, J.; PSZCZOKOWSKA, A.; KULIK, T. Effect of fungal infection and the application of the biological agent EM 1 on the rate of photosynthesis and transpiration in pea (*Pisum sativum* L.) leaves. **Polish Journal of Natural Sciences**, v. 23, n. 1, p. 35-47, 2008.

PALM, C. A.; GACHEGO, C.; DELVE, R.; CADISCH, G.; GILLER, K. E. Organic inputs for soil fertility management in tropical agroecosystems: Application of an organic resource database. **Agriculture, Ecosystems and Environment**, v. 83, n. 1-2, p. 27-42, 2001.

PIGOZZO, A. T. J.; LENZI, E.; LUCA JÚNIOR, J.; SCAPIM, C. A.; VIDIGAL FILHO, P. S.; COSTA, A. C. S. Reação do solo e disponibilidade de micronutrientes, em solo de textura média, tratado com lodo de esgoto e cultivado com milho. **Acta Scientiarum. Agronomy**, v. 30, n. 4, p. 569-579, 2008.

SAHAIN, M. F. M.; EL-MOTTY, E. Z. A.; EL-SHIEKH, M. H.; HAGAGG, L. F. Effect of some biostimulant on growth and fruiting of Anna apple trees in newly reclaimed areas. **Research Journal of Agriculture and Biological Sciences**, v. 3, n. 5, p. 422-429, 2007.

SAS-Statistical Analisys System. **SAS® 9.1.2 qualification tools user's guide**. Cary: SAS Institute, 2004.

SCHENCK ZU SCHWEINSBERG-MICKAN, M.; MÜLLER, T. Impact of effective microorganisms and other biofertilizers on soil microbial characteristics, organic-matter decomposition, and plant growth. **Journal of Plant Nutrition and Soil Science**, v. 172, n. 5, p. 704-712, 2009.

SENEVIRATNE, G. Litter quality and nitrogen release in tropical agriculture: a synthesis. **Biology and Fertility of Soils**, v. 31, n. 1, p. 60-64, 2000.

SHEN, J.; YUAN, L.; ZHANG, J.; LI, H.; BAI, Z.; CHEN, X.; ZHANG, W.; ZHANG, F. Phosphorus Dynamics: From Soil to Plant. **Plant Physiology**, v. 156, n. 3, p. 997-1005, 2011.

SONG, M.; JIANG, J.; XU, X.; SHI, P. Correlation between CO_2 efflux and net nitrogen mineralization and its response to external C or N supply in an alpine meadow soil. **Pedosphere**, v. 21, n. 5, p. 666-675, 2011.

TEDESCO, M. J.; GIANELLO, C.; BISSAN, C. A.; BOHNEN, H.; VOLKWEISS, S. J. **Análise de solo, plantas e outros materiais**. Porto Alegre: UFRGS, 1995.

TOMÉ JÚNIOR, J. B. **Manual para interpretação de análise de solo**. Guaíba: Agropecuária, Brasil, 1997.

USDA-Unites States Department of Agriculture. **Soil taxonomy**. A basic system of soil classification for making and interpreting soil survey. Washington D.C.: Soil Survey Staff, 1999.

VAN VLIET, P. C. J.; BLOEM, J.; GOEDE, R. G. M. Microbial diversity, nitrogen loss and grass production after

addition of effective micro-organisms® (EM) to slurry manure. **Applied Soil Ecology**, v. 32, n. 2, p. 188-198, 2006.
ZYDLIK P.; ZYDLIK Z. Impact of biological effective microorganisms (EM) preparations on some physic-chemical properties of soil and the vegetative growth of apple-tree rootstocks. **Nauka Przyroda Technologie**, v. 2, n. 1, p. 1-8, 2008.

22

Soil penetration resistance mapping quality: effect of the number of subsamples

Osmar Henrique de Castro Pias[1*], Maurício Roberto Cherubin[2], Claudir José Basso[3], Antônio Luis Santi[3], José Paulo Molin[2] and Cimélio Bayer[1]

*[1]Departamento de Ciência do Solo, Universidade Federal do Rio Grande do Sul, Av. Bento Gonçalves, 7712, 91540-000, Porto Alegre, Rio Grande do Sul, Brazil. [2]Escola Superior de Agricultura "Luiz de Queiroz", Universidade de São Paulo, Piracicaba, São Paulo, Brazil. [3]Universidade Federal de Santa Maria, Frederico Westphalen, Rio grande do Sul, Brazil. *Author for correspondence. E-mail: henriquepias@yahoo.com.br*

ABSTRACT. There is no consensus in the literature regarding how many subsamples are needed to perform accurate on-farm soil penetration resistance (SPR) mapping. Therefore, the objective of this study was to define the number of subsamples per sampling point needed to quantify the SPR. The experiment was performed in a 4.7 ha area and employed a 50 × 50 m grid system (18 sampling points). The SPR was evaluated using a digital penetrometer in two different years with 1, 2, 3, 4, 5, 6, 9, 12, and 15 subsamples per sampling point. The SPR maps produced with increasing numbers of subsamples were compared to the reference maps (15 subsamples) using the relative deviation coefficient and Pearson´s linear correlation. A reduction in the number of subsamples promoted an increase in the variability of the SPR data. Generally, the results from this study suggest the use of at least four subsamples per sampling point to achieve SPR maps with a coefficient of relative deviation less than 10% (30% maximum error per point around the mean) and significant correlation with the reference maps (15 subsamples).

Keywords: soil compaction, sampling protocol, precision agriculture, microvariability.

Introduction

Soil penetration resistance (SPR) is a measurement utilized to quantify the mechanical impedance of the soil for plant root growth (Bengough, Mckenzie, Hallet, & Valentine, 2011). Therefore, SPR is considered one of the main parameters for the diagnosis of the levels of soil compaction and determination of the most restrictive soil layers for root growth (Girardello et al., 2014). This tool has been widely used by researchers and service providers because it is rapid and easily used in the field compared to other more conventional methods, such as soil bulk density (Molin, Dias, & Carbonera, 2012).

The SPR is usually quantified using traditional sampling methods, which include the collection of various subsamples (*e.g.*, 12-15) in a random manner along the field and are considered independent samples (Tavares-Filho & Ribon, 2008; Storck et al., 2016). As a result, a mean SPR value is obtained for and applied to the total sampled area. Since the introduction of precision agriculture in Brazil in the early 2000s, systematic sampling protocols to assess the SPR have been widely applied in commercial fields. This method considers the spatial dependence among the sampling points and consequently the spatial variability of the SPR within the area (Molin et al., 2012). The objective of this methodological change is to identify the spatial variability (horizontal and vertical) of the SPR in the sampled area, which enables the creation of thematic maps to guide site-specific management of compacted subareas and soil layers in the field (Girardello et al., 2014).

Measurements of SPR values are highly influenced by diverse intrinsic (*e.g.*, soil moisture, texture and structure) and extrinsic (*e.g.*, management system) soil factors. As consequence, high coefficients of variation are usually observed (Beutler et al., 2007; Storck et al., 2016). Therefore, to correctly determine the spatial variability of the SPR, it is crucial to establish an adequate density of sampling points per area and select the number of subsamples that best represents each sampling point. Regarding the sampling density, studies have demonstrated that ideal sampling grids are approximately 50 × 50 m (*i.e.*, four samples per ha) (Cherubin, Santi, Basso, Eitelwein, & Vian, 2011) or 30 × 30 m (*i.e.*, more than 10 samples per ha) (Debiasi, Franchini, Oliveira, & Machado 2012). However, no study has defined the sampling point and its possible influence on the mapping of this variable.

In traditional sampling methodology, between 12 and 15 subsamples must be collected to compose a mean SPR value with maximum errors varying between 5 and 15% (Tavares Filho & Ribon, 2008; Molin et al., 2012). Georeferenced sampling normally uses many sampling points; therefore, a larger number of subsamples per sampling point increases the sampling cost and makes this method less attractive to farmers (Molin et al., 2012). There is no consensus in the literature concerning the number of subsamples used to collect the SPR data. Instead, reports vary with the use of one (Souza et al., 2006; Marasca et al., 2011), two (Debiasi et al., 2012), three (Tormena, Barbosa, Costa, & Gonçalves, 2002; Cherubin et al., 2011), five (Silva, Passos, & Beltrão, 2009) and ten subsamples (Secco, Reinert, Reichert, & Silva, 2009; Girardello et al., 2014) per sampling point. The use of an insufficient number of subsamples may result in inaccurate data collection, which generates recommendations for unnecessary interventions.

In this study, we tested the hypothesis that an insufficient number of subsamples per sampling point affected the representative of the assessment and the accuracy of the generated SPR thematic maps. The aim of this study was to evaluate the impact of the number of subsamples per sampling point on the quality of the SPR mapping and to determine the number of subsamples necessary to generate thematic maps with adequate accuracy for on-farm precision agriculture in crop production systems based on no-till farming.

Material and methods

Description of the study area

This on-farm study was conducted in an agricultural area near Palmeira das Missões city in southern Brazil (latitude 28°72′62″ S and longitude 69°14′34″ W), with a mean altitude of 600 m. The relief of the area is smoothly undulating, and the soil presents a clay texture (636 g kg^{-1} of clay, 316 g kg^{-1} of silt and 48 g kg^{-1} of sand content) and is classified as Rhodic Acrudox according to Soil Taxonomy (Soil Survey Staff, 2014) and "Latossolo Vermelho distrófico" according to the Brazilian System of Soil Classification (Santos et al., 2013). The area has been cultivated under a no-tillage cropping system without machinery traffic control since 1997 (*i.e.*, 15 years when the study was performed), including crop succession with wheat in the winter season and soybean or eventually corn in the summer season.

Determination of the soil penetration resistance

The study was performed in 2012 (year I) and reproduced in 2013 (year II). In both years, the data were collected in May after the soybean harvest. The 4.7 ha agricultural area was georeferenced and divided into a regular quadrangular sampling grid of 50 × 50 m to yield 18 sampling points (Figure 1). The SPR was determined down to a 0.30 m depth using a portable digital penetrometer (PenetroLOG®, model PLG 1020, Falker Automação, Porto Alegre, Rio Grande do Sul State, Brazil) with a cone diameter of 12.83 mm. The rod was inserted into the soil at a constant speed close to 20 mm s^{-1}. When the insertion speed surpassed 30 mm s^{-1}, the equipment registered an error and the measurement was remade. Fifteen subsamples were collected from each sampling point following the inter-row position of the previous crop within a radius of 3 m around the georeferenced point. The SPR evaluations were performed two days after a heavy rain. The whole procedure took only one working day in both years.

The water content of the soil at the moment of SPR evaluation was determined using the gravimetric method (Embrapa, 1997) with disturbed soil samples collected from the 0.00-0.15 m layer at three points (points 2, 9 and 16) (Figure 1). The average soil moisture was 310 and 330 g kg^{-1} for years I and II, respectively.

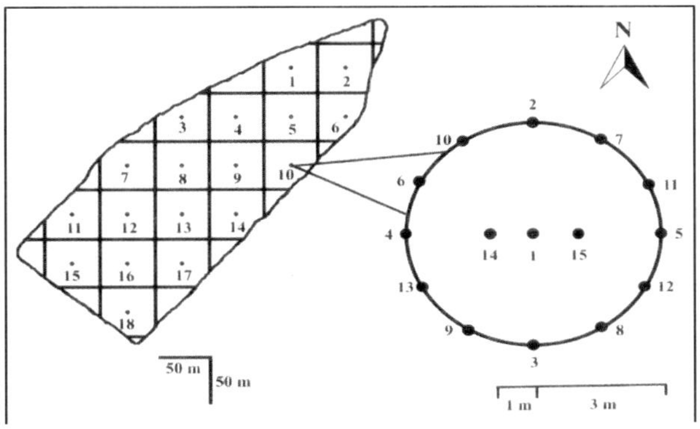

Figure 1. Schematic representation of the sampling grid (50 x 50 m) used in this study area highlighting the distribution of the 15 subsamples of soil penetration resistance in each sampling point in Palmeira das Missões (RS), southern Brazil.

Mathematical and statistical analyses

The datasets from each sampling point (18 points) and soil layer (*i.e.*, 0.00-0.05, 0.05-0.10, 0.10-0.15, 0.15-0.20, 0.20-0.25 and 0.25-0.30 m) composed of 15 subsamples were organized into a spreadsheet and subjected to outlier analysis. Any values that fell outside of the range of two standard deviations from the mean were considered outliers. Subsequently, the determination of the optimum number of subsamples was performed based on Equation 1 as proposed by Petersen and Calvin (1965):

$$n = \frac{t_\alpha^2 x\, S^2}{D^2} \qquad (1)$$

where *n* is the number of subsamples, t is the value from the distribution table for the function of the level of significance (α) and the degrees of freedom used to estimate the sample variance, S is the sample standard deviation of the mean (15 subsamples per sampling point) and D is the result of the SPR mean at each sampling point divided by the percentage variation enabled around the mean. The level of significance used was 0.05%, and the optimum number of subsamples for each sampling point was determined considering maximum errors of 10, 20 and 30% around the mean.

Soil penetration resistance mapping

The SPR maps were produced using the CR-Campeiro 7 software (Department of Geomatics, Federal University of Santa Maria, Rio Grande do Sul State, Brazil) by considering different numbers of subsamples (*i.e.*, 1, 2, 3, 4, 5, 6, 9, 12, and 15) per sampling point. A specific map was produced for each soil layer studied down to 0.30 m (*i.e.*, 0.00-0.05, 0.05-0.10, 0.10-0.15, 0.15-0.20, 0.20-0.25, and 0.25-0.30 m). The reduced number of points ($n < 50$ points) limited the use of the geostatistical approach to verify the existence of spatial dependence among the samples (Webster & Oliver, 2007). Therefore, the maps were produced using the interpolator inverse square distance as broadly recommended in the literature (*e.g.*, Coelho, Souza, Uribe-Opazo, & Pinheiro Neto, 2009; Wu, Hung, & Patton, 2013).

The SPR data from each soil layer considering the different numbers of subsamples were subjected to a descriptive statistical analysis to obtain the positional means (minimum, mean and maximum) and dispersion (coefficients of variation–CV, %). The CV values were used to classify the variability of the data into low (CV < 12%), medium (CV= 12 to 62%) and high (CV > 62%) as proposed by Warrick and Nielsen (1980). The normality hypothesis was tested and confirmed using the W test ($p \geq 0.05$) (Shapiro & Wilk, 1965), and thus no data transformation was necessary. The data analysis was completed using the statistical package SAS (SAS, 2010).

Analysis of the soil penetration resistance mapping quality

Two parameters were used to evaluate the effect of the number of subsamples on the accuracy of the thematic maps: Pearson's linear correlation coefficient ($p \leq 0.05$) and the relative deviation coefficient (RDC, %) (Coelho et al., 2009). The mean SPR values from 15 subsamples per sampling point were considered as a reference (standard) for comparison with the other maps produced using different numbers of subsamples (*i.e.,* 1, 2, 3, 4, 5, 6, 9, and 12).

The RDC, which is expressed as an absolute value, shows the dissimilarity between the two maps as demonstrated by the differences between the interpolated points of each map. The RDC was determined using Equation 2, which was adapted from the equations applied by Coelho et al. (2009) and Cherubin et al. (2015).

$$RDC = \sum \left[(SPRj - SPRref) / SPRref \right] x\ (100 / n) \qquad (2)$$

where n is the number of sampling points (18), SPRref is the soil penetration resistance value at point *i* (reference value obtained using 15 subsamples per sampling point), and SPRj is the soil penetration resistance value at point i determined using different numbers of subsamples (*i.e.,* 1, 2, 3, 4, 5, 6, 9, and 12).

Results and discussion

Number of subsamples per sampling point

After the preliminary analysis to detect outliers, 2.7% of the raw data was removed. This procedure is fundamental when the SPR is measured using portable penetrometers with manual operation because the roughness of the soil surface (Catania et al., 2013) and the variation in the speed of the rod going into the soil profile can influence the results (Valadão Jr, Biachini, Valadão, & Rosa, 2014).

The optimum number of subsamples per sampling point was similar in both years studied, indicating good consistency of the results (Figure2). The surface layer (0.00-0.05 m) required a larger number of subsamples to adequately represent the SPR of the sampling point. Therefore, 30 and 35 subsamples were required from each sampling in years I and II, respectively, to maintain 75% of the sampling points with a maximum error of 10%. If all of the sampling points (18) achieve this accuracy (*i.e.*, a maximum error of 10%), 44 subsamples need to be collected. The higher microvariability observed in the surface soil layer (radius 3 m) is due to various factors, including the soil and crop management practices, effects of plant roots, wet-dry cycles, and the potential surface sealing that is commonly observed in no-tillage systems, especially when little straw is present (Cherubin et al., 2011; Silva, Bianchi & Cunha, 2016). High variation of microvariability of the SPR data was observed between sampling points (Figure 2). Thus, for some points in the surface layer, less than 10 subsamples were sufficient to obtain values with a maximum deviation of less than 10%.

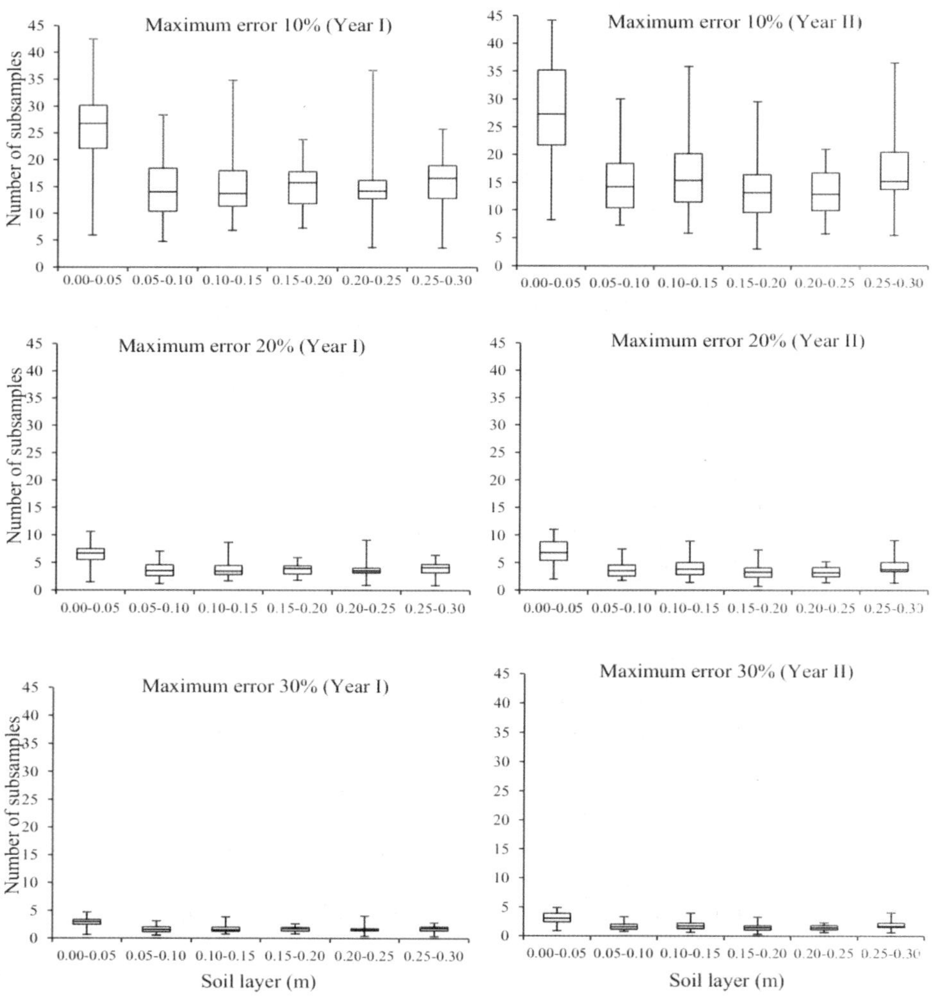

Figure 2. Number of subsamples per sampling point required to obtain soil penetration resistance (SPR) values with maximum errors of 10, 20 and 30% around the mean, for different soil layers in two years in Palmeira das Missões (RS), southern Brazil.

For the surface soil layer in both years, 11 and 5 subsamples per sampling point were sufficient to obtain SPR values with maximum errors of 20 and 30% around the mean, respectively. However, the surface layer of the soil under a no-tillage system is periodically disturbed during the opening of the sowing row, thereby minimizing possible physical restrictions (*i.e.*, high SPR values) for plant root growth (Moreira et al., 2016). Thus, when the SPR is determined, the major interest of the technician is an evaluation of the compaction state of the subsuperficial layers of the soil (below 0.05 m), where the higher SPR levels restrict the growth of roots at depth and may limit crop development, mainly due to water stress (Tormena et al., 2002; Cardoso et al., 2006). In a study of the effect of high SPR values on soybeans, Cardoso et al. (2006) observed that impediments to root growth in the subsurface caused the root system to concentrate in the soil surface layer (0.00-0.05 m), which was the zone that retained the lowest water content, thereby negatively influencing nutrient absorption.

For the deeper soil layers (0.05-0.30 m), the optimum numbers of subsamples per sampling point were similar. At these depths, the collection of 15 and 18 subsamples was necessary to obtain a maximum error of 10% around the mean for 75% of the sampling points for years I and II, respectively. However, 35 samples needed to be collected for all sampling points to attain this accuracy level, demonstrating that the high SPR microvariability among sampling point sob served in the surface soil layer persisted in the deeper soil layers. Only 8 and 4 samples would be required if we allowed maximum variations of 20 and 30% around the mean for the layers between 0.05 and 0.30 m in all sampling points, respectively. However, for 75% of the sampling points, only 5 (20% error) and 2 (30% error) subsamples were sufficient.

The use of reduced numbers of subsamples decreases the operational cost of field SPR sampling (less time-consuming). However, the low accuracy of the measurement (*i.e.*, a higher level of error) can make the results less reliable and even technically unviable depending on the goal of the assessment (Tavares-Filho & Ribon, 2008). For example, suppose that the SPR in a soil layer at one specific point is 3.5 MPa and a 30% maximum error is allowed; in this scenario, the result of the evaluation could be between 2.5 and 4.55 MPa. These values encompass SPR values that are considered adequate for root growth or are highly restrictive. Allowing a 20% maximum error results in values that vary from 2.8 to 4.2, and allowing a maximum error of 10% results in a range of 3.15 to 3.8, which does not generate large differences in terms of soil management decisions. Thus, an increase in the robustness of sampling will increase the accuracy of the information and in turn support management decisions that prevent unnecessary soil disturbances to alleviate compaction and its deleterious effects on soil ecosystem services. Nevertheless, the increased operational costs involved in more intensive soil sampling should always be considered to ensure that the evaluation is financially feasible for the farmer. Based on these factors, the decision concerning the number of subsamples that should be taken in an SPR evaluation depends on the accuracy required by the farmer/consultant (goals of the assessment) and the capacity for investments.

Descriptive statistics of the SPR sampling point data set

The highest mean SPR values (close to 3 MPa) were obtained from the soil layers below 0.15 m in depth in year I (Table 1) and the 0.10-0.15 and 0.15-20 m layers in year II (Table 2). Soil compaction in layers below the action zone of seeder disks and shank openers has been frequently detected in soils under the no-tillage system (Cardoso et al., 2006; Debiasi, Levien, Trein, Conte, & Kamimura, 2010; Cherubin et al., 2011; Moreira et al., 2016). The absence of soil disturbance, the low diversified cropping system and especially the systematic traffic of heavy machinery under soil moisture conditions favorable for compaction have been proposed as the main causes associated with soil compaction in no-tillage areas (Tormena et al., 2002; Debiasi et al., 2010).

The analyses of the mean SPRs from the whole area (18 sampling points), from the different soil layers and over the two year period showed that the number of subsamples did not largely influence the results. Therefore, if the objective of the SPR evaluation was to obtain a general diagnosis of the area (traditional sampling), the number of subsamples would not influence the results. This finding corroborates the results obtained by Tavares Filho and Ribon (2008) and Molin et al. (2012), which indicate that 12-15 subsamples are sufficient to obtain satisfactory results using conventional sampling.

However, the number of subsamples had an elevated influence on the amplitude of the SPR values (*i.e.*, the range between the minimum and maximum values), resulting in a reduction of the amplitude of the data with an increase in the number of subsamples.

A lower amplitude of the SPR values between the sampling points indicates that the value obtained for each sampling point when a higher number of subsamples is used more accurately represents the SPR mean since the microvariability of the area is better considered.

Table 1. Descriptive statistics of soil penetration resistance (SPR, MPa) in the soil profiles obtained with different numbers of subsamples (1-15) per sampling point in year I in Palmeira das Missões (RS), southern Brazil.

Statistical	Number of subsamples								
parameters	1	2	3	4	5	6	9	12	15
				0.00–0.05 (m) (year I)					
Minimum	1.19	1.27	1.57	1.58	1.62	1.55	1.74	1.78	1.81
Mean	2.05	2.09	2.14	2.11	2.08	2.10	2.18	2.20	2.21
Maximum	3.36	3.06	2.83	2.76	2.69	2.79	2.87	2.84	2.78
CV (%)[§]	28.66	23.19	15.05	14.18	14.70	15.08	13.43	13.07	13.60
				0.05–0.10 (m) (year I)					
Minimum	1.62	1.68	1.85	1.83	2.02	2.17	2.11	2.22	2.28
Mean	2.66	2.55	2.56	2.47	2.52	2.58	2.58	2.62	2.64
Maximum	3.58	3.55	3.22	3.04	2.95	2.92	2.98	3.06	3.05
CV (%)	21.56	19.42	14.25	13.96	12.08	9.91	10.15	9.64	8.69
				0.10–0.15 (m) (year I)					
Minimum	2.03	2.05	2.24	2.18	2.25	2.09	2.29	2.41	2.44
Mean	2.79	2.79	2.78	2.74	2.76	2.77	2.77	2.81	2.83
Maximum	4.19	3.96	3.79	3.52	3.36	3.32	3.12	3.10	3.27
CV (%)	22.06	16.31	14.78	12.85	11.12	10.93	8.12	7.23	7.07
				0.15–0.20 (m) (year I)					
Minimum	1.70	2.05	2.22	2.27	2.28	2.31	2.43	2.63	2.65
Mean	2.93	2.87	2.92	2.93	2.94	2.93	2.97	3.02	3.04
Maximum	4.76	3.84	3.68	3.77	3.55	3.53	3.30	3.38	3.40
CV (%)	27.86	17.44	14.95	13.51	12.67	12.05	9.25	7.85	7.48
				0.20–0.25 (m) (year I)					
Minimum	1.57	1.59	1.86	2.35	2.43	2.52	2.65	2.65	2.66
Mean	3.02	3.09	3.09	3.09	3.09	3.10	3.14	3.15	3.17
Maximum	4.37	4.30	3.80	3.74	3.71	3.71	3.82	3.61	3.64
CV (%)	25.36	20.38	16.19	13.05	12.14	10.9	9.23	8.29	8.47
				0.25–0.30 (m) (year I)					
Minimum	1.40	1.94	2.14	2.31	2.34	2.43	2.61	2.58	2.60
Mean	3.12	3.10	3.05	3.12	3.13	3.13	3.17	3.18	3.17
Maximum	4.86	4.61	3.92	4.09	3.89	3.77	3.77	3.81	3.78
CV (%)	29.04	21.77	17.34	16.2	16.04	14.27	11.38	10.01	9.96

[§]CV = coefficient of variation.

Table 2. Descriptive statistics of soil penetration resistance (SPR, MPa) in the soil profiles obtained with different numbers of subsamples (1-15) per sampling point in year II in Palmeira das Missões (RS), southern Brazil.

Statistical	Number of subsamples								
parameters	1	2	3	4	5	6	9	12	15
				0.00–0.05 (m) (year II)					
Minimum	1.25	1.25	1.41	1.35	1.46	1.42	1.42	1.61	1.67
Mean	2.02	2.04	1.99	1.97	2.00	2.07	2.12	2.11	2.12
Maximum	3.08	3.88	2.47	2.46	2.55	2.55	2.48	2.53	2.53
CV (%)[§]	22.48	22.66	16.68	15.72	15.74	15.81	14.27	13.66	12.09
				0.05–0.10 (m) (year II)					
Minimum	1.51	1.76	1.93	2.21	2.10	2.27	2.38	2.38	2.46
Mean	2.60	2.63	2.70	2.69	2.71	2.75	2.75	2.72	2.74
Maximum	3.65	3.46	3.29	3.15	3.29	3.25	3.13	3.05	3.03
CV (%)	24.13	20.91	15.30	11.90	12.39	10.97	7.74	7.31	5.79
				0.10–0.15 (m) (year II)					
Minimum	1.77	2.14	2.38	2.40	2.42	2.38	2.49	2.43	2.48
Mean	2.82	2.88	2.91	2.90	2.91	2.92	2.92	2.92	2.92
Maximum	3.93	4.20	3.49	3.47	3.53	3.46	3.30	3.22	3.17
CV (%)	22.73	18.47	11.47	11.79	10.96	9.82	8.08	7.75	6.60
				0.15–0.20 (m) (year II)					
Minimum	1.77	2.14	2.38	2.40	2.42	2.38	2.49	2.43	2.48
Mean	2.82	2.88	2.91	2.90	2.91	2.92	2.92	2.92	2.92
Maximum	3.93	4.20	3.49	3.47	3.53	3.46	3.30	3.22	3.17
CV (%)	16.52	11.73	8.55	7.95	8.75	8.18	8.40	7.59	7.16
				0.20–0.25 (m) (year II)					
Minimum	1.83	2.09	2.20	2.17	2.18	2.25	2.25	2.24	2.26
Mean	2.83	2.75	2.63	2.63	2.63	2.65	2.65	2.65	2.68
Maximum	3.80	3.38	3.07	3.06	3.11	3.08	3.05	3.11	3.18
CV (%)	16.9	12.63	11.14	10.45	11.24	10.69	9.48	9.31	8.65
				0.25–0.30 (m) (year II)					
Minimum	1.53	1.99	1.98	2.02	2.03	2.16	2.21	2.16	2.21
Mean	2.69	2.69	2.64	2.64	2.64	2.69	2.68	2.67	2.68
Maximum	3.68	3.53	3.39	3.42	3.67	3.53	3.29	3.28	3.19
CV (%)	21.96	17.87	16.74	15.51	15.73	13.80	10.23	10.76	9.76

[§]CV = coefficient of variation.

Moreover, the errors resulting from possible sampling faults are diluted (Molin et al., 2012). For example, the maximum SPR value observed in year I using 15 subsamples was 3.27 MPa (0.10-0.15 m layer), whereas when using only one subsample the maximum at this location was 4.19 MPa. This evidence is important and emphasizes the need to correctly choose the number of subsamples when localized intervention in the field is guided by the spatial variability in the SPR as proposed by Girardello et al. (2014). The utilization of an insufficient number of subsamples can incorrectly indicate a need to conduct interventions in areas, which makes this type of management technically and economically inefficient (Tavares Filho & Ribon, 2008; Molin et al., 2012)

Generally, the data show CV values classified as low or medium. The coefficient of variation values were classified with low variation (< 12%) when more than six subsamples were used (Warrick & Nielsen, 1980). The exception was the samples collected from the 0.00-0.05 m soil layer, which were classified as medium (12 < CV < 62%), where only one subsample from years I and II reached values of 29 and 23%, respectively. Independent of the year of the study, a reduction in the number of subsamples resulted in an increase in the CV values (Tables 1 and 2). The observation of higher CV values is an indication of the existence of higher spatial variability of the attribute in that area (Oliveira et al., 2015), which requires the utilization of sampling plans that use a larger number of samples to faithfully reproduce the spatial variability at that location (Siqueira et al., 2014). Although we did not investigate different sampling grid sizes in this study, the results obtained indicate that using a higher number of subsamples per sampling point is an SPR mapping strategy to uses less dense sampling grids. This finding need to be confirmed in future studies.

Quality of the thematic SPR maps in the function of the number of subsamples

In the surface layer of the soil (0.00-0.05 m), we found a significant correlation with the reference maps (15 subsamples) when at least six and four subsamples were used in years I and II, respectively (Table 3).For the other soil layers, three (year I) and four (year II) subsamples were sufficient to obtain a significant correlation with the reference maps. For all of the layers measured, a reduction in the correlation coefficient was observed with a reduction in the number of subsamples collected. These results indicated that the maps obtained using lower numbers of subsamples per sampling point presented higher deviations in their estimates, thereby reducing the reliability of the information (Cherubin et al., 2015).

In both years, the SPR maps for the surface soil layer presented the lowest correlations with the reference maps (15 subsamples). This result is due to higher variation in the SPR values in this soil layer, as was previously discussed for the item numbers of subsamples per sampling point. Tavares Filho and Ribon (2008) compared no-tillage and conventional tillage systems, and Storck et al. (2016) studied an integrated crop-livestock system; both studies also found that the 0.00-0.10 m layer presented the highest variations and consequently needed a larger number of subsamples than any other soil layer to achieve a good level of reliability. However, considering that the surface layer of the soil under a no-tillage system is periodically disturbed during crop sowing, which minimizes possible physical restrictions to plant root growth, the decision of the subsample numbers per sampling point to provide penetration resistance measurements should be based on the microvariability of this parameter in the deeper soils layers (Moreira et al., 2016).

The RDC results (Figure3) were similar to the results obtained for the correlation analysis for the two study years and all of the soil layers measured, with a correlation of -0.88 between the RDC and Pearson´s correlation. A high correlation (r = 0.96) between the RDC and the Kappa index, which is another procedure used to evaluate the similarity between thematic maps, has been shown in the literature (Bazzi, Souza, Uribe Opazo, Nóbrega, & Neto, 2008). Independent of the soil layer measured, there was an increase in the deviation (sampling errors) in the maps with a reduction in the number of subsamples. The highest RDC values were found in the 0.00-0.05 m layer in year I, with a deviation of 21%, and in the 0.00-0.05 and 0.05-0.10 m layers in year II, with a deviation of 19%. The use of RDC analysis to compare SPR maps has not been documented in the literature, which characterizes this study as pioneering in this area. However, this coefficient has been used with success for other variables, such as grain yield (Bazzi et al., 2008; Coelho et al., 2009) and soil chemical attributes (Cherubin et al., 2015).

To obtain RDC values less than 10% in the surface soil layer (0.00-0.05 m), 9 and 6 subsamples were necessary for years I and II, respectively. This number decreased to 4 subsamples for the deeper soil layers in both years. Because the RDC is calculated from the mean difference in the modulus of the interpolated values in relation to the reference map (Coelho et al., 2009), no RDC value is considered optimum, and the choice of the acceptable deviation coefficient depends on the degree of reliability desired by the researcher. In this study, an RDC of 10% was considered a suitable value that could guide the interpretation of the results, as suggested by Bazzi et al. (2008).

Table 3. Correlation between the soil penetration resistance (SPR) maps obtained with different numbers of subsamples (1-12) per sampling point and the reference maps obtained with 15 subsamples in two years in Palmeira das Missões (RS), southern Brazil.

Number of subsamples	Soil layer (m)						Mean
	0.00-0.05	0.05-0.10	0.10-0.15	0.15-0.20	0.20-0.25	0.25-0.30	
	Year I						
1	0.42^{ns}	0.63*	0.42^{ns}	0.62*	0.30^{ns}	0.43^{ns}	0.47
2	0.44^{ns}	0.49*	0.49*	0.59*	0.41^{ns}	0.43^{ns}	0.49
3	0.46^{ns}	0.58*	0.58*	0.65*	0.54*	0.68*	0.58
4	0.46^{ns}	0.70*	0.64*	0.68*	0.67*	0.72*	0.65
5	0.44^{ns}	0.74*	0.77*	0.79*	0.78*	0.81*	0.72
6	0.61*	0.80*	0.75*	0.85*	0.86*	0.89*	0.79
9	0.87*	0.95*	0.86*	0.94*	0.92*	0.96*	0.92
12	0.95*	0.95*	0.93*	0.96*	0.97*	0.98*	0.96
	Year II						
1	0.16^{ns}	0.16^{ns}	0.26^{ns}	0.46^{ns}	0.16^{ns}	0.46^{ns}	0.29
2	0.28^{ns}	0.17^{ns}	0.30^{ns}	0.42^{ns}	0.43^{ns}	0.37^{ns}	0.33
3	0.45^{ns}	0.39^{ns}	0.45^{ns}	0.50*	0.51*	0.54*	0.47
4	0.64*	0.51*	0.52*	0.65*	0.63*	0.63*	0.60
5	0.69*	0.56*	0.57*	0.83*	0.74*	0.76*	0.69
6	0.77*	0.64*	0.69*	0.86*	0.86*	0.83*	0.78
9	0.85*	0.89*	0.80*	0.85*	0.94*	0.90*	0.87
12	0.97*	0.95*	0.91*	0.92*	0.97*	0.97*	0.95

*and ns= significant and non-significant Pearson's correlation coefficients ($p < 0.05$), respectively; n= 18 points.

The SPR maps produced considering the different numbers of subsamples in years I and II are shown in Figure 4. Generally, we observed an increase in the dissimilarity in relation to the reference map (15 subsamples) with a reduction in the number of subsamples, resulting in a loss of accuracy in the maps that consequently could lead to erroneous interpretations concerning the real physical conditions of the soil. This loss of accuracy can be visualized by the increase in amplitude of the observed SPR values, with the maps produced with a smaller number of subsamples using more SPR classes (*viz.*, colors). Therefore, the utilization of an insufficient number of subsamples overestimates the variability of the data and can indicate a necessity for denser sampling grids (*viz.*, more sampling points), which increases the sampling cost.

In the literature, divergent opinions exist regarding the SPR value that should be considered the critical limit for plant root growth. These values vary according to the characteristics of the soil, management practices and crops. Traditionally, SPR values between 2.0 and 2.5 MPa (Taylor, Robertson & Parker, 1966) are considered the critical limits for root growth. However, various studies have shown that plants tolerate higher SPR values (up to 3 MPa) in areas with no-tillage systems (Secco et al., 2009; Girardello et al., 2014; Moraes, Debiasi, Carlesso, Franchini, & Silva, 2014), probably due to the better soil structure and the greater presence of continuous biopores (Moraes et al., 2014). Independent of the critical limit considered, subareas with SPR values considered restrictive for root growth were detected using the mapping strategy. In this sense, this study could help farmers, consultants and researchers with decision-making regarding the sampling procedure that should be used for SPR evaluations in agricultural soils.

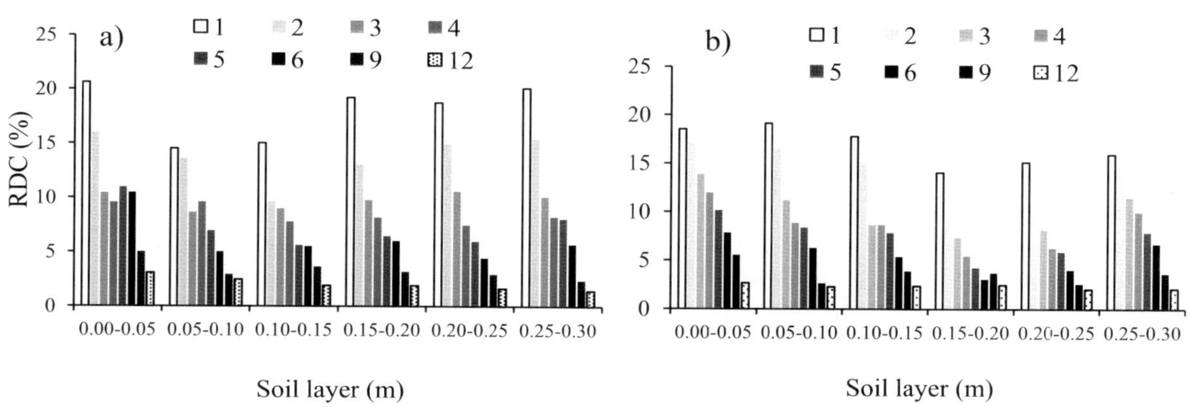

Figure 3. Relative deviation coefficient (RDC%) between soil penetration resistance maps (SPR, MPa) obtained with different numbers of subsamples (1-12) per sampling point and the reference maps (15 subsamples) for different soil layers in years I (a) and II (b) in Palmeira das Missões (RS), southern Brazil.

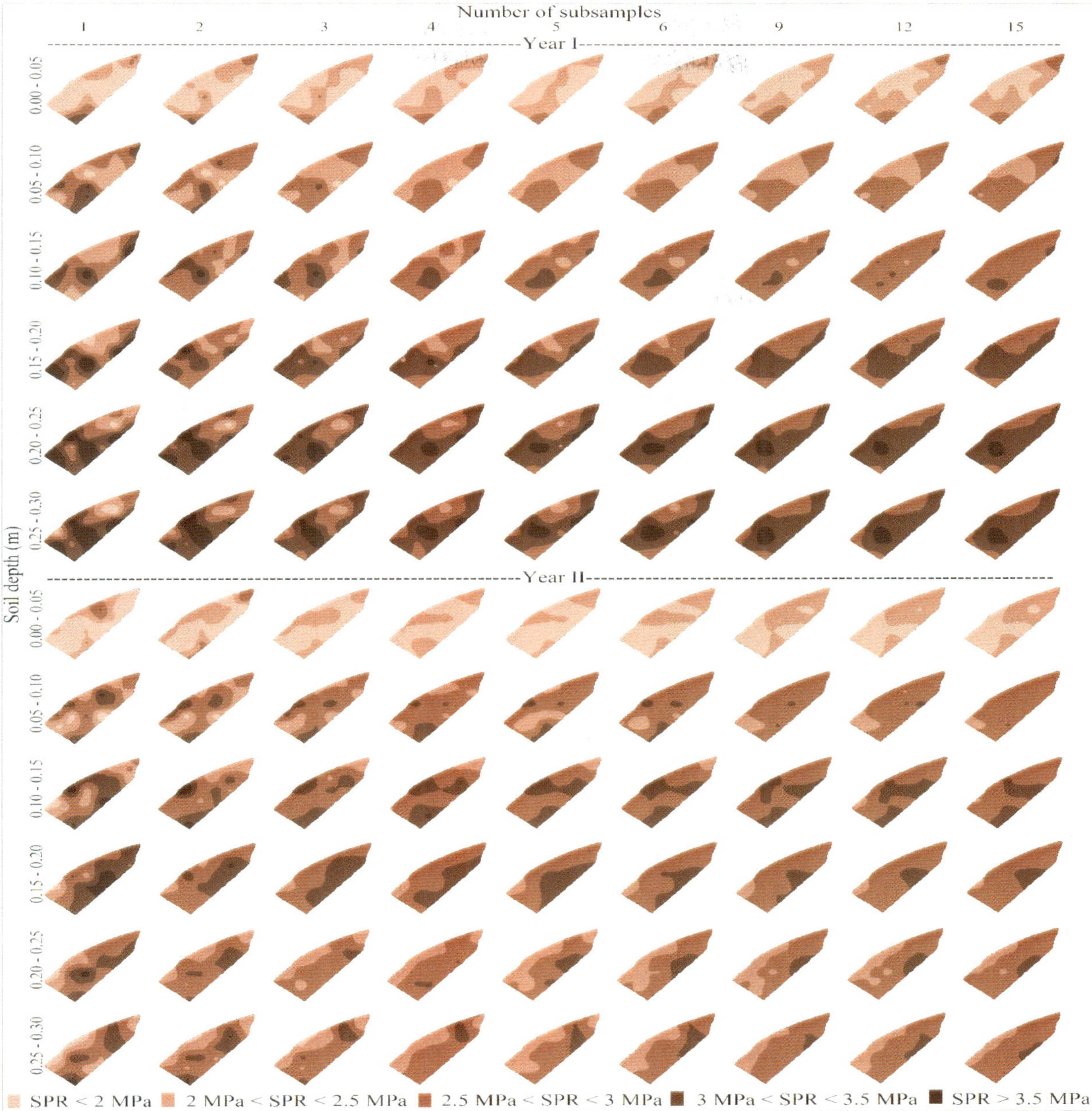

Figure 4. Thematic maps of soil penetration resistance (SPR) obtained by considering different numbers of subsamples (1-12) per sampling point and the reference maps (15 subsamples) for the different soil layers in two years in Palmeira das Missões (RS), southern Brazil.

Conclusion

The number of subsamples used to obtain the soil penetration resistance that properly represents a sampling point depends on the level of error tolerated in the mapping, with a higher number of subsamples resulting in more accurate maps.

A reduction in the number of subsamples promotes an increase in the variability of soil penetration resistance data. Generally, this study suggests that at least four subsamples per sampling point achieves soil penetration resistance maps with a coefficient of relative deviation less than 10% (30% maximum error per point around the mean) and significant correlation with the reference maps (15 subsamples).

References

Bazzi, C. L., Souza, E. G., Uribe Opazo, M. A., Nóbrega, L. H. P., & Neto, R. P. (2008). Influência da distância entre passadas de colhedora equipada com monitor de colheita na precisão dos mapas de produtividade na cultura do milho. *Engenharia Agrícola, 28*(2), 355-63.

Bengough, A. E., Mckenzie, B. M., Hallet, P. D., & Valentine, T. A. (2011). Root elongation, water stress, and mechanical impedance: A review of limiting stress and beneficial root tip trails. *Journal of Experimental Botany, 62*(1), 59-68.

Beutler, N. A., Cebturion, J. F., Centurion, M. A. P. C., Leonel, C. L., São João, A. C. E., & Freddi, O. S. (2007). Intervalo hídrico ótimo no monitoramento da compactação e da qualidade física de um Latossolo Vermelho cultivado com soja. *Revista Brasileira de Ciência do Solo, 31*(6), 1223-1232.

Cardoso, E. G., Zotarelli, L., Piccinin, J. L., Torres, E., Saraiva, O. F., & Guimarães, M. F. (2006). Sistema radicular da soja em função da compactação do solo no sistema plantio direto. *Pesquisa Agropecuária Brasileira, 41*(3), 493-501.

Catania, P., Vallone, P., Pipitone, F., Argento, G. F., Spartà, G., & Laudicina, V. A. (2013). Soil management effect on soil penetration resistance in the vineyard. *Journal of Agricultural Engineering, 44*(2), 344-347.

Cherubin, M. R., Santi, A. L., Basso, C. J., Eitelwein, M. T., & Vian, A. L. (2011). Variabilidade da resistência do solo à penetração em função da dimensão da malha amostral. *Revista Plantio Direto, 125*(5), 4-9.

Cherubin, M. R., Santi, A. L., Eitelwein, M. T., Amado, T. J. C., Simon, D. H., & Damian, J. M. (2015). Dimensão da malha amostral para caracterização da variabilidade espacial de fósforo e potássio em Latossolo Vermelho. *Pesquisa Agropecuária Brasileira, 50*(2), 168-177.

Coelho, E. C., Souza, E. G., Uribe-Opazo, M. A., & Pinheiro Neto, R. (2009). Influência da densidade amostral e do tipo de interpolador na elaboração de mapas temáticos. *Acta Scientiarum Agronomy, 31*(1), 165-174.

Debiasi, H., Levien, R., Trein, C. R., Conte, O., & Kamimura, K. M. (2010). Produtividade de soja e milho após coberturas de inverno e descompactação mecânica do solo. *Pesquisa Agropecuária Brasileira, 45*(6), 603-612.

Debiasi, H., Franchini, J. C., Oliveira, F. A., &Machado, T. M. (2012). Ajuste de grades amostrais para o mapeamento da resistência à penetração de um Latossolo Bruno. In: Inamasu, B. Y., Naime, J. M., Resende, A. V., & Bassoi, L. H. (Ed.). *Agricultura de Precisão: Um novo olhar* (p. 138-42). São Carlos: Embrapa CNPSo.

Empresa Brasileira de Pesquisa Agropecuária [Embrapa]. (1997). *Manual de métodos de análise de solo* (2a ed.). Rio de Janeiro, RJ: Embrapa.

Girardello, V. C., Amado, T. J. C., Santi, A. L., Cherubin, M. R., Kunz, J., & Teixeira, T. G. (2014). Resistência à penetração, eficiência de escarificadores mecânicos e produtividade da soja em Latossolo argiloso manejado sob plantio direto de longa duração. *Revista Brasileira de Ciência do Solo, 38*(4), 1234-1244.

Marasca, I., Oliveira, C. A. A., Guimarães, E. C., Cunha, J. P. A. R., Assis, R. L., Perin, A., & Menezes, L. A. S. (2011). Variabilidade espacial da resistência do solo à penetração e teor de água em sistema de plantio direto na cultura da soja. *Bioscience Journal, 27*(2), 239-246.

Molin, J. P., Dias, C. T. S., & Carbonera, L. (2012). Estudos com penetrometria: Novos equipamentos e amostragem correta. *Revista Brasileira de Engenharia Agrícola e Ambiental, 16*(5), 584-590.

Moraes, M. T., Debiasi, H., Carlesso, R., Franchini, J. C., & Silva, V. R. (2014). Critical limits of soil penetration resistance in a Rhodic Eutrudox. *Revista Brasileira de Ciência do Solo, 38*(1), 288-298.

Moreira, W. H., Tormena, C. A., Karlen, D. L., Silva, A. P., Keller, T., & Betioli Jr, E. (2016). Seasonal changes in soil physical properties under long-term no-tillage. *Soil Tillage Research, 160*(1), 53-64.

Oliveira, I. A., Junior, J. M., Campos, M. C. C., Aquino, R. E., Freitas, L., Siqueira, D. S., & Cunha, J. M. (2015). Variabilidade espacial e densidade amostral da suscetibilidade magnética e dos atributos de Argissolos da Região de Manicoré, AM. *Revista Brasileira de Ciência do Solo, 39*(3), 668-681.

Petersen, R. G., & Calvin, L. D. (1965). Sampling. In Black, C. A. (Ed.). *Methods of soil analysis: Physical and mineralogical properties, including statistics of measurement and sampling* (p. 54-71). Madison: American Society of Agronomy.

Santos, H. G., Jacomine, P. K. T., Anjos, L. H. C., Oliveira, V. A., Lumbreras, J. F., Coelho, M. R., Almeida, J. Á., Cunha, T. J. F., & Oliveira, J. B. (2013). *Sistema Brasileiro de Classificação de Solos* (3a ed.). Brasília, DF: Embrapa Solos.

Secco, D., Reinert, D. J., Reichert, J. M., & Silva, V. R. (2009). Atributos físicos e rendimento de grãos de trigo, soja e milho em dois Latossolos compactados e escarificados. *Ciência Rural, 39*(1), 58-64.

Shapiro, S. S., & Wilk, M. B. (1965). An analysis of variance test for normality: complete samples. *Biometrika, 52*(3/4), 591-611.

Silva, J. M., Passos, A. L. R., & Beltrão, F. A, S. (2009). Análise espacial da densidade, umidade e resistência do solo à penetração sob sistemas de cultivo. *Engenharia Ambiental, 6*(3), 103-118.

Silva, W. M., Bianchini, A., & Cunha, C. A. (2016). Modeling and correction of soil penetration resistance for variations in soil moisture and soil bulk density. *Engenharia Agrícola, 36*(3), 449-459.

Siqueira, D. S. J., Junior, M., Pereira, G. T., Barbosa, R. S., Teixeira, D. B., & Peluco, R. G. (2014). Sampling density and proportion for the characterization of the variability of Oxisol attributes on different materials. *Geoderma, 232-234*(1), 172-182.

Soil Survey Staff. (2014). *Keys to soil taxonomy* (12th ed.). Washington, DC: USDA-Natural Resources Conservation Service.

Souza, Z. M., Campos, M. C. C., Cavalcante, I. H. L., Marques JR, J., Cesarin, L. G., & Souza, S. R. (2006). Dependência espacial da resistência do solo à penetração e do teor de água do solo sob cultivo continuo de cana-de açúcar. *Ciência Rural*, *36*(1), 128-134.

Statistical Analysis System [SAS]. (2010). *Sas/Stat user's guide: statistics, version 9.1.3*. Cary, NC: SAS.

Storck, L., Kobata S. G. K., Brum, B., Soares, A. B., Modolo, A. J., & Assmann, T. S. (2016). Sampling plan for using a motorized penetrometer in soil compaction evaluation. *Revista Brasileira de Engenharia Agrícola e Ambiental*, *20*(3), 250-255.

Tavares-Filho, J., & Ribon, A. A. (2008). Resistência do solo à penetração em resposta ao número de amostras e tipo de amostragem. *Revista Brasileira de Ciência do Solo*, *32*(2), 487-494.

Taylor, H. M., Robertson, G. M., & Parker, J. J. (1966). Soil strength root penetration for mediun to coarse textured soil materials. *Soil Science Society American Journal*, *102*(1), 18-22.

Tormena, C. A., Barbosa, M. C., Costa, A. C. S., & Gonçalves, A. C. A. (2002). Densidade, porosidade e resistência à penetração em Latossolo cultivado sob diferentes sistemas de preparo de solo. *Scientia Agrícola*, *591*(4), 795-801.

Valadão Jr, D. D., Biachini, A., Valadão F. C. A., & Rosa, R. P. (2014). Penetration resistance according to penetration rate, cone base size and different soil conditions. *Bragantia*, *73*(2), 171-177.

Warrick, A. W., & Nielsen, D. R. (1980). Spatial variability of soil physical properties in the field. In Hillel, D. (Ed.), *Applications of soil physics* (p. 319-44). New York: Academic Press.

Webster, R., & Oliver, M. A. (2007). *Geoestatitic for Enviromental Scientists* (2nd ed.). Chichester, England: John Wiley & Sons Ltd.

Wu, Y. H., Hung, M. C., & Patton, J. (2013).Assessment and visualization of spatial interpolation of soil pH values in farmland. *Precision Agriculture*, *14*(6), 565-585.

Reducing conditions on barium absorption in rice plants cultured in $BaSO_4$-enriched soil

Marcio Osvaldo Lima Magalhães[1*], Nelson Moura Brasil do Amaral Sobrinho[1], Everaldo Zonta[1], Alfredo Tolón Becerra[2], Xavier Bolívar Lastra-Bravo[2] and Izabella Bezerra Coutinho[1]

*[1]Departamento de Solos, Universidade Federal Rural do Rio de Janeiro, BR-465, km 7, 23890-000, Seropédica, Rio de Janeiro, Brazil. [2]Rural Engineering Department, University of Almería, Almería, Spain. *Author for correspondence. E-mail: marciomagalhaes@gmail.com*

ABSTRACT. To evaluate the possible solubilization of barium sulfate in soils under reducing conditions and its effects on barium bioavailability, an *Oryza sativa* pot trial was established. Increasing barium doses and two redox potential conditions were evaluated. The geochemical fractionation data demonstrated that reducing conditions led to an increase in the levels of more labile forms of barium and a reduction in more stable forms. Furthermore, higher doses of barium were found to have a negative impact on grain production. The highest levels of barium accumulation in the leaves, roots, and grains were observed with the highest barium dose under reducing conditions. These results demonstrate that reducing conditions increased barium bioavailability and absorption by rice plants.

Keywords: contamination, barite, *Oryza sativa*.

Introduction

Barium is present in small quantities in igneous rocks such as feldspars and micas. The primary source, however, is the mineral barite ($BaSO_4$), which is widely used in the petroleum industry as a drilling fluid component due to its high density (4.2 g cm^{-3}) (FAM et al., 2003). Barite ($BaSO_4$) often is added as a weighting agent to drilling muds to counteract pressure in the geologic formations being drilled, preventing a blowout (NEFF, 2008). During the drilling of oil wells for oil prospecting and production, the drilling fluid is mixed with ground rock, thus releasing its component substances (POZEBON et al., 2005). Consequently, these component substances greatly influence the concentration of barium and other substances in oil well residues (MELTON et al., 2000).

Soils in flood areas exhibit alterations in elemental composition and microbial metabolism that trigger a series of physical, chemical, and biological transformations. These changes subsequently produce soil characteristics that are profoundly distinct from the initial conditions (SEYBOLD et al., 2002). These new chemical environments are ecologically relevant because the change redox states affecting the mobility and bioavailability of elements present in the soil.

The barium contained in barite is relatively immobile, and its low water solubility (2.47 mg L^{-1} at 25°C) reduces its bioavailability. However, the solubility of barium sulfate can be improved under highly reducing environments (MAGALHÃES et al., 2011a; MONNIN et al., 2001), reflecting the fact that sulfate reduction occurs at Eh values of -250 mV (KHANAL; HUANG, 2003). Previous studies have shown that barite can serve as a sulfate source for the respiration of anaerobic bacteria (BALDI et al., 1996). When these wastes are disposed into poorly drained soil, the release of the most toxic form of barium, Ba^{+2}, into the environment (USEPA, 2005) may contaminate bodies of water and introduce the element into the food chain (MAGALHÃES et al., 2011b).

Because barium is present in most soils, low, non-toxic levels of barium, approximately 4 to 50 mg kg^{-1}, are commonly found in plants (CHAUDHRY et al., 1977). Nogueira et al. (2010), for example, found no signs of barium toxicity in corn plants with barium concentrations ranging from 90 to 106 mg kg^{-1}. In a hydroponic trial conducted with soy plants, however, Suwa et al. (2008) observed signs of toxicity, namely, high concentrations of barium in the leaves (4,970 mg kg^{-1}) and reductions in dry mass, following treatment with 5 mM barium.

The aim of the present study was to test the effects of oxidizing and reducing conditions on the bioavailability of barium and its absorption by rice plants (*Oryza sativa*).

Material and methods

The soil samples used were collected from an Oxisol, at a depth of 0-20 cm, in the State of Rio de Janeiro, Brazil. The soil characteristics were as follows: barium: 223 mg kg^{-1} (ISO, 1995), pH in H_2O: 5.8; sand: 272 g kg^{-1}; silt: 132 g kg^{-1}; clay: 596 g kg^{-1}; Corg.: 10,7 g kg^{-1} and a $CEC_{(pH\ 7.0)}$: 97.0 $mmol_c$ dm^{-3} (EMBRAPA, 1997).

Using barium sulfate ($BaSO_4$) as a source, barium doses were applied according to the values proposed by the 420 CONAMA Resolution (BRASIL, 2009). The treatments were as follows: soil without barium application (control); 100 mg barium kg^{-1} of dry soil (dose 1); 300 mg barium kg^{-1} of dry soil (dose 2); and 3,000 mg barium kg^{-1} of dry soil (dose 3).

The pots were filled with 5 dm^3 of soil, and the soil was homogenized after the barium doses were added. The moisture conditions were 70% of field capacity (oxidizing conditions) and saturation (reducing conditions). To maintain the reducing conditions, a 7 cm layer of water was applied to the soil surface. During the incubation period, drainage was impeded for all treatments, and the pots were covered to prevent water loss by evaporation. A total of 32 experimental units were prepared in a factorial design (4 x 2) with four barium levels, two moisture levels, and four replicates.

During the incubation period, pH and Eh values were determined 2 hours after flooding on days 2, 4, and 6. Following the first week of incubation, the assessment was performed on a weekly basis for the saturated soil until stabilization of the redox potential required for the reduction of sulfate was obtained (i.e., Eh values of -250 mV). The Eh values of the soil under oxidizing and reducing conditions were directly measured using a specific electrode. Combined ORP analyzer electrodes were used for measuring Eh, and half-cell pH analyzer electrodes were used for measuring pH with the pH/Ion 450M® Analyzer. Thirty days after a redox potential value of -250 mV was obtained, rice seedlings (*Oryza sativa* var. Bico Ganga) were transplanted to the pots. The choice for rice was due to the fact that it is a culture that adapts to both dryland and flooding conditions.

Following the incubation period, the barium was geochemically fractionated for all experimental treatments according to the method recommended by the European Communities Bureau of Reference (BCR) and adopted by Sahuquillo et al. (1999). The geochemical fractions were as follows: the acid-soluble fraction, extracted with 0.11 mol L^{-1} acetic acid (F1); the fraction bound to Fe and Mn oxides, extracted with 0.1 mol L^{-1} hydroxylamine hydrochloride (F2); the fraction bound to organic matter, extracted with 8.8 mol L^{-1} hydrogen peroxide and 1.0 mol L^{-1} ammonium acetate (F3); and the residual fraction, which represents the aqua regia extracts from all other fractions (F4).

Ten days after germination, seedlings homogeneous in size and vigor were selected, and two plants were transplanted into each pot. A parceled fertilization method was used, with 80 kg N ha^{-1}, 40 kg P ha^{-1} (single dose), and 40 kg K ha^{-1} applied at the time of planting. After 40 days, surface fertilization was performed by applying 40 kg N ha^{-1} and 40 kg K ha^{-1}. The plants were collected at the end of the experiment, after approximately 4 months of growth, corresponding to the growth cycle of the cultivar. The collected plants were separated into shoots, roots, and grains and then rinsed in deionized water. The plant samples were oven-dried at 70°C until a constant weight was obtained. The digestion material was obtained by grinding the shoots, roots, and grains in a Willey-type mill using a 2 mm mesh. Nitroperchloric digestion was used to determine the barium concentrations, at a ratio of 6:1, according to the method described by Tedesco et al. (1995). The total barium content in the shoots, roots, and grains were calculated using the determined concentration and the amount of dry matter produced.

The levels of barium in the soil and plant extracts were quantified using a plasma emission spectrometer (ICP-OES; Perkin Elmer, model OPTIMA 3000), with Detection Limit = 0.036 mg kg^{-1}

and Quantification Limit = 0.36 mg kg^{-1} for barium. To validate the determination of the pseudo total content of barium in the soil and plants, the following reference materials were used: NIST SRM 2709a (San Joaquin soil) and SRM 1573a (tomato leaves), which had barium concentrations of 979 ± 28 mg kg^{-1} (95% recovery) and 63 mg kg^{-1} (93% recovery), respectively. These values are within the range classified as normal by the National Institute of Standards and Technology (NIST) for soil and plant samples. The oxidation-reduction variations were assessed by analysis of variance with F-test ($\rho < 0.05$), and the mean values of soil redox potential conditions compared by Tukey test ($\rho < 0.05$). Analyses were performed using the SAS/STAT 9.2® software (SAS Institute Inc., Cary, NC, USA). An error bar was also used to highlight possible differences among barium contents for the same redox potential condition.

Results and discussion

Figure 1A shows that the soil Eh values at 70% field capacity ranged from ± 350 to 450 mV, which is within the range for oxidized soil (800-300 mV) (CAMARGO et al., 2001). The pH values also remained constant, ranging from 5.8 to 6.0. When saturation conditions in the soil were combined with prolonged flooding (Figure 1B), a reduction in Eh values and subsequent stabilization below -250 mV were observed, indicative of highly reducing conditions. According to the literature (KHANAL; HUANG, 2003; JONES; INGLE JR., 2005; MAGALHÃES et al., 2011b), the reduction of sulfate to sulfide by sulfate-reducing bacteria (SRB) in the soil may occur at this Eh value. At the onset of flooding, Eh values of -250 mV and pH values below 6 were observed. As saturation continued, Eh values decreased, and pH values increased until stabilizing at an approximately neutral value.

The observed increase in pH is characteristic of soils under flooding conditions and reflects the accumulation of protons (GONÇALVES et al., 2008; LIMA et al., 2005; VEPRASKAS; FAULKNER, 2001). As described by Seybold et al. (2002), Eh and pH levels reflect the activity of protons and electrons in the soil because the excess of one causes a deficit in the other. According to Camargo et al. (1999), the increase in pH after submergence depends not only on the ratio of OH^- and H^+ consumption but also on the ratio of H^+ and electron consumption. The change in pH necessarily depends on two conditions: a well-developed reduction process and a sufficient amount of reduced Fe (SEYBOLD et al., 2002; PONNAMPERUMA, 1978). The measured Eh and pH values in the present study indicate that the incubation period was sufficient to promote these conditions and thus favored the reduction of oxidized soil components, particularly sulfate.

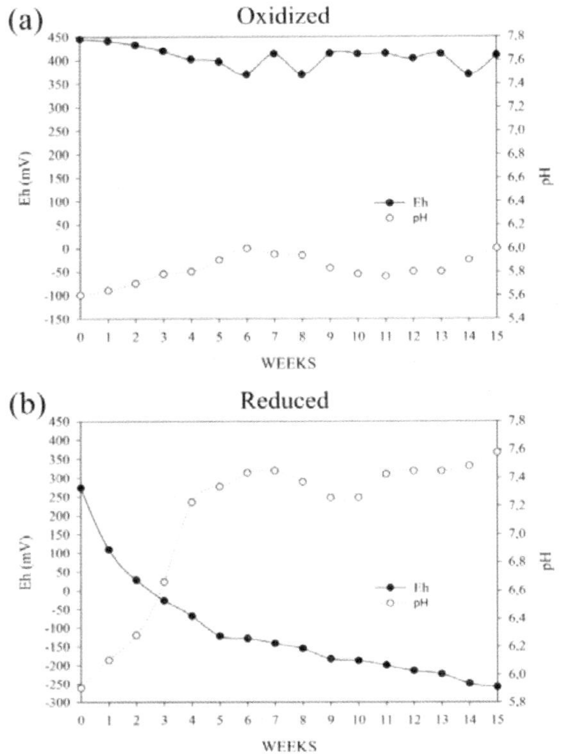

Figure 1. Redox potential values (Eh) for two condidition (oxidized and reduced) and the relationship between Eh and pH after weeks of flooding.

Figure 2 shows the mean barium distributions in the different geochemical fractions obtained by BCR scheme. Shows that under reducing conditions in the soil (-250 mV), the barium concentration in the acid soluble fraction (F1) increased significantly. The increases observed for the different treatments under reducing conditions were as follows: 11 mg kg^{-1} for the control, 25 mg kg^{-1} for dose 1, 87 mg kg^{-1} for dose 2, and 135 mg kg^{-1} for dose 3. Higher treatment doses yielded greater increases in the barium concentration of the F1 fraction. This fraction contains barium in solution that is electrostatically adsorbed on the surface of the soil colloids (exchangeable) and precipitated in the form of carbonates (FILGUEIRAS et al., 2004; MARIN et al., 1997).

The increased barium concentration in the F1 fraction under saturated conditions was accompanied by decreased barium concentrations in the oxide-bound (F2) and residual (F4) fractions. The binding of heavy metals with oxides does not guarantee surface immobilization because these elements can be released by oxide reduction in this fraction, resulting in a negative impact on the soil biota (CHLOPECKA, 1996; RODRIGUEZ et al., 2009). No significant differences were observed for the barium concentration in the organic matter fraction (F3). These results demonstrate the low affinity of barium for organic matter, because this complex is considered limited by some authors (BODEK et al., 1988; DANG et al., 2002). Used geochemical fractionation and observed that only 1% of the total barium was present in the fraction bound to organic matter (SMEDA; ZYRNICKI, 2002). The results of the geochemical fractionation in the present study demonstrate that reducing conditions increased barium concentrations in the most labile fraction (F1) and reduced the amount of barium associated with more stable fractions (F2 and F4). The increase observed in F1, which was promoted by increased $BaSO_4$ solubility, may result in increased barium bioavailability (ALBERTA ENVIRONMENT, 2009; MAGALHÃES et al., 2011b).

For both of the redox potential conditions, no significant differences were observed in the shoot, root, and total dry matter production of control plants and rice plants treated with arying barium doses (Figure 3). However, differences in grain mass were observed in plants that received the highest barium dose and plants without $BaSO_4$ application (control). Relative to the control, the highest barium dose resulted in decreases of approximately 18% and 38% in grain yield under oxidizing and reducing conditions, respectively. The findings of the present study therefore indicate that the highest barium dose (3,000 mg kg^{-1}) negatively affected grain yield. In the literature, there is limited information about the influence of barium on plant metabolism. However, in a study that utilized *Glycine max* in a nutrient solution, Suwa et al. (2008) reported similar results: high barium doses affected plant development, particularly yield, demonstrating the phytotoxic effect of this element. Llugany et al. (2000) observed phytotoxicity symptoms in *Phaseolus vulgaris* plants even at low barium concentrations.

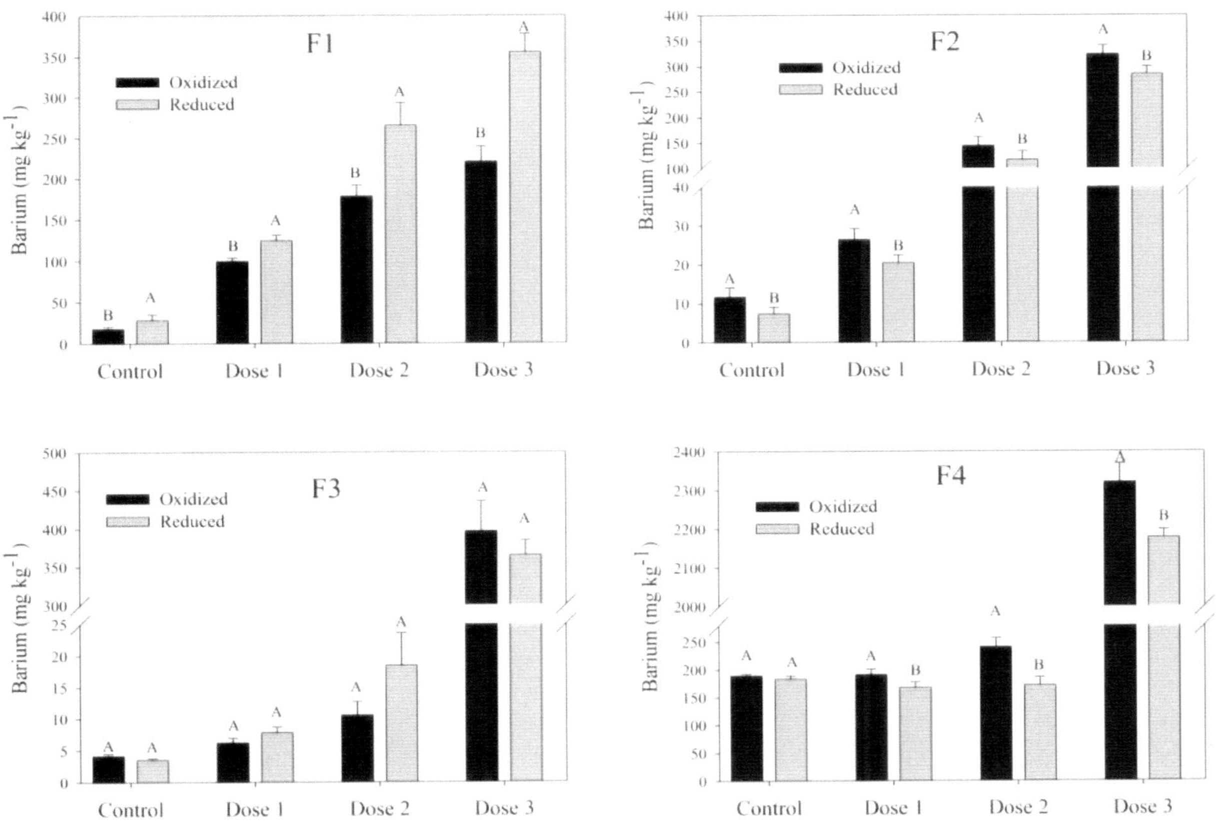

Figure 2. Barium distribution in different geochemical fractions as a function of dose and redox potential (oxidized and reduced). F1: soluble acid fraction; F2: fraction bound to iron and manganese oxide; F3: fraction bound to organic matter and sulfides; F4: residual fraction. Letters (for each barium content) indicate a significant difference ($p < 0.05$). Control: no barium sulfate application; Dose 1: 100 mg kg^{-1}; Dose 2: 300 mg kg^{-1}; Dose 3: 3,000 mg kg^{-1}.

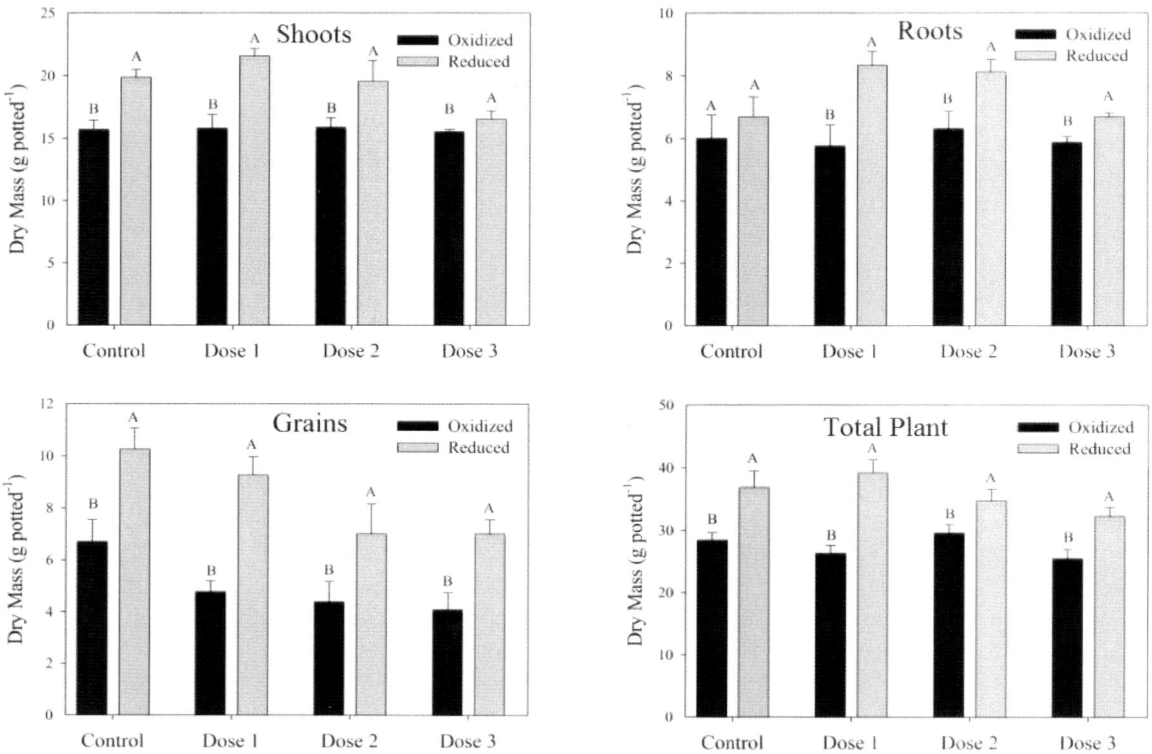

Figure 3. Dry mass production of rice plants as a function of barium dose and redox potential (oxidized and reduced). Letters (for each barium content) indicate a significant difference ($p < 0.05$). Control: no barium sulfate application; Dose 1: 100 mg kg^{-1}; Dose 2: 300 mg kg^{-1}; Dose 3: 3,000 mg kg^{-1}.

In the present study, the biomass of rice plants was significantly different under the two soil redox conditions, with a greater biomass observed in reduced soils than in oxidized soils. Although reducing conditions yielded greater grain mass at all doses, this increase was lowest at the highest dose, and significant differences were observed between doses.

There is no reference data describing the toxicity of barium to rice plants grown in nutrient solution. Although there are no reports of the effects caused by the toxicity of barium, they may be associated to the deficiency of other elements due to competition, for instance, competition with calcium for root absorption, due to its similarity (NOGUEIRA et al., 2010).

According to the data obtained, higher moisture levels (saturation) resulted in higher dry matter production for all of the evaluated parameters. Thus, saturation positively influenced the development of plants. Several studies have shown that the highest rice yields are obtained under flood conditions. For example, Patel et al. (2010) observed a mean reduction of 27% in plant yield when plants were cultivated under oxidizing conditions compared to flooded cultivation (reducing conditions).

Among the different doses treatments, no significant differences in barium content were observed in the shoots of rice plants cultivated in oxidized soil (Figure 4), and the mean concentration was 180 mg kg^{-1}. However, there were differences in barium absorption by rice plants under reducing conditions, with the lowest concentrations observed in the control (150 mg kg^{-1}) and the highest concentrations (620 mg kg^{-1}, approximately 4 times higher) observed in plants that received the highest barium dose (dose 3). The effects of doses 1 and 2 were not significantly different. A comparison of the barium content on plant shoots under different redox potential revealed no significant differences in the control (background levels of barium). For plants treated with $BaSO_4$, however, reducing conditions resulted in the highest barium concentrations. At the highest dose, reducing conditions resulted in a 3-fold increase in the barium content of plant shoots. Absorption in plant roots was also dependent on the dose under both soil redox conditions. The following barium concentrations were measured in plants cultivated in oxidized soil: 61 mg kg^{-1} for the control, 222 mg kg^{-1} for dose 1, 508 mg kg^{-1} for dose 2, and 826 mg kg^{-1} for dose 3. There was a 13-fold increase in the barium concentration in plants that received the highest dose compared to the control. Furthermore, the significant differences observed in reduced soil were greater than those in oxidized soil. Under reducing conditions, there was a 36-fold difference between the control and the highest dose. Analysis of the barium content in the grains revealed higher concentrations in plants cultured under reducing conditions compared to oxidizing conditions, especially at higher doses.

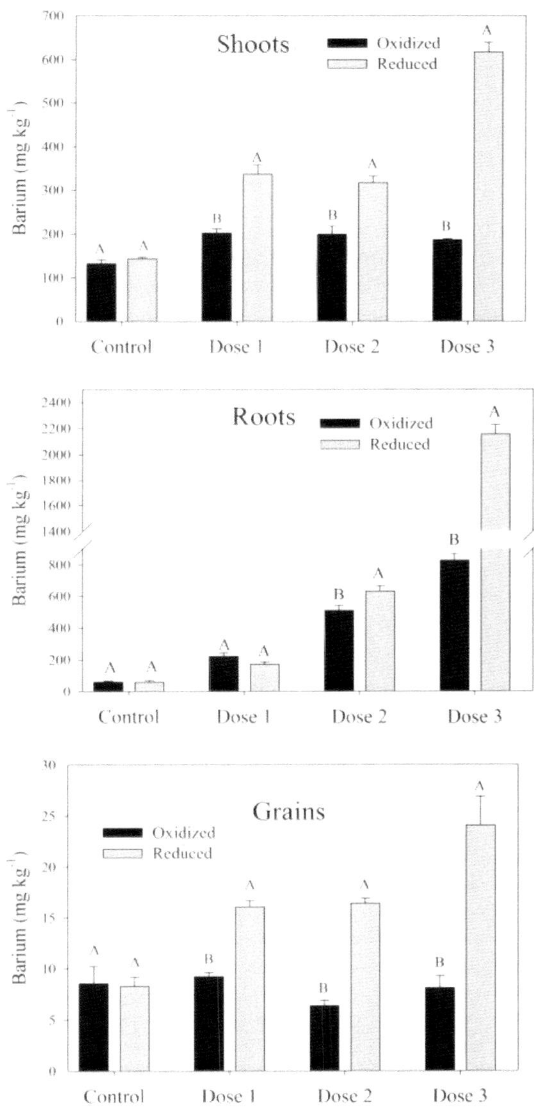

Figure 4. Barium concentration in leaves, roots, and grains of rice plants as a function of barium dose and redox potential (oxidized and reduced). Letters (for each barium content) indicate a significant difference ($p < 0.05$). Control: no barium sulfate application; Dose 1: 100 mg kg^{-1}; Dose 2: 300 mg kg^{-1}; Dose 3: 3,000 mg kg^{-1}.

The levels of barium that can be considered normal or toxic for plants are not well established (SUWA et al., 2008). According to Nogueira et al. (2010), barium is commonly found in plant tissues but not at levels that can be considered toxic, and the values of barium absorption they observed in *Zea mays* plants ranged from 90 to 106 mg kg^{-1}. These levels were not found to influence plant development or produce toxicity symptoms. In a *Glycine max* hydroponic trial, Suwa et al. (2008) observed high concentrations of barium in the leaves (4,970 mg kg^{-1}) following a treatment dose of 5 mM barium. These levels led to decreased dry mass production, which is one indicator of phytotoxicity.

Another trend observed in the present study was increased translocation of barium from the roots to the leaves under low dosage concentrations. In contrast, higher barium concentrations in the soil (doses 2 and 3) decreased translocation, resulting in greater barium immobilization in the roots. The ability of plants to immobilize metal in the roots is one of the mechanisms some plant species have evolved to tolerate heavy metals present in the soil (MAGALHÃES et al., 2011a; SANTOS et al., 2007).

The first barrier against the entering of heavy metals, especially regarding root level, is the immobilization of heavy metals in the cell wall and by extracellular carbohydrates, such as mucilage and callose (FRITZ, 2007; KARTEL et al., 1999), avoiding the presence of free ions in root tissues and, consequently, the translocation of ions to shoots, thereby reducing phytotoxicity. Pectins and histidine stand out for immobilizing heavy metals in the cell wall (LEITA et al., 1996).

Among the doses treatments, no significant differences were observed with respect to the amount of barium accumulated in the shoots and grains under oxidizing conditions (Figure 5). Under reducing conditions, however, plants cultured at the highest dose exhibited greater accumulation relative to the control. In the roots, there was a significant difference between the control and the dose treatments for both redox potential. An assessment of the total barium accumulated in the roots under oxidizing conditions revealed that the only significant difference was between the highest dose and the control. Under reducing conditions, the effect for the highest dose was significantly different from all other dosage treatments and the control. For natural concentrations of barium in the soil (control), there was no significant difference in barium accumulation as a function of redox conditions (oxidized and reduced). For the dose treatments, however, reducing conditions resulted in higher barium accumulation, except in the roots, where a significant difference was only observed at the highest dose. The accumulation increase in plants grown under reducing conditions was approximately 3-fold for the shoots, roots, filled grains, and accumulated total at the highest dose. Except in the control, differences were observed in the amount of accumulated barium between the two soil redox conditions, with greater accumulation observed under reducing conditions.

These findings demonstrate that highly reducing conditions improve barium sulfate solubility. As demonstrated by the geochemical fractionation (Figure 2), barium bioavailability increased, enabling absorption and accumulation of this element in plants.

It is worth noting that the increase in barium accumulation in the grains was approximately 300%, this increases the probability of introducing barium in to the food chain because rice grains are a staple food in many countries.

According to the United Nations Food and Agriculture Organization, the consumption of rice around the world amounts to an average 84.8 kg person^{-1} year^{-1}, which is equivalent to an average daily consumption per person of approximately 230 g. The maximum daily ingestion of barium per kg of body mass suggested by the United States Environmental Protection Agency (USEPA, 2005) is 0.2 mg kg^{-1} bw day^{-1}. Based on this information, a person weighing 70 kg may ingest a maximum of 14 mg of barium per day. If this person consumes an average of 230 g day^{-1} of rice grains produced under the most critical conditions presented in this study (extremely reduced soil with the highest amount of barium), this person would ingest a total of 5.5 mg of barium, or approximately 40% of the maximal value of barium. This value may be considered high because barium is found in many food groups and in water (USEPA 2005; YSART et al., 1999). It is also important to compare the values obtained in this study with the values measured in plants grown in soil with natural barium concentrations (control). Under such conditions, the daily ingestion of barium from rice would be 1.8 mg, or 13% of the reference dose.

Pearson correlation analysis was performed for the barium extracted in the acid soluble fraction (F1) and the barium absorbed and accumulated in different plant parts for the two redox potential conditions (Table 1). High and significant correlations were observed for all variables, except for barium concentration and accumulation in the grains under oxidizing conditions. The correlation of the barium concentration in F1 with barium concentration in the leaves under reducing conditions was greater (r = 0.87) than the correlation under oxidizing conditions (r = 0.54). A similar pattern was observed in the roots, with the highest correlation always under reducing conditions. There was no correlation between the barium concentration in F1 and the concentration and accumulation of barium in the grains under oxidizing conditions. However, a high correlation (r = 0.86) was observed for these parameters under reducing conditions.

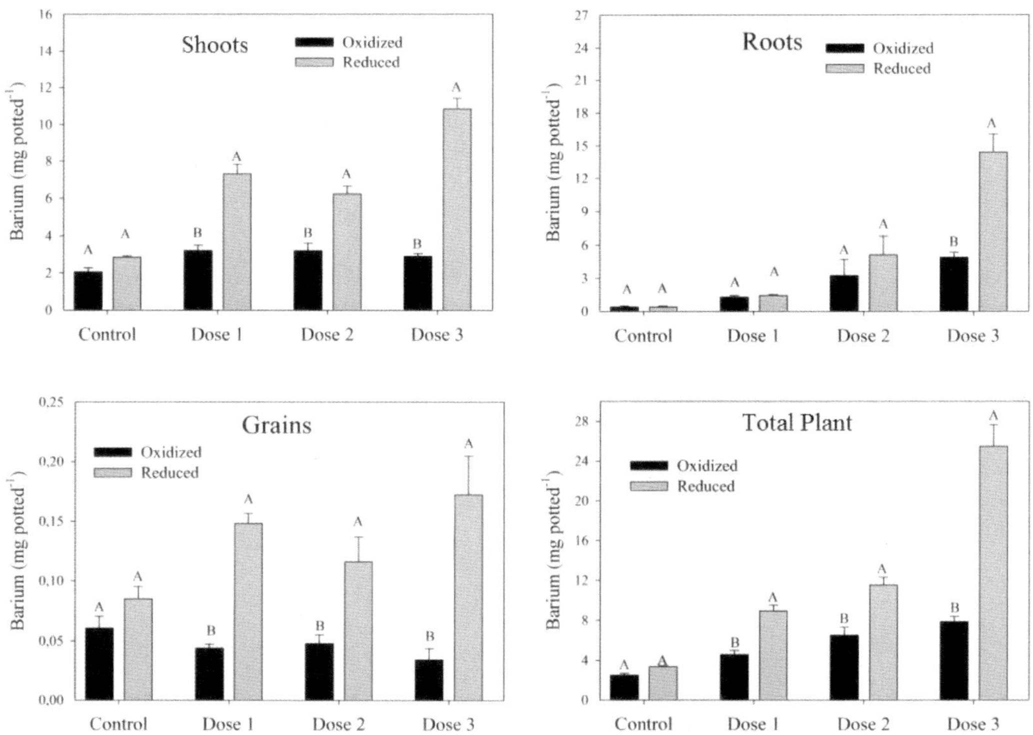

Figure 5. Barium uptake in leaves, roots, grains, and whole rice plants as a function of barium dose and redox potential (oxidized and reduced). Letters (for each barium content) indicate a significant difference ($p < 0.05$). Control: no barium sulfate application; Dose 1: 100 mg kg^{-1}; Dose 2: 300 mg kg^{-1}; Dose 3: 3,000 mg kg^{-1}.

In general, higher correlations were observed under reducing conditions than under oxidizing conditions. These results are consistent with the higher levels of barium found in the F1 fraction under reducing conditions, which promoted the absorption and accumulation of barium in the plants.

Table 1. Pearson correlation coefficients for the barium extracted in the acid soluble fraction (F1) and the barium absorbed and accumulated in different plant parts, as a function of the redox potential.

Correlation		Oxidized	Reduced
Concentration	Shoot	0,54*	0,87**
	Root	0,73**	0,87**
	Grain	-0,23	0,86**
Accumulated	Shoot	0,51*	0,83**
	Root	0,81**	0,88**
	Grain	-0,34	0,51*
	Total Plant	0,89**	0,89**

**significant at 1%; *significant at 5%.

Conclusion

Reducing conditions led to increased levels of barium in the most labile fraction.

The highest dose of barium resulted in a decrease in grain yield.

Saturation promoted the absorption and accumulation of barium in plants, increasing the risk of the introduction of this element into the food chain.

The highest correlation between the soluble acid fraction and the barium accumulation in plants was observed under reducing conditions.

References

ALBERTA ENVIRONMENT. **Soil remediation guidelines for barite**: environmental health and human health. Ottawa: Alberta, 2009.

BALDI, F.; PEPI, M.; BURRINI, D.; KNIEWALD, G.; SCALI, D.; LANCIOTTI, E. Dissolution of barium from barite in sewage sludges and cultures of *Desulfovibrio desulfuricans*. **Applied and Environmental Microbiology**, v. 62, n. 7, p. 2398-2404, 1996.

BODEK, L.; LYMAN, W. J.; REEHI, W. F.; ROSENBLATT, D. H. **Environmental inorganic chemistry**: properties, processes, estimation methods. New York: Pergamon Press, 1988.

BRASIL. Ministério do Meio Ambiente, Conselho Nacional de Meio Ambiente. Resolução n.º 420, de 28 de dezembro de 2009. **Diário Oficial da República Federativa do Brasil**. Brasília, 2009.

CAMARGO, F. N. O.; ARAUJO, G. S.; ZONTA, E. Alterações eletroquímicas de solos alagados. **Ciência Rural**, v. 29, n. 1, p. 171-180, 1999.

CAMARGO, O. A.; ALLEONI, L. R. F.; CASAGRANDE, J. C. Reações dos micronutrientes e elementos tóxicos no solo. In: FERREIRA, M. E.; CRUZ, M. C. P.; RAIJ, B. V.; ABREU, C. A. (Ed.). **Micronutrientes e elementos tóxicos na agricultura**. São Paulo: Potafós, 2001. p. 89-117.

CHAUDHRY, F. M.; WALLACE, A.; MUELLER, R. T. Barium toxicity in plants. **Communications in Soil Science and Plant Analysis**, v. 8, n. 9, p. 795-797, 1977.

CHLOPECKA, A. Assessment of form of Cd, Zn and Pb in contaminated calcareous and gleyed soils in southwest Poland. **Science of the Total Environment**, v. 188, n. 2, p. 253-262, 1996.

DANG, Z.; LIU, C.; HAIGH, M. J. Mobility of heavy metals associated with the natural weathering of coal mine spoils. **Environmental Pollution**, v. 118, n. 3, p. 419-426, 2002.

EMBRAPA-Empresa Brasileira de Pesquisa Agropecuária. **Manual de métodos de análises de solos**. Rio de Janeiro: Centro Nacional de Pesquisa de Solos, 1997.

FAM, M. A.; DUSSEAULT, M. B.; FOOKS, J. C. Drilling in mudrocks: rock behavior issues. **Journal of Petroleum Science and Engineering**, v. 38, n. 3, p. 155-166, 2003.

FILGUEIRAS, A. V.; LAVILLA, I.; BENDICHO, C. Evaluation of distribution, mobility and binding behaviour of heavy metals in surficial sediments of Louro River (Galicia, Spain) using chemometric analysis: a case study. **Science of the Total Environment**, v. 330, n. 1, p. 115-129, 2004.

FRITZ, E. Measurement of cation exchange capacity (CEC) of plant cellwalls by X-ray microanalysis (EDX) in the transmission electron microscope. **Microscopy and Microanalysis**, v. 13, n. 4, p. 233-244, 2007.

GONÇALVES, G. K.; SOUSA, R. O.; VAHL, L. C.; BORTOLON, L. Solubilização dos fosfatos naturais Patos de Minas e Arad em dois solos alagados. **Revista Brasileira de Ciência do Solo**, v. 32, n. 5, p. 2157-2164, 2008.

ISO-International Organisation for Standardisation. **ISO 11466**. Soil quality - extraction of trace elements soluble in aqua regia. Geneva: International Organisation for Standardisation, 1995.

JONES, B. D.; INGLE JR., J. D. Evaluation of redox indicators for determining sulfate-reducing and dechlorinating conditions. **Water Research**, v. 39, n. 18, p. 4343-4354, 2005.

KARTEL, M. T.; KUPCHIK, L. A.; VEISOV, B. K. Evaluation of pectin binding of heavymetal ions in aqueous solutions. **Chemosphere**, v. 38, n. 11, p. 2591-2596, 1999.

KHANAL, S. K.; HUANG, J. C. ORP-based oxygenation for sulfide control in anaerobic treatment of high-sulfate wastewater. **Water Research**, v. 37, n. 9, p. 2053-2062, 2003.

LEITA, L.; DE NOBILI, M.; CESCO, S.; MONDINI, C. Analysis of intercellular cadmium forms in roots and leaves of bush bean. **Journal of Plant Nutrition**, v. 19, n. 3, p. 527-533, 1996.

LIMA, E. N.; MELLO, J. W. V.; SCHAEFER, C. E. G. R.; KER, J. C. Dinâmica da mobilização de elementos em solos da Amazônia submetidos á inundação. **Acta Amazônica**, v. 35, n. 3, p. 317-330, 2005.

LLUGANY, M.; POSCHENRIEDER, C.; BARCELO, L. Assessment of barium toxicity in bush beans. **Archives of Environmental Contamination and Toxicology**, v. 39, n. 4, p. 440-444, 2000.

MAGALHÃES, M. O. L.; AMARAL SOBRINHO, N. M.; MAZUR, N. Use of industrial waste associated with the remedy *Eucalyptus urophylla* of soil contaminated with cadmium and zinc. **Ciência Florestal**, v. 21, n. 2, p. 253-261, 2011a.

MAGALHÃES, M. O. L.; AMARAL SOBRINHO, N. M.; ZONTA, E.; LIMA, L. S.; PAIVA, F. S. D. Barium mobilit in soil treated with barium sulfate under conditions of oxidation and reduction. **Química Nova**, v. 34, n. 9, p. 1544-1549, 2011b.

MARIN, B.; VALLADON, M.; POLVE, M.; MONACO, A. Reproducibility testing of a sequential extraction scheme for the determination of trace metal speciation in a marine reference sediment by inductively coupled plasma-mass spectrometry. **Analytica Chimica Acta**, v. 342, n. 2, p. 91-112. 1997.

MELTON, H. R.; SMITH, J. P.; MARTIN, C. R.; NEDWED, T. J.; MAIRS, H. L.; RAUGHT, D. L. **Offshore discharge of drilling fluids and cuttings** – a scientific perspective on public policy. Rio de Janeiro: IBP, 2000.

MONNIN, C.; DUPRE, C. G.; ELDERFIELD, H.; MOTTL, M. M. Barium geochemistry in sediment pore waters and formation waters of the oceanic crust on the eastern flank of Juan de Fuca Ridge (ODP Leg 168). **Geochemistry, Geophysics, Geosystems**, v. 2, n. 1, p. 15-21. 2001.

NEFF, J. M. Estimation of bioavailability of metals from drilling mud barite. **Integrated Environmental Assessment and Management**, v. 4, n. 2, p. 184-193, 2008.

NOGUEIRA, T. A. R.; DEMELO, W. J.; FONSECA, I. M.; MARQUES, M. O.; HE, Z. Bariun uptake by maize plants as effected by sewage sludge in long-term fild study. **Journal of Hazardous Materials**, v. 181, n. 1, p. 1148-1157, 2010.

PATEL, D. P.; DAS, A.; MUNDA, G. C.; GHOSH, P. K.; BORDOLOI, J. S.; KUMAR, M. Evaluation of yield and physiological attributes of high-yielding rice varieties under aerobic and flood-irrigated management practices in mid-hills ecosystem. **Agricultural Water Management**, v. 97, n. 9, p. 1269-1276, 2010.

PONNAMPERUMA, F. N. Electrochemical changes in submerged soils and the growth of rice. In: IRRI. **Soils and rice**. Los Baños: IRRI, 1978. p. 421-441.

POZEBON, D.; LIMA, E. C.; MAIA, S. M.; FACHEL, J. M. G. Heavy metals contribution of non-aqueous fluids used in offshore oil drilling. **Fuel**, v. 84, n. 1, p. 53-61, 2005.

RODRIGUEZ, L.; RUIZ, E.; ALONSO-AZCARATE, J.; RINCON, J. Heavy metal distribution and chemical speciation in tailings and soils around a Pb–Zn mine in Spain. **Journal of Environmental Management**, v. 90, n. 2, p. 1106-1116, 2009.

SAHUQUILLO, A.; LÓPEZ-SÁNCHES, J. F.; RUBIO, R.; RAURET, G.; THOMAS, R. P.; DAVIDSON, C. M.; URE, A. Use of certified reference material for extractable trace metals to asses sources of uncertainty in the BCR three-stage sequencial extraction procedure. **Analytica Chemical Acta**, v. 382, n. 3, p. 317-327, 1999.

SANTOS, F. S.; MAGALHAES, M. O. L.; MAZUR, N.; AMARAL SOBRINHO, N. M. B. Chemical amendment and phytoestabilization of industrial residue contaminated with Zn and Cd. **Scientia Agricola**, v. 64, n. 5, p. 506-512, 2007.

SEYBOLD, C. A.; MERSIE, W.; HUANG, J.; MCNAMEE, C. Soil redox, pH, temperature, and water-table patterns of a freshwater tidal wetland. **Wetlands**, v. 22, n. 1, p. 149-158, 2002.

SMEDA, A.; ZYRNICKI, W. Application of sequential extraction and the ICP-AES method for study of the partitioning of metals in fly ashes. **Microchemical Journal**, v. 72, n. 1, p. 9-16, 2002.

SUWA, R.; JAYACHANDRAN, K.; NGUYEN, N. T.; BOULENOUAR, A.; FUJITA, K.; SANEOKA, H. Barium toxicity effects in soybean plants. **Archives of Environmental Contamination and Toxicology**, v. 55, n. 3, p. 397-403, 2008.

TEDESCO, M. J.; GIANELLO, C.; BISSANI, C. A.; BOHNEN, H.; VOLKWEISS, S. J. **Análise de solo, plantas e outros materiais**. Porto Alegre: UFRGS, 1995.

USEPA-United States Environmental Protection Agency. **Toxicological review of barium and compounds**. Revised version of previous report dated 1999. In support of summary information on the Integrated Risk Information System (IRIS). Washington, D.C.: U.S. Environmental Protection Agency, 2005.

VEPRASKAS, M. J.; FAULKNER, S. P. Redox chemistry of hydric soils. In RICHARDSON, J. L.; VEPRASKAS, M. J. (Ed.). **Wetland soils**: genesis, hydrology, landscapes, and classification. Boca Raton: Lewis Publishers, 2001. p. 85-106.

YSART, G.; MILLER, P.; CREWS, H. Dietary exposure estimates of 30 elements from the UK total diet study. **Food Additives Contaminants**, v. 16, n. 9, p. 391-403, 1999.

24

Chemical attributes of a savannah Typic Hapludox soil under management systems

Cid Tacaoca Muraishi[1], Marlene Cristina Alves[1*], Aparecido da Silva Júnior[1] and Zigomar Menezes de Souza[2]

*[1]Departamento de Fitossanidade, Faculdade de Engenharia, Universidade Estadual Paulista "Júlio de Mesquita Filho", Av. Brasil, 56, Cx.Postal 31, 15385-000, Ilha Solteira, São Paulo, Brazil. [2]Departamento de Água e Solo, Faculdade de Engenharia Agrícola, Universidade Estadual de Campinas, Campinas, São Paulo, Brazil. *Author for correspondence. E-mail: mcalves@agr.feis.unesp.br*

ABSTRACT. This study was conducted at the Experimental Station belonging to UNESP Engineering University, Ilha Solteira Campus, based in Selvíria, Mato Grosso do Sul State, Brazil, with the aim of evaluating chemical alterations in an Oxisol after being managed for two years with organic and/or chemical fertilization and different tillage systems during the agricultural years of 2004/2005 and 2005/2006. The treatments were: conventional-tillage; chisel tillage and no-tillage; the fertilization treatments were: control (no fertilization); chemical fertilization (300 kg ha^{-1} from the 20-00-20); organic fertilization (cattle manure - 20 Mg ha^{-1}); organic + ½ the recommended chemical fertilization for the used crop; 20 and 30 Mg ha^{-1} of sewage sludge. Soybean was used in the first year and sorghum in the next year, evaluating the soil chemical attributes in four layers. The soil chemical attributes were changed in the first year; the organic fertilization, sewage sludge and organic+chemical fertilization were efficient to change the chemical attributes; the sewage sludge was more efficient in soil P recuperation and, the no-tillage system contributed to soil K increase.

Keywords: cattle manure, sewage sludge, soil management, no-tillage.

Introduction

Savannah soils usually provide low yields due to the high Al saturation, low contents of most of the mineral nutrients that are essential to the plants development, low organic matter content, leading to low CEC and high phosphorus fixation (FAQUIN et al., 2000). The savannah is equivalent to a quarter of the national territory. It is estimated that most of this area can be used for agricultural purposes. One of the challenges of establishing crops in savannah areas is the fertilization management of the economically important cultures under the no-tillage system, especially when it comes to nitrogen, which is largely required by these crops.

The agricultural machinery which is available to prepare the soil leads to alterations in the soil chemical, physical and biological attributes. Each machine prepares the soil in a different way and alters these attributes differently as a consequence (VÁZQUEZ et al., 2006; TORMENA et al., 2008). The intensity in which the soil is revolved and the crops residues are incorporated causes changes in the organic matter content, in the cation exchange capacity, in the pH, in the ion dynamic and in the soil aggregation (TRANNIN et al., 2008).

The no-tillage system is based on crop rotation and is characterized by growing in a soil which is both unprepared and covered by straw for unlimited time. This system usually demands the use of no-tillage seeders to cut the straw, open a small furrow to place the fertilizer and the seeds, not to mention all the conservationist procedures which allow good quality straw to be kept in a proper amount in order to improve the sustainability of the whole ecosystem.

In order to minimize the river pollution, the domestic and industrial wastes have been treated and the outcome is caked sewage sludge. The handling and the final disposal of the sewage sludge are the process most important phases. The agricultural use of sewage sludge as nutrients and organic matter sources is an alternative method for the final disposal of this waste from which some benefits can be gained (ALVES et al., 2006).

The agricultural use of sewage sludge has become a sensible and promising way of disposing it once it is a good nutrient source, especially regarding phosphorus and nitrogen (BINDER et al., 2002). Phosphorus is considered one of the hindering factors to agricultural yields, due to the adsorption and/or fixation processes in soil colloids (CORRÊA et al., 2005). Sewage sludge usually contains low potassium content due to the obtaining process, the opposite of what happens to phosphorus, which is present in a high content and, most importantly, approximately 80% of it is available in the first application year (SIMONETE et al., 2003). Another important factor is the sewage sludge decomposition in the soil which allows a better uptake of the nutrients by the plants as they are slowly released via the organic matter mineralization process (DEBOSZ et al., 2002).

According to Borges and Coutinho (2004), after the sewage sludge application, there was an increase in the organic matter content and in the pH of the soil without correctives. Nascimento et al. (2004), evaluating the chemical changes in a soil under corn and beans crops where sewage sludge was used, observed an increase in the organic matter, total nitrogen, phosphorus, sodium, calcium and magnesium contents though potassium and sodium contents were not altered.

The manure incorporation in the agricultural soils has been done since the beginning of the agriculture. Until half of the 19th century, when the advance of the fertilizer industry took place, the use of manure was the main nutrient source for the plants. Galvão et al. (2008) mentioned that the manure composition varies and is influenced by many factors, such as animal species, breed, age, feed and the way the raw material (manure) is treated, and that cattle manure has 57.1% OM, 1.67% N, 0.86% P_2O_5, 1.37% K_2O and C/N ratio of 32/1.

For Sharpley et al. (2004) and Souto et al. (2005), the benefits of using manure can be ranked as follows: improvement of the soil physical attributes and of the nutrient supply; increase in the organic matter content, leading to better water infiltration as well as an increase in the cation exchange capacity.

Silva et al. (2004), when studying the manure use for corn crops observed that the pH, the exchangeable calcium and magnesium contents and the organic matter content were higher for depths of 0.00-0.20 m than for depths of 0.20-0.40 m. There was an increase in the phosphorus content and there was no difference in the potassium content. Silva et al. (2007), quoted by the same author, also observed linear increases in the potassium and phosphorus contents, with an increase of the manure dose, but he also observed increased pH and Mg and organic matter contents.

Manure was widely used in the past; however, with the development of the chemical fertilizers, the interest in the organic ones decreased. Currently, due to the environmental concern, the interest in the use of manure, that is, sustainable agriculture has increased. Its chemical function is shown by its ability of interacting with metals, oxides and metal hydroxides and of forming organic-metallic compounds acting as N, P and S stock. In acidic soils under savannah, the organic matter effect on the availability of the applied phosphorus has a residual characteristic (MOREIRA et al., 2001; ANDRADE et al., 2002).

Based on what was mentioned, this research work had the objective of studying the changes in the chemical attributes of a Typic Hapludox in a savannah in Mato Grosso do Sul State, after two years under different tillage systems and managements with organic and/or chemical fertilization.

Material and methods

The experiment was conducted at the Experimental Station belonging to UNESP Engineering University, Ilha Solteira Campus, based in Selvíria, Mato Grosso do Sul State, Brazil. The experimental area has 51°22'W and 20°22'S as geographic coordinates and 335 m altitude. The climate is Aw, defined as humid tropical with rains in the summer and drought in the winter, temperature of 24.5°C, precipitation of 1,232 mm and average relative moisture of 64.8%. The studied area soil was classified as Latossolo Vermelho distrófico, according to Brazilian Soil Classification System (DEMATTÊ, 1980; EMBRAPA, 2006), loam texture, moderately plain to slightly ondulated landscape, corresponding to Typic Hapludox (SOIL SURVEY STAFF, 1975).

The area had been under pasture with *Brachiaria decumbens* and intensive grazing for twenty years. During this period, the soil was not corrected or fertilized. No renewal, recovering or maintenance treatment was previously performed in the studied pasture. In November 2001, cotton and corn were grown under no-tillage and conventional tillage systems (harrow and leveling bars). The soil correction was done in order to increase the percent base saturation up to 70%. The chemical fertilization was performed according to the results from the soil chemical analysis and 300 kg ha^{-1} of the 20-0-20 formula was used for both crops (cotton and corn). From April 2002 to October 2003 the area remained fallow, which made the *Brachiaria* grow back again spontaneously.

In October 2003, a soil sampling was performed in order to have another chemical characterization of the experimental area. Composed samples from 20 simple samples were collected from three different soil layers: 0.00-0.05; 0.05-0.10 and 0.10-0.20 m. The analysis results can be observed in Table 1 and were used to determine the P and K amounts to be used for sowing soy according to Mascarenhas and Tanara's recommendations (1996).

The used experimental design was the randomized blocks, in a (3 x 6) factorial design, with 3 tillage systems (conventional – harrow and leveling bars; chisel tillage – subsoil plow and light harrow and, no-tillage) and 6 fertilization treatments (control – not fertilized; chemical fertilization – 300 kg ha^{-1} of the 20-0-20 formula; organic fertilization – cattle manure 20 Mg ha^{-1}; organic fertilization – cattle 20 Mg ha^{-1} + ½ the recommended chemical fertilization; 20 Mg ha^{-1} of sewage sludge and, 30 Mg ha^{-1} of sewage sludge), with 4 blocks. The treatments were applied in 2003 and there were no other applications in the following years.

A hoe was used to establish the organic treatments in order to spread the material on the soil surface; after that, the disc harrow was used. This operation was necessary since the sewage sludge must be incorporated to the soil for sanitary reasons.

On December 27, 2003, IAC 19 soy was sowed with a 0.45-m distance between lines and 20 seeds per meter. In the 2004/2005 yield, foraging sorghum (hybrid BR 700) was sowed with 0.45-m between lines on December 20, using 8.0 kg ha^{-1}. Sorghum was chosen due to its resistance to the drought.

The sewage sludge, predominantly from domestic waste with 85.42% moisture, was obtained from SANEAR, in Araçatuba, Sao Paulo State. The heavy metals content is low and for some elements it is none. The used cattle manure came from the cattle confinement area on a farm belonging to Damha Animal Nutrition Group. Table 2 presents the chemical attributes of the used sewage sludge and cattle manure, according to the proposed by Sarruge and Haag (1974).

The soil samplings were performed after the soy and sorghum harvest. A composed sample from each plot was collected from 5 simple samples from the layers 0.00-0.05, 0.05-0.10 and 0.10-0.20 m. The soil chemical analyses were performed for P, O.M., pH, K, Ca, Mg, H+Al, Al; SB, CEC, and %BS were calculated according to Raij and Quaggio's (1983) methodology.

A variance analysis and the means were compared using Tukey test at 5% probability.

Table 1. Soil chemical characterization of the experimental area before the experiment establishment in October 2003.

Layer (m)	Resin P (mg dm^{-3})	O.M. (g dm^{-3})	pH ($CaCl_2$)	K	Ca	Mg	H+Al	Al	SB	CEC	BS
				(mmol$_c$ dm^{-3})							%
0.00-0.05	10	18	5.6	0.8	20	7	15	0	27.6	42.6	65
0.05-0.10	13	17	5.3	0.2	15	6	16	0	21.0	37.0	57
0.10-0.20	5	15	5.0	0.2	10	5	18	1	16.0	34.0	47

O.M.= organic matter; SB= bases sum; CEC= exchange cation ; BS= bases saturation.

Table 2. Chemical attributes of sewage sludge and cattle manure.

O.M.	N	P	K	Ca	Mg	S	B	Cu	Fe	Mn	Zn	Moisture
(g dm^{-3})	g kg^{-1}						(mg kg^{-1})					kg kg^{-1}
Sewage sludge												
200	71	19	15	11	3	8	16.4	160.1	960.6	115.7	583.5	0.85
Cattle manure												
252	10	5	15	15	3	2	1.0	12.0	7217.0	99.0	51.0	0.33

Results and discussion

According to Table 3, it was observed significance for P only in the depth of 0.05-0.10 m, where the treatments with sewage sludge provided better results; the same was observed by Galdos et al. (2004) in the first years of their experiment. The sewage sludge application works as a correcting fertilization, increasing the initial low contents to very high ones. However, its application in order to increase soil P up to very high contents must be carefully performed in areas where there is surface drainage, as it could lead to the contamination of water courses (TRANNIN et al., 2005). According to Silva et al. (2007) and Sharpley et al. (2004), continuous applications of organic compounds to the soil can probably cause an accumulation of nutrients and their vertical migration as a consequence.

As for the pH, it was observed that the sewage sludge treatments provided the lowest numbers not differing from the chemical and organic fertilization treatments (Table 3). According to Andrade et al. (2002), the organic matter biodegradation can also cause transient acidification in the soil, due to the production of organic acids. According to Oliveira et al. (2002), pH alterations in the soil, caused by the sewage sludge incorporation, are directly related to the type of treatment given to the waste; therefore, only alkaline sewage sludge tends to elevate soil pH, regardless of the nature and organic changes of these materials. Galvão et al. (2008), when studying the nutrient accumulation in sandy soils fertilized with cattle manure, found contrary results, that is, the organic fertilization provided a pH increase.

Differences among the treatments concerning the organic matter were not observed in the studied depths (Table 3). This is due to the waste degradation rate in the soil which, according to Melo et al. (2004)

The time of degradation of organic matter from sewage sludge is very short compared to other organic materials. The results weren´t the same as observed by De Maria et al. (2007), that is, for the authors there was increase in the organic-C content even when applying high doses of the waste.

In all the studied layers, we observed that the use of sewage sludge led to lower K contents when compared to the chemical fertilization, due to the plants absorption and leaching (Table 3). Similar results were observed by Rocha et al. (2004) when studying eucalyptus; K contents decreased 36 months after the application. Trannin et al. (2005), evaluating the use of an industrial biosolid for corn crops, observed low K contents, which can be justified by the Na increase in the soil and plant, considering that this element has antagonistic effect on K due to the competition among these ions for the absorption sites.

As for the potential acidity, it was observed that the sewage sludge use increased it (Table 3) in all soil layers, which can be due to higher organic matter mineralization (COLODRO; ESPINDOLA, 2006). Similar results were observed by Galdos et al. (2004) as the experiments with corn on a soil treated with sewage sludge for two years showed an acidifying effect of the waste, decreasing pH in the 0.00-0.05 m layer from 5.8 to 5.0 and increasing the potential acidity; there were no changes for pH or potential acidity in other depths.

Table 3. Soil chemical attributes regarding treatments and the studied soil layers. Selvíria, Mato Grosso do Sul State, 2004/2005 and 2005/2006 seasons.

Treatments	Resin P ($mg\ dm^{-3}$)	O.M. ($g\ dm^{-3}$)	pH ($CaCl_2$)	K	Ca	Mg	H+Al	Al	SB	CEC	BS
				($mmol_c\ dm^{-3}$)							%
Layer 0.00-0.05 m											
Control	7.9 c	19.40	5.3 ab	1.1 c	20.1	8.3	16.9 ab	0.8	29.5	40.4	63
Chemical F.	12.0 c	19.80	5.4 ab	1.3 abc	21.7	7.7	16.5 ab	0.6	41.3	47.5	62
Organic F.	9.7 c	18.70	5.3 ab	1.4 ab	18.5	8.2	16.4 ab	0.6	36.8	44.7	59
Org. + chem.	14.8 bc	18.50	5.5 a	1.5 a	22.5	7.1	15.4 b	0.5	41.8	46.7	64
20 Mg ha^{-1*}	21.2 ab	18.80	5.1 b	1.1 bc	19.1	7.8	17.9 a	0.7	36.1	46.1	59
30 Mg ha^{-1*}	23.0 a	19.40	5.1 b	1.3 abc	21.1	8.6	18.1 a	0.8	42.0	48.4	59
Layer 0.05-0.10 m											
Control	11.9 b	17.40	5.1 b	0.7 cd	18.9	7.0	17.5 ab	0.9	26.7	44.0	59
Chemical F.	11.0 b	17.10	5.3 ab	0.9 abc	22.1	7.3	16.8 ab	0.8	30.3	47.2	60
Organic F.	12.4 b	16.80	5.3 ab	1.0 ab	18.6	7.5	17.0 ab	0.9	27.1	45.2	58
Org. + chem.	12.8 b	17.20	5.6 a	1.1 a	23.9	6.8	15.0 b	0.8	31.8	46.9	64
20 Mg ha^{-1}	22.7 a	16.10	5.1 b	0.6 d	19.6	6.4	18.3 a	0.7	26.9	45.0	57
30 Mg ha^{-1}	23.7 a	17.20	5.0 b	0.8 bcd	19.1	6.8	18.8 a	1.1	26.8	45.6	54
Layer 0.10-0.20 m											
Control	8.2	13.50	5.1	0.5 c	16.5	6.3	16.8	0.8	23.2	40.6	55
Chemical F.	8.2	14.20	5.3	0.8 abc	17.1	6.5	17.7	0.7	24.4	41.0	57
Organic F.	8.8	12.90	5.2	0.9 ab	15.8	6.5	16.5	0.7	23.3	39.8	56
Org. + chem.	8.9	13.40	5.4	1.1 a	18.3	6.9	16.0	0.8	26.4	42.3	60
20 Mg ha^{-1}	11.4	14.20	5.1	0.6 bc	15.2	5.7	17.9	0.9	21.6	39.4	53
30 Mg ha^{-1}	11.7	12.90	5.2	0.7 bc	17.7	5.9	16.8	0.8	22.8	39.6	54

Means followed by the same letter in the column do not statistically differ from each other according to Tukey test at 5% probability. *Sewage sludge.

In Table 4, it is possible to observe that for the nutrients which showed significance, the highest results were obtained in the year of 2004. As for the effect of the interaction sampling year x fertilization (Table 5), significance was only observed for the 0.00-0.05 m layer when the years were compared for the treatments with sewage sludge and, in 2005, the results were higher, which was similar to the observed by Galdos et al. (2004) as, according to him, there was an improvement in the chemical quality of the soil caused by the application of sewage sludge only in the second year after using the waste. Concerning the fertilization, significance was only observed in 2005, when the treatments with sewage sludge showed the best results.

For the 0.10-0.20 m layer, the year of 2004 presented the best results and when the fertilizations were compared again, the treatments with sewage sludge showed the best results with highest values for P soil content (Table 4). Similar results were presented by Barbosa et al. (2002), when, after applying sewage sludge for two years in a row, it was observed that the treatments with the waste led to significant improvement of the soil chemical attributes studied. Colodro and Espindola (2006) also observed an increase in the P soil content in an area which had been treated with biosolid for two years. Although the continuous biosolid application increased the total P soil content, approximately 40% of which was part of the organic P being very stable in the soil and available to the plants.

In regard to magnesium, there was a decrease in its content throughout the year (Table 4), which was similar to the observed by Rocha et al. (2004); the authors noticed a decrease in the Mg content after 32 months from the application day due to the low Mg content in sewage sludge. However, Trannin et al. (2008) evaluated the chemical and physical attributes of a soil treated with industrial sewage sludge where corn was being grown and observed a significant increase in the magnesium soil content, due to the waste characteristics.

There was a pH decrease and increases in the potential acidity (H+Al) and in the exchangeable Al and, therefore, an increase in the Al saturation (Table 3) as a consequence of the sewage sludge application. It can be partly explained by the fact that this waste is produced without adding limestone, having therefore low correcting efficiency. The organic N mineralization and consequent nitrification could also have contributed to the soil acidification. Similar results were observed by Galdos et al. (2004), Trannin et al. (2005) and Colodro and Espindola (2006).

For the effect of the interaction sampling year x tillage concerning K contents (Table 6), it was observed that the significance only occurred for the no-tillage system. Being the year of 2005 the period when the best results were observed, the no-tillage system was proved to be the best when compared to the others. Nevertheless, the results were different from the observed by Trannin et al. (2005), who observed a decrease in K contents, due to the application of sewage sludge, as there was an increase in Na contents in the soil and also in the plants. Therefore, this increase could have been provided by the no-tillage system, once it leads to a better nutrient cycling in the soil where straw from the previous crop is preserved.

Table 4. Soil chemical attributes for different soil layers regarding the years. Selvíria, 2004/2005 and 2005/2006 seasons.

Treatments	Resin P ($mg\ dm^{-3}$)	O.M ($g\ dm^{-3}$)	pH ($CaCl_2$)	K	Ca	Mg	H+Al	Al	SB	CEC	BS
				($mmol_c\ dm^{-3}$)							%
						Layer 0.05-0.10 m					
2004	12.2 a	19.0	5.4 a	1.7 a	29.3 a	9.1 a	17.5 a	1.0 a	39.3 a	55.7 a	70.3 a
2005	17.3 b	19.2	5.2 b	0.9 b	11.8 b	6.8 b	16.3 b	0.4 b	20.1 b	38.6 b	52.8 b
						Layer 0.05-0.10 m					
2004	17.8 a	18.2 a	5.4 a	0.7 b	28.8 a	8.8 a	16.8	1.1 a	38.4 a	55.1 a	69.2 a
2005	13.7 b	15.7 b	5.1 b	0.9 a	11.9 b	5.2 b	17.6	0.7 b	18.2 b	36.2 b	49.1 b
						Layer 0.10-0.20 m					
2004	11.2 a	14.60 a	5.2	0.9 a	22.7 a	8.0 a	17.3	1.0 a	31.6 a	48.9 a	64.6 a
2005	7.9 b	12.40 b	5.1	0.7 b	10.8 b	4.6 b	16.1	0.7 b	15.6 b	32.1 b	47.3 b

Means followed by the same letter in the column do not statistically differ from each other according to Tukey test at 5% probability.

Table 5. Effect of the interaction between sampling year x fertilization for P contents in different soil layers. Selvíria, 2004/2005 and 2005/2006 seasons.

			Layer 0.00-0.05 m			
Years	Control	Chemical F.	Org. F.	Org. + Chem.	20 Mg ha^{-1}*	30 Mg ha^{-1}*
2004	6.6	11.3	10.3	16.0	15.0 b	14.0 b
2005	9.1 B	12.6 B	9.0 B	13.7 B	27.0 a A	32.1 a A
			Layer 0.10-0.20 m			
2004	11.3 a	9.6	12.6 a	12.0 a	12.2	12.5
2005	5.1 b C	6.6 BC	4.8 b C	5.9 b C	11.3 AB	10.3 A

Means followed by the different letters, capital letters in the line and minuscule letters in the column, differ from each other according to Tukey test at 5% probability. *Sewage sludge.

Table 6. Effect of the interaction between sampling year x tillage for potassium contents in the 0.05-0.10 m layer. Selvíria, 2004/2005 and 2005/2006 seasons.

	Layer 0.05-0.10 m		
Year	Conventional	Chisel Tillage	No-tillage
2004	0.8 a	0.8 a	0.8 b
2005	0.8 aB	0.8 aB	1.2 aA

Means followed by different letters, capital letters in the line and minuscule letters in the column, differ from each other according to Tukey test at 5% probability.

Conclusion

The chemical attributes of Typic Hapludox were modified in the first year after fertilizations. Manuring, sewage sludge fertilization and the combination manure + chemical fertilization were efficient in modifying the chemical attributes of the studied soil. Sewage sludge was the most efficient treatment for soil P content and the no-tillage system contributed for the increase of the soil K content.

References

ALVES, M. C.; GONZALEZ, A. P.; COLODRO, G.; PERECIN JÚNIOR, H.; VÁQUEZ, E. V. Influence of Biosolids Rate on Chemical Properties of an Oxisol in São Paulo, Brazil. **Communications in Soil Science and Plant Analysis**, v. 37, n. 15-20, p. 2481-2493, 2006.

ANDRADE, A.T.; FERNANDES, L. A.; FAQUIN, V. Organic residue, limestone, gypsum, and phosphorus adsorption by lowland soils. **Scientia Agricola**, v. 59, n. 2, p. 349-355, 2002.

BARBOSA, G. M. C.; TAVARES FILHO, J.; FONSECA, I. C. B. Propriedades químicas de um Latossolo Vermelho eutroférrico após aplicação por dois anos consecutivos de lodo de esgoto. **Acta Scientiarum. Agronomy**, v. 24, n. 5, p. 1501-1505, 2002.

BINDER, D. L.; DOBERMANN, A.; SANDER, D. H.; CASSMAN, K. G. Biosolids as nitrogen source for irrigated maize and rainfed sorghum. **Soil Science Society of America Journal**, v. 66, n. 2, p. 531-543, 2002.

BORGES, M. R.; COUTINHO, E. L. M. Metais pesados do solo após aplicação de biossólido: II - Disponibilidade. **Revista Brasileira de Ciência do Solo**, v. 28, n. 3, p. 557-568, 2004.

COLODRO, G.; ESPINDOLA, C. R. Alterações na fertilidade de um latossolo degradado em resposta à aplicação de lodo de esgoto. **Acta Scientiarum. Agronomy**, v. 28, n. 1, p. 1-5, 2006.

CORRÊA, R. M.; NASCIMENTO, C. W. A.; SOUZA, S. K. S.; FREIRE, F. J.; SILVA, G. B. Gafsa rock phosphate and triple superphosphate for dry matter production and P uptake by corn. **Scientia Agricola**, v. 62, n. 2, p. 159-164, 2005.

DEBOSZ, K.; PETERSEN, S. O.; KURE, L. K.; AMBUS, P. Evaluating effects of sewage sludge and household compost on soil physical, chemical and microbiological properties. **Applied Soil and Ecology**, v. 19, n. 3, p. 237-248, 2002.

DE MARIA, I. C.; KOCSSI, M. A.; DECHEN, S. C. F. Agregação do solo em área que recebeu lodo de esgoto. **Bragantia**, v. 66, n. 2, p. 291-298, 2007.

DEMATTÊ, J. L. I. **Levantamento detalhado dos solos do campus experimental de Ilha Solteira**. Piracicaba: Esalq/USP, 1980. 114 p. (Mimeografado).

EMBRAPA-Empresa Brasileira de Pesquisa Agropecuária. Ministério da Agricultura, Pecuária e Abastecimento. **Sistema Brasileiro de Classificação de Solos**. 2. ed. Rio de Janeiro: Embrapa Solo, 2006.

FAQUIN, V.; LIMA, D. V.; FURTINI NETO, A. E.; MORAIS, A. R.; CURI, N.; HIGA, N. T. Nutrição mineral do braquiarão e da soja cultivados em Latossolos sob Cerrado da região de Cuiabá-MT. **Ciência e Agrotecnologia**, v. 24, n. 1, p. 110-117, 2000.

GALDOS, M. V.; DE MARIA, I. C.; CAMARGO, O. A. Atributos químicos e produção de milho em um latossolo vermelho eutroférrico tratado com lodo de esgoto. **Revista Brasileira de Ciência do Solo**, v. 23, n. 3, p. 569-577, 2004.

GALVÃO, S. R. S.; SALCEDO, I. H.; OLIVEIRA, F. F. Acumulação de nutrientes em solos arenosos adubados com esterco bovino. **Pesquisa Agropecuária Brasileira**, v. 43, n. 1, p. 99-105, 2008.

MASCARENHAS, H. A. A.; TANARA, R. T. Crescimento em vasos, de cultivares de soja e de trigo em função da saturação de alumínio. **Scientia Agricola**, v. 52, n. 2, p. 257-262, 1996.

MELO, V. P.; BEUTLER, A. N.; SOUZA, Z. M.; CENTURION, J. F.; MELO, W. J. Atributos físicos de Latossolos adubados durante cinco anos com biossólido. **Pesquisa Agropecuária Brasileira**, v. 39, n. 1, p. 67-72, 2004.

MOREIRA, S. G.; KIEHL, J. C.; PROCHNOW, L. I.; PAULETTI, V. Calagem em sistema de semeadura direta e efeitos sobre a acidez do solo, disponibilidade de nutrientes e produtividade de milho e soja. **Revista Brasileira de Ciência do Solo**, v. 25, n. 1, p. 71-81, 2001.

NASCIMENTO, C. W. A.; BARROS, D. A. S.; MELO, E. E. C.; OLIVEIRA, A. B. Alterações químicas em solos e crescimento de milho e feijoeiro após aplicação de lodo de esgoto. **Revista Brasileira Ciência do Solo**, v. 28, n. 2, p. 385-392, 2004.

OLIVEIRA, F. C.; MATTAZZO, M. E.; MARCIANO, C. R.; ROSSETTO, R. Efeitos de aplicações sucessivas de lodo de esgoto em um Latossolo Amarelo distrófico cultivado com cana-de-açúcar. **Revista Brasileira de Ciência do Solo**, v. 26, n. 2, p. 505-519, 2002.

RAIJ, B. V.; QUAGGIO, J. A. **Métodos de análises de solo para fins de fertilidade**. Campinas: Instituto Agronômico de Campinas, 1983. (Boletim Técnico n. 81).

ROCHA, G. N.; GOLÇALVES, J. L. M.; MOURA, I. M. Mudanças da fertilidade do solo e crescimento de um povoamento de Eucalyptus grandis fertilizado com biossólido. **Revista Brasileira de Ciência do Solo**, v. 28, n. 4, p. 623-639, 2004.

SARRUGE, J. R.; HAAG, H. P. **Análises químicas em plantas**. Piracicaba: Esalq, 1974. 56p. (Mimeografado).

SHARPLEY, A. N., MCDOWELL, R. W.; KLEINMAN, P. J. A. Amounts, agricultural drainage: Historical perspective and current research. forms, and solubility of phosphorus in soils receiving manure. **Soil Science Society of America Journal**, v. 68, n. 8, p. 2048-2057, 2004.

SILVA, T. O.; MENEZES, R. S. C.; TIESSEN, H.; SAMPAIO, E. V. S. B.; SALCEDO, I. H.; SILVEIRA, L. M. Adubação orgânica da batata com esterco e/ou *Crotalaria juncea*. I - Produtividade vegetal e estoque de nutrientes no solo em longo prazo. **Revista Brasileira de Ciência do Solo**, v. 31, n. 1, p. 39-49, 2007.

SILVA, J.; SILVA, P. S. L.; OLIVEIRA, M.; SILVA, K. M. B. Efeito de esterco bovino sobre os rendimentos de espigas verdes e de grãos de milho. **Horticultura Brasileira**, v. 22, n. 2, p. 326-331, 2004.

SIMONETE, M. A.; KIEHL, J. C.; ANDRADE, C. A.; TEIXEIRA, C. F. A. Efeito do lodo de esgoto em um Argissolo e no crescimento e nutrição de milho. **Pesquisa Agropecuária Brasileira**, v. 38, n. 10 p. 1187-1195, 2003.

SOIL SURVEY STAFF. **Soil Taxonomy**. A basic system of soil classification for making and interpreting soil surveys. USDA Agric. Handbook 436, 1975.

SOUTO, P. C.; SOUTO, J. S.; SANTOS, R. V.; ARAÚJO, G. T.; SOUTO, L. S. Decomposição de estercos dispostos em diferentes profundidades em área degradada no semi-árido da Paraíba. **Revista Brasileira de Ciência do Solo**, v. 29, n. 1, p. 125-130, 2005.

TORMENA, C. A.; SILVA, A. P.; IMHOFF, S. D. C.; DEXTER, A. R. Quantification of the soil physical quality of a tropical oxisol using the S index. **Scientia Agricola**, v. 65, n. 1, p. 56-60, 2008.

TRANNIN, I. C. B.; SIQUEIRA, J. O.; MOREIRA, F. M. S. Avaliação agronômica de um biossólido industrial para a cultura do milho. **Pesquisa Agropecuária Brasileira**, v. 40, n. 3, p. 261-269, 2005.

TRANNIN, I. C. B.; SIQUEIRA, J. O.; MOREIRA, F. M. S. Atributos químicos e físicos de um solo tratado com biossólido industrial e cultivado com milho. **Revista Brasileira de Engenharia Agrícola e Ambiental**, v. 12, n. 3, p. 223-230, 2008.

VÁQUEZ, E. V.; MIRANDA, J. G. V.; ALVES, M. C.; GONZALEZ, A. P. Effect of tillage on fractal indices describing soil surface microrelief of a Brazilian Alfisol. **Geoderma**, v. 134, p. 428-439, 2006.

25

The effects of industrial waste from enzyme production on pasture growth and soil chemical properties

André Faé Giostri[1*], Luis Fernando Montrucchio Bond[1], Antônio Carlos Vargas Motta[1], Ida Chapaval Pimentel[2], Paulo César de Faccio Carvalho[3] and Jair Alves Dionísio[1]

*[1]Programa de Pós-graduação em Ciência do Solo, Universidade Federal do Paraná, Rua dos Funcionários, 1540, 80035-050, Curitiba, Paraná, Brazil. [2]Setor de Ciências Biológicas, Departamento de Patologia Básica, Centro Politécnico, Universidade Federal do Paraná, Curitiba, Paraná, Brazil. [3]Programa de Pós-graduação em Produção Vegetal, Universidade Federal do Paraná, Curitiba, Paraná, Brazil. *Author for correspondence. E-mail: andregiostri@hotmail.com*

ABSTRACT. Farmers have used liquid waste from the enzyme industry as fertilizer. To determine the impact of liquid waste from enzymes (LWEs) on soil properties, plant growth and nutrient content in pasture, an experiment was conducted with five different rates of LWE application (0, 45, 90, 135 and 180 m^3 ha^{-1}) to a Cambisol in August 2006. In August 2007, 60 m^3 ha^{-1} of LWE was applied in every treatment. Soil was sampled on four different dates and at four different depths (0-10, 10-20, 20-40 and 40-60 cm) to verify changes in its chemical properties. Soil was improved in terms of acidity and availability of Ca^{2+}, Mg^{2+}, K^+ and P in the 0-10 cm layer, indicating that the effect of LWE is corrective concerning these attributes. There was also improvement in pasture quality through increased levels of N, P and K in plants during the first year of waste application. The use of larger doses of LWE led to increases in productivity (42.6 kg dry weight m^{-3} residues). These results indicate that the application of LWE is feasible for the fertilization of this grassland containing low-fertility acidic Cambisol.

Keywords: soil acidity, forage, carbon, fertilizater.

Introduction

Industrial waste has been used in agricultural applications as a source of nutrients, such as vinasse in sugarcane areas (CANELLAS et al., 2003). Long-term use was shown to be feasible when low concentrations of heavy metals and other organic contaminants were present in the waste (CAVALLET et al., 2006).

Among the soil systems to which industrial waste has been applied, pastures are the most appropriate because they provide easy access to farm equipment year round and are found throughout the Brazilian territory. Pasture ecosystems occupy 173 million hectares of land in Brazil, of which 5.7 million hectares are natural pastures located in Paraná State (IBGE, 2010). In most Brazilian pastures, the soil has low natural fertility; it is necessary to constantly replace nutrients in these soils. Pasture productivity and quality decrease over time when no nutrient replacement takes place. This reduction causes a subsequent decrease in the performance of pasture-raised cattle (COSTA et al., 2009).

In many regions of the world, pasture fertilization occurs through the use of organic waste that is usually generated by confined animals. This type of fertilization is a cheap way for nutrients to be returned to the field and can improve pasture productivity (POWER; SUTTON, 1996). The indiscriminate use of organic waste can increase nutrient levels in the soil to the point of causing environmental problems (SHARPLEY et al., 1993; KINGERY et al., 1994; POWER; SUTTON, 1996). It is necessary to consider local legislation when applying organic waste as a fertilizer; many countries, such as Brazil, prohibit the application of any residue on pastures that contains animal source products (BRASIL, 2001).

It has become more common in developing countries over the last decade for waste from industries that use grain and other materials from fields as an alternative to using animal, waste. These industries can generate large amounts of waste. Returning these industrial wastes to the field might provide a good nutrient source for pastures and has an environmental appeal. Waste derived from microbial fermentation processes that yield industrial enzymes contains N, P, K, Ca, Mg and micronutrients (CAVALETT et al., 2006). Because LWE (liquid waste from enzymes) is neutralized with Ca and Mg oxides, it can increase soil pH and other associated soil chemical properties (CAVALETT et al., 2006).

Application of LWE is an effective way to dispose of industrial waste and may increase the productivity of pastures via a beneficial effect on soil nutrient status.

Using LWE as a fertilizer is an effective way to dispose of industrial waste and may increase the productivity of pastures via a beneficial effect on soil nutrient status.

An experiment was performed to determine the impact of LWE on soil properties and plant growth and nutrition in pasture.

Material and methods

Location

The experiment was carried out in the municipality of Fazenda Rio Grande, in the Metropolitan Region of Curitiba. This municipality is located on the First Plateau of Paraná with the coordinates 24°16'56.71" south and 51°28'32.91 west.

According to the Köppen classification, the climate of this region is classified as Cfb. The climate is subtropical mesothermal and humid with no dry season and cool summers. The temperature in the coldest months ranges from -3 to 18°C; in the hottest months, temperature ranges from 10 to 34°C. The annual rainfall is 1600-1800 mm. The relative air humidity is 70-80% (IAPAR, 2000).

Experimental design

The area exhibits an undulated relief at a mean altitude of 900 m (IAPAR, 1994). It is covered with non-fertilized native field, where grass stands are predominant as an extensive pasture system to raise cattle and horse.

This experiment was carried out over 2 years in an area of nearly 2 hectares, where hermarthria (*Hemarthria altissima*) grass seedlings were introduced in rows. The experimental design was a random block with four blocks. Each block included five plots (7 x 15 m) corresponding to different waste application rates, totaling 20 plots.

The treatments comprised five different rates of LWE application: 0, 45, 90, 135 and 180 m^3 ha^{-1}. The maximum rate recommended by the company that produces this LWE (Novozymes Latin America Ltda) is 90 m^3 ha^{-1}. Therefore this maximum rate and two rates above the maximum rate (135 and 180 m^3 ha^{-1}) were used to test for possible negative effects of high rates of application. LWE was applied manually using water cans on August 3–4, 2006. The waste was transferred to a 1 m^3 reservoir and stirred continually to keep the product in suspension before spreading over each plot. The plots were subdivided into sections (5 per plot) during the application to ensure a homogenous application of the LWE (with each section receiving an equal portion of the LWE application). For the highest application rate, half of the application volume was applied to each plot, and the other half was applied only after the first application had completely infiltrated. This procedure was performed to minimize surface runoff.

A second 60 m^3 application was carried out on August 15, 2007. A 5 m^3 tank with a fan type spreader was used to apply liquid waste to the pasture surface.

Soil

The soil in the study area was a Cambisol (Cambissolo Háplico distrófico; EMBRAPA, 2006). Soil samples were collected from the 0-20 cm layer of the experimental area on July 11, 2006. The results of the soil analysis are shown in Table 1.

Table 1. Chemical attributes of the soil in the study area at a depth of 0-20 cm before the application of LWE.

pH	Al^{3+}	(H + Al)	Ca^{2+}	Mg^{2+}	K^+	Na^+	P	C	Clay
$CaCl_2$	------------------		$cmol_c$ dm^{-3}	------------------			mg dm^{-3}	g dm^{-3}	g kg^{-1}
4.15	1.29	12.2	1.32	0.65	0.11	0.06	2.4	34.1	350

Characterization of Industrial LWE

The LWE biomass used in this study is a biological waste consisting of a sub-product of phytase enzyme production and is free of animal residues. The production company used a specific microorganism to produce the phytase enzyme. Before field application, the waste underwent a process that used hydrated lime [$Ca(OH)_2$ + $Mg(OH)_2$] to inactivate enzymes and microorganisms. After a 24-hour period, a batch of waste was stirred, sampled and analyzed. LWE was only cleared for agricultural use after a negative microbiological growth test, which was verified by the quality control service of the production company. The chemical characteristics of the LWE used were described by APHA (1998) and are shown in Table 2.

Table 2. Chemical characteristics of LWE applied on a native pasture in Fazenda Rio Grande, Paraná State, Brazil.

Chemical characteristic	Value
Ca(1)	173.9 g kg^{-1}
Mg(1)	106.6 g kg^{-1}
Total P(1)	44.3 g kg^{-1}
K(1)	0.3 g kg^{-1}
pH $CaCl_2$(2)	12.88
Ammonium(1)	33.6 g kg^{-1}
Nitric N(1)	0.33 g kg^{-1}
Total N(1)	65.1 g kg^{-1}
Organic C(1)	34.6 g kg^{-1}
Total C(1)	189.6 g kg^{-1}
C/N ratio	2.91
Total solid (110°C)	126.9 %

(1)Dried base (65°C) and (2)wet basis .

Sampling, analysis and pasture productivity

Soil samples were collected at depths of 0-10, 10-20, 20-40 and 40-60 cm on days 34, 132, 216 and 336 after the first application of LWE. Soil was collected at each plot with a Dutch auger at 15 points to produce a composite sample. The soil was air-dried and sifted through a 2 mm sieve. Chemical analysis of the soil was performed at the UFPR fertility laboratory, according to the methodology described by Marques and Motta (2003) to determine the following: pH $CaCl_2$, pH by Shoemaker, McLean and Pratt (SMP), exchangeable elements by KCl 1 M (Al^{3+}, Ca^{2+}, Mg^{2+}), available nutrients by Mehlich I (K^+ and P) and potential acidity by Ca acetate 0.5 M - pH 7.0 (H+Al). The sum of bases (SB) was calculated from the analytical results. The organic C content was determined by the Walkley-Black method adapted for colorimetric measurements by Raij et al. (1987).

Pasture dry matter was assessed at 58, 128, 233, 496 and 540 days after the first application of LWE. During autumn and winter, no assessment was conducted because the forage production in both seasons was very low. Green leaf samples were collected from an area of 0.25 m^2 at four random points per plot, totaling 1 m^2. At each point, all of the material was cut from the plant at a height of 2 cm. Soon thereafter, all plants of the plot were cut at the same height, and the cut material was removed from the plot. The sampled material was weighed to obtain a wet mass, rinsed under running water and deionized water and dried in an oven at 65°C to obtain the dry mass. Chemical analysis was performed for C, N, P, K, Ca, and Mg in the plant tissue according to the methodology described by Reissman and Martins (2007). Nitrogen and carbon contents were determined by dry combustion using a VARIO EL III - Elementary® apparatus. Plant samples were subjected to dry digestion in porcelain crucibles for P, Ca, Mg, and K analysis. Phosphorus content was determined colorimetrically using an UV/VIS spectrophotometer. Calcium, Mg, Fe, Mn, Zn, and Cu contents were determined by atomic absorption spectrophotometry. Potassium content was determined by flame spectrophotometry.

Statistical analysis

The data from the five treatments in four randomized blocks were examined via regression analysis, in which a curve trend was observed and where the significance cutoff was $p < 0.05$. Statistical analysis was performed using the ASSISTAT® 7.5 beta statistical program. The data are presented as the mean of 4 replicates.

Results and discussion

Soil analysis

The application of LWE resulted in lower soil acidity, higher pH ($CaCl_2$) and decreased Al^{3+} (Figure 1). The increases in soil pH were linear overall for all treatments up to 60 cm deep; the most significant effects occurred close to the surface. This finding demonstrates that LWE has an impact on soil acidity, both at the surface and at greater depths (Figure 1).

It is unusual to obtain a change in soil acidity below a depth of 20 cm and in a short period of time by application of a product at the soil surface. The effect of LWE on soil acidity below a depth of 20 cm could be associated with two potential processes: 1 - the occurrence of a large number of biopores when soil is not tilled (GEOHRING et al., 2001); and 2 - the addition of organic compounds with LWE, which facilitates the penetration of corrective compounds (PAVAN, 1999). Petrere and Anghinoni (2001) reported that the effect of lime applied at the soil surface reached a lower soil layer in a native pasture compared to no-tillage and traditional cropping systems.

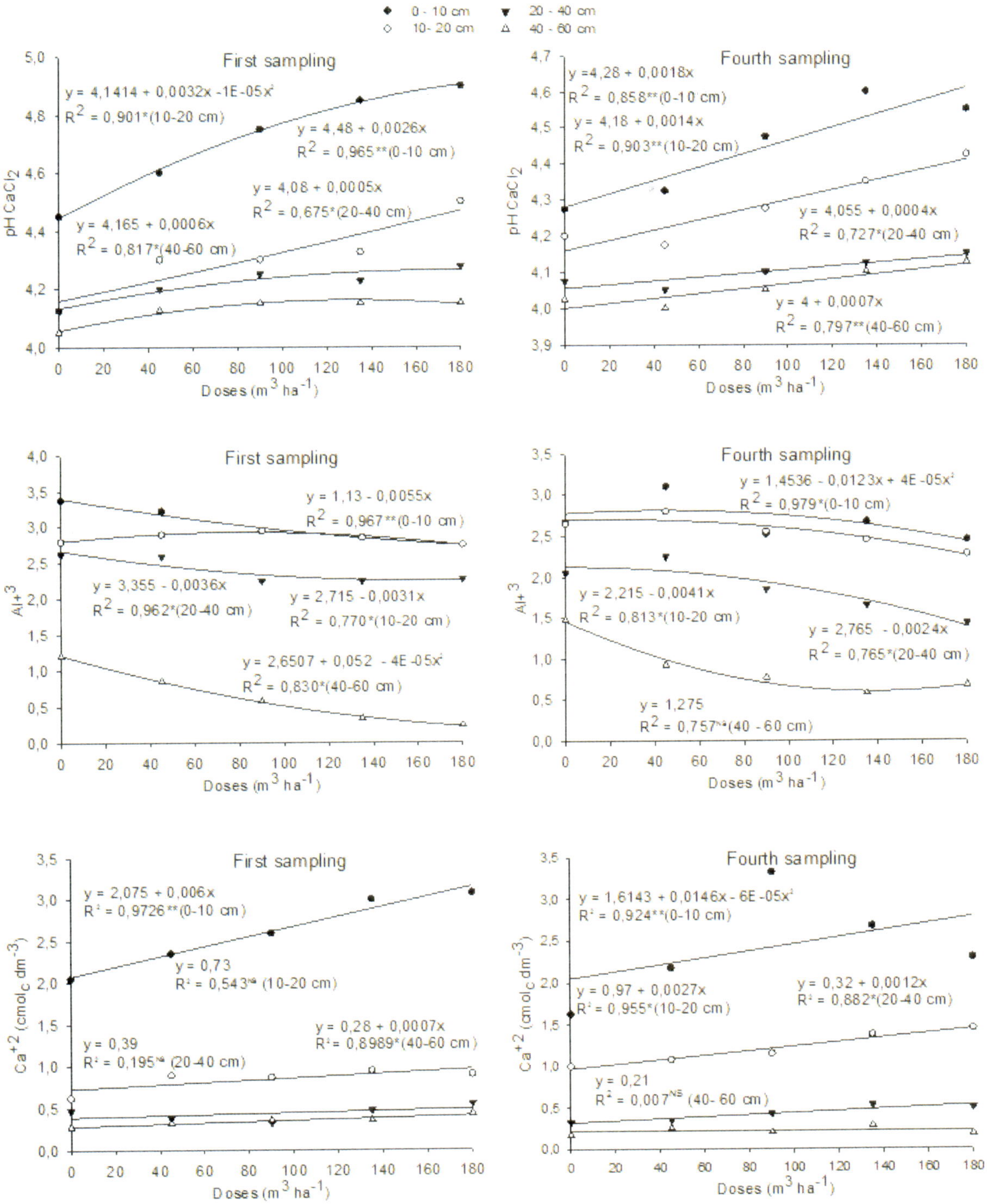

Figure 1. The pH ($CaCl_2$) and concentrations of Al^{3+} and Ca^{2+} at different depths in soil treated with different doses of LWE in the municipality of Fazenda Rio Grande, Paraná State, Brazil. *($p < 0.001$), **($p < 0.05$) and NS (non-significant). The data represent the mean of 4 replications.

The soil in this experiment had a high buffering capacity, which can be observed by a high value of (H + Al) (Table 1); the changes obtained in this soil were therefore considerable (Figure 1). The increase in soil pH may explain the increase in dry matter productivity (Figure 3). Native grasses have been shown to exhibit a low-to-zero response to lime application in terms of yield in conditions similar to this experiment (MACEDO et al., 1979).

Concentrations of Ca^{2+} (Figure 1) and Mg^{2+} (Figure 2) in the soil increased in response to LWE addition, especially at 10 cm.

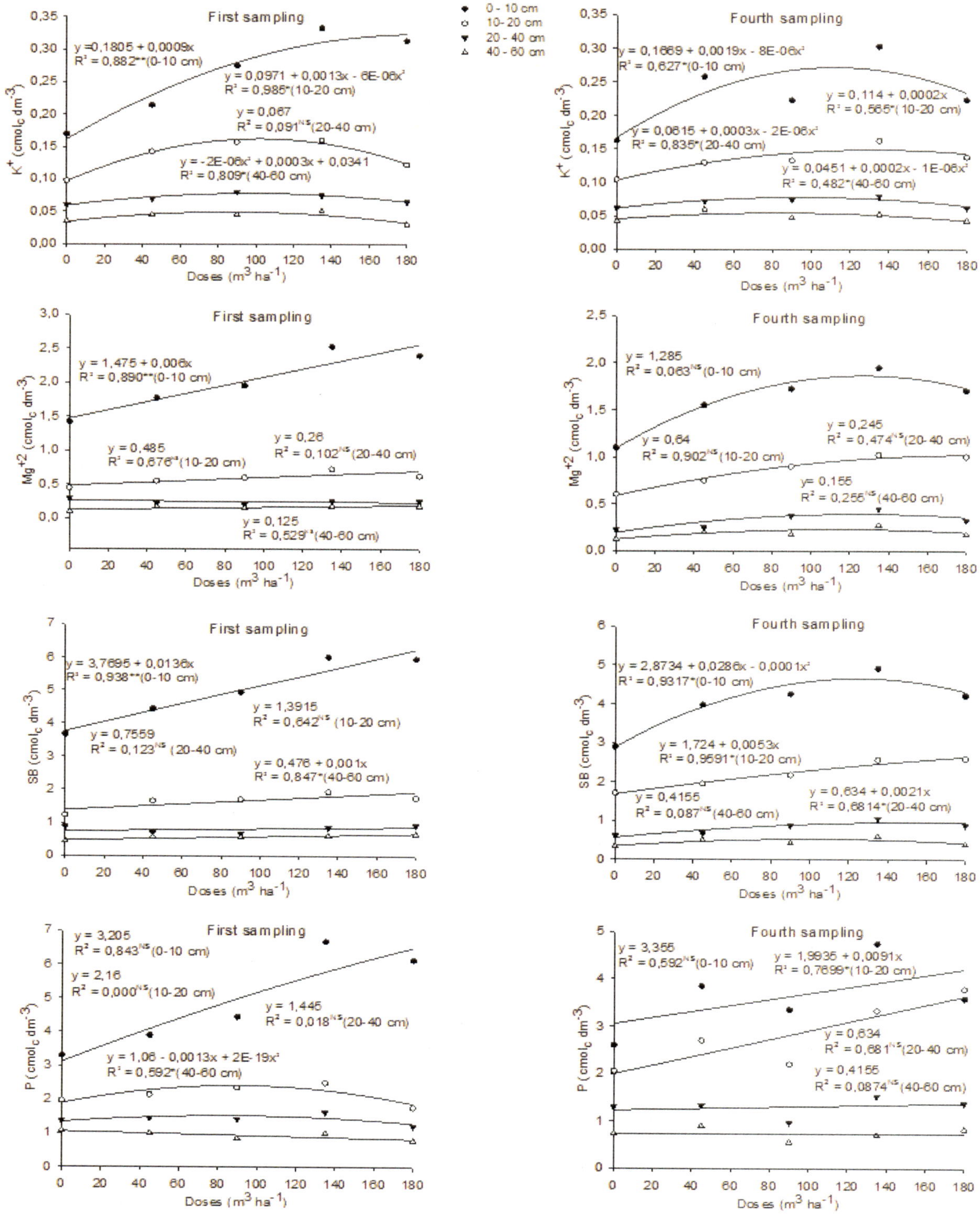

Figure 2. Concentrations of Mg^{2+}, K^+, P and BS at different depths in soil treated with different doses of LWE in the municipality of Fazenda Rio Grande, Paraná State, Brazil; *($p < 0.001$), **($p < 0.05$) and NS (non-significant).

This result was expected because these elements were abundant in the waste; these elements are also associated with higher pH and lower Al^{3+} in the soil. Potassium (K^+) levels had changed substantially by the first sample collection; its soil concentration increased almost two-fold at 0-10 cm depth (Figure 2), demonstrating that K^+ can be made readily available using LWE. These results also confirm that a major surface application of the LWE yielded a positive response at a depth of 60 cm. Work at Tocantis State performed by Araújo et al. (2007) also showed that K^+ concentration in the soil increased when different sources of organic matter were used for fertilization; this nutrient is released by organic compounds, especially cattle manure.

The increase of Ca^{2+} in the soil was associated with a decrease of the Ca in the leaf tissue of the first cut. Values of Ca remained the same in the other cuts, suggesting that an interaction with K helped to decrease Ca absorption by plants.

The phosphorous (P) concentration in the soil was consistently higher in surface layers and decreased as the assessed depth increased (Figure 2). This response was expected because the high amount of P in the LWE used in this study. Additionally, P exhibits low soil mobility and concentrates within the first soil layer. Cavallet et al. (2006) also observed an increase in P concentration in the superficial layers when LWE was used in a fertilization experiment involving annual crops in a no-tillage system.

The sum of bases increased due to LWE application, especially in the 0-10 cm depth. The sum of bases was higher when collections were conducted closer to the date of LWE application (Figure 2). This result was expected, as the sum of bases yielded by LWE is a direct effect. Decreases in Al^{3+} concentrations are the result of the formation of hydroxides of Al when pH increased.

The sum of bases increased down to the 40 cm layer,suggesting that these bases could leach downwards over time when continuously applied to the same area; this was also observed by Kingery et al. (1994) when organic waste was applied to soil. There is a strong association between variations in soil pH and the sum of bases, suggesting that biopores and the application of organic residue can affect the results observed here.

Plant tissue

The C concentration in the pasture was between 417.5 and 428.7 g kg^{-1}, which was close to the value of 400 g kg^{-1} found by Mazza et al. (2009) for Mombasa grass (*Panicum maximum* Jacq.) in the region of the First Plateau of Paraná. The C concentration did not vary with the use of LWE (Table 3), even when gains were made in productivity (Table 3).

Although the C concentration in the pasture remained unchanged, C production increased with LWE use (Table 3) along with an increase of dry matter production (Figure 3).

Table 3. Regression analysis of concentration data for N, P, K, Ca, Mg and C and the C/N ratio in plant tissues in the first and second years of the study. Data were collected after applications of LWE in natural pastures with hermarthria grass in the municipality of Fazenda Rio Grande, Paraná State, Brazil.

Treatment(g kg^{-1})	First year			Second year		
	Cutting	Equation	R^2	Cutting	Equation	R^2
C	1	$\hat{y} = \bar{y} = 422.3$	NS	4	$\hat{y} = \bar{y} = 420.7$	NS
	2	$\hat{y} = \bar{y} = 426.6$	NS	5	$\hat{y} = \bar{y} = 423.8$	NS
	3	$\hat{y} = \bar{y} = 423.0$	NS			
N	1	$y = 2.15+0.005x$	0.92**	4	$\hat{y} = \bar{y} = 2.06$	NS
	2	$\hat{y} = 1.58+0.007x-0.00003x^2$	0.89*	5	$\hat{y} = \bar{y} = 1.73$	NS
	3	$\hat{y} = \bar{y}$ 1.25	NS			
C/N ratio[(1)]	1	$\hat{y} = 19.65-0.0315x$	0.86**	4	$\hat{y} = \bar{y} = 20.97$	NS
	2	$\hat{y} = \bar{y} = 26.87-0.0955x+0.00045x^2$	0.91*	5	$\hat{y} = \bar{y} = 24.96$	NS
	3	$\hat{y} = \bar{y} = 34.26$	NS			
K	1	$\hat{y} = 14.87+0.1725x-0.00079x^2$	0.92**	4	$\hat{y} = \bar{y} = 22.71$	NS
	2	$\hat{y} = 11.74+0.0305x$	0.90**	5	$y = 15.9+0.09x-0.00052x^2$	0.92*
	3	$\hat{y} = \bar{y} = 7.91$	NS			
P	1	$\hat{y} = 2.06+0.0156x-0.00006x^2$	0.96*	4	$\hat{y} = \bar{y} = 2.24$	NS
	2	$\hat{y} = 1.68+0.0148x-0.00006x^2$	0.87*	5	$\hat{y} = \bar{y} = 2.07$	NS
	3	$\hat{y} = 1.50+0.0021x$	0.82*			
Ca	1	$\hat{y} = \bar{y} = 2.32-0.007x$	0.61*	4	$\hat{y} = \bar{y} = 2.00$	NS
	2	$\hat{y} = \bar{y} = 0.95$	NS	5	$\hat{y} = \bar{y} = 2.73$	NS
	3	$\hat{y} = \bar{y} = 0.86$	NS			
Mg				4	$\hat{y} = \bar{y} = 2.13$	NS
				5	$\hat{y} = \bar{y} = 2.59$	NS

[(1)]Dimensionless. y - Nutrient concentrations or relationships of nutrients in plants. x - LWE rates (m^{-3} ha^{-1}); *($p < 0.001$), **($p < 0.05$) and NS (non-significant).

Each m^3 of LWE applied yielded 20 kg ha^{-1} of C in the aerial part of the plants; this yield was observed in both years. This increase in C sequestration could have a positive effect on the maintenance of organic matter in the soil (COSTA et al., 2009).

It was observed that the concentration of N in the plant tissue had increased at the first sampling, though the LWE effect decreased over time (Table 3). Such increases greatly affect pasture quality, as N content is associated with protein concentration and animal performance (SENGER et al., 1996). At the first cutting, the concentration of N in plant tissue showed a linear increase in response to LWE application (Table 3); this was most likely due to the ready availability of N as ammonium in LWE (Table 2). The organic fraction of N in LWE was found in easily mineralized compounds; the C/N relationship is considerably lower than 20/1 (Table 2). Increases in productivity and the concentration of N in plant tissue occurred simultaneously, with a visible change in the leaves in the form of an intense green color. Alvim et al. (1999) and Mazza et al. (2009) also observed a linear response due to increasing rates of nitrogen fertilization.

Quadratic behavior was observed in the concentration of N in the plant in response to LWE application at the second cutting. No change in the concentration of N in the plants was found at the third cutting (Table 3). The decrease in LWE influence on N concentration over time could be associated with N leaching. The leaching process should be high for nitrate, which is directly related to pH. Nitrificant bacteria cannot survive at the low pH observed in native conditions. The incremental change in soil pH caused by LWE application could boost nitrate loss. These results are in contrast with the strong linear increase in dry matter production, suggesting a probable effect of N dilution.

An inverse relationship was found between LWE application rates and the C/N ratio in pasture (Table 3). This result was expected because the C concentration did not vary and the concentration of N in the plants increased when LWE was applied during the initial phases of the trial.

The concentration of P in the plants showed a quadratic response to the LWE rates applied in the first and second cuttings, reaching a maximum value close to a rate of 130 m^3 ha^{-1} (Table 3). In these cuttings, the concentration of P in the pasture when LWE was applied reached values that can be considered high. The concentration of P at the third cutting showed a linear response to LWE, though at lower concentrations when compared with the other cuts (Table 3). In the fourth and fifth cuttings, there was no difference in the P concentration at different LWE applications. This result was not expected; P is characterized as being a nutrient with a high residual effect.

The potassium concentration in plant tissue at the first cutting was found to show a quadric response, reaching a maximum concentration at an LWE application rate of 109 m^3 ha^{-1}. There was a linear increment in the concentration of K in the plants at the second cutting, while no changes were found at the third and fourth cuttings. The K concentration showed a quadratic response to LWE application again at the fifth cutting (Table 3).

The P and K concentrations obtained from the plant tissue are within adequate levels for grasses according to Raij et al. (1996) at 1.0-3.0 g kg^{-1} and 12-25 g kg^{-1}, respectively. The only exceptions to this pattern were found for K in the second and third cuttings, which the concentrations were below 12 g kg^{-1}. In the second colection LWE was not applied. A low K concentration was observed in the pasture for all treatments at the third collection, though no symptoms associated with deficiency were observed.

That shows that treatment should be applied at smaller amount for K to be supplied to the pasture, given the mobility of that element in the soil, and as such it can be easily lost. The decrease in the concentrations of K in plant tissue was greatest between the second and third collection, when a deficient concentration was observed.

The loss of K in plant tissue is not exclusively caused by the loss of the residual effect of fertilization but may also be due to a dilution effect, because the third sampling yielded the highest production of dry matter for the studied period. Seasonal conditions played an important role in K variations as well, as shown by Senger et al. (1996), who verified that higher concentrations of P, Mg and K occurred in the springtime compared to other seasons.

The concentration of Ca in the plants showed an inverse relationship to LWE application rates (Table 3) applied at the first cutting, despite the large amount of LWE applied. This finding could be due to an antagonist effect of one or more nutrients that were added to LWE.

The concentration of Ca in the plant tissue varied from low in the first year to adequate in the second year, with values within the 2-6 g kg^{-1} range being considered adequate for grasses according to Raij et al. (1996).

Although the Ca concentration was low according to the concentrations established for pastures, there was no sign of associated nutritional deficiency in the pasture (i.e., no sign of retarded growth or a lighter green color than is usual). This result suggests that these levels were sufficient for the conditions at hand. Even when lower or unchanged Ca concentrations were observed when LWE was applied at increasing rates, increase of the productivity was observed; in this case, Ca concentrations seemed to have little effect on production.

Magnesium concentrations were only measured at the fourth and fifth collection. The concentration of Mg was only a concern after low Ca concentrations to be observed. Balance between Ca and Mg is important for plant and animal nutrition. With this information, it was possible to establish the best way to correct the soil composition; the treatment for extremely low Mg is gypsum, the treatment for low Mg is calcitic limestone and high Mg can be treated with dolomitic limestone.

The concentration of Mg in the plants was not influenced by the amount of LWE applied. The measured levels of Mg in all treatments, starting with the fourth collection, were within the adequate range of 1.5-4.0 g kg^{-1} (Table 3) according to Raij et al. (1987). Senger et al. (1996) observed similar Mg concentrations in dry matter in a natural pasture that varied from 1.2-2.5 g kg^{-1}; however, Silva et al. (1996) and Senger et al. (1996) reported lower concentrations that varied between 1.0-1.4 and between 0.6-1.5 g kg^{-1}, respectively.

The data confirm that the applied waste was responsible for the improvement in production and pasture quality. This improvement was due to increases in protein and nutrient content, especially in the cuttings conducted after (or closer to) LWE applications.

The results presented in Table 4 indicate that the amounts of N, P and K extracted by the pasture increased, especially in the cuttings conducted in the first year. These results were expected; isolated or combined effects of increases in productivity and the N, P and K concentrations in the pasture were observed during this period. Despite successive decreases in the concentrations of N, P and K in plant tissue in the first, second and third cuttings (Table 3), the highest concentrations of these nutrients were found in the first cut and were due to higher productivity (Figure 3).

Pasture productivity

When no LWE was applied, the cumulative productivity values for natural pasture with hermarthria grass were 4,552.4 and 4,627 kg ha^{-1} of dry matter for the first and second years, respectively. A similar production potential of approximately 4000 kg ha^{-1} $year^{-1}$ has been reported for native pastures in southern Brazil (DURIGON et al., 2002; CORREA et al., 2006; GATIBONI et al., 2000). When natural pastures with hermartria grass were fertilized with LWE, their cumulative productivity in the first year reached a maximum value of 11642 kg ha^{-1} of dry matter at the highest rate of LWE addition. The productivity increases observed in this work are similar to those obtained by Durigon et al. (2002) in native pastures during the spring-summer period in the State of Rio Grande do Sul; in that study, an application rate of 40 m^3 ha^{-1} of liquid pig manure led to an increase from 3,900.0 to 10,300.0 kg ha^{-1} of dry matter.

Although the obtained productivities were much higher than normally indicated for native conditions, productivities were lower than the 23,100.0 kg ha^{-1} $year^{-1}$ of dry matter reported by Alvim et al. (1999); in that study, a Tifton 85 pasture of selected pasture grass in Minas Gerais was

Table 4. Nutrient and C concentrations measured in the first and second year from natural pastures with hermarthria grass treated with different rates of LWE, in the municipality of Fazenda Rio Grande, Paraná State, Brazil.

Treatment (g ha^{-1})	First			Second		
	Cutting	Equation	R^2	Cutting	Equation	R^2
C	1	$\hat{y} = 90.22+1.561x-0.0057x^2$	0.98**	4	$\hat{y} = \bar{y} = 926.46$	NS
	2	$\hat{y} = 438.93+9.88x$	0.99**	5	$\hat{y} = \bar{y} = 1199.79$	NS
	3	$\hat{y} = 1376.19+7.708x$	0.89**			
N	1	$\hat{y} = 4.26+0.117x-0.0004x^2$	0.99**	4	$\hat{y} = \bar{y} = 45.87$	NS
	2	$\hat{y} = 18.14+0.4481x$	0.97**	5	$\hat{y} = \bar{y} = 48.47$	NS
	3	$\hat{y}= 37.12+0.2804x$	0.92**			
K	1	$\hat{y} = 3.15+0.1271x-0.0005x^2$	0.99**	4	$\hat{y} = \bar{y} = 50.44$	NS
	2	$\hat{y} = 10.38+0.4176x$	0.99**	5	$\hat{y} = \bar{y} = 51.41$	NS
	3	$\hat{y} = 25.48+0.1458x$	0.93**			
P	1	$\hat{y} = 0.43+0.0142x-5E-05x^2$	0.99**	4	$\hat{y} = \bar{y} = 4.76$	NS
	2	$\hat{y} = 2.12+0.0567x$	0.92**	5	$\hat{y} = \bar{y} = 5.85$	NS
	3	$\hat{y} = 4.68+0.0415x$	0.95**			
Ca	1	$\bar{y}= \hat{y} = 0.60$	NS	4	$\hat{y} = \bar{y} = 4.47$	NS
	2	$\bar{y}= \hat{y} = 2.26$	NS	5	$\hat{y} = \bar{y} = 7.56$	NS
	3	$\bar{y}= \hat{y} = 4.32$	NS			
Mg				4	$\hat{y} = \bar{y} = 4.69$	NS
				5	$\hat{y} = \bar{y} = 6.20+0.0125x$	0.59*

y - Nutrients or relationships of nutrients in plants. x - LWE rates (m^{-3} ha^{-1}); *($p < 0.001$), **($p < 0.05$) and NS (Non-significant).

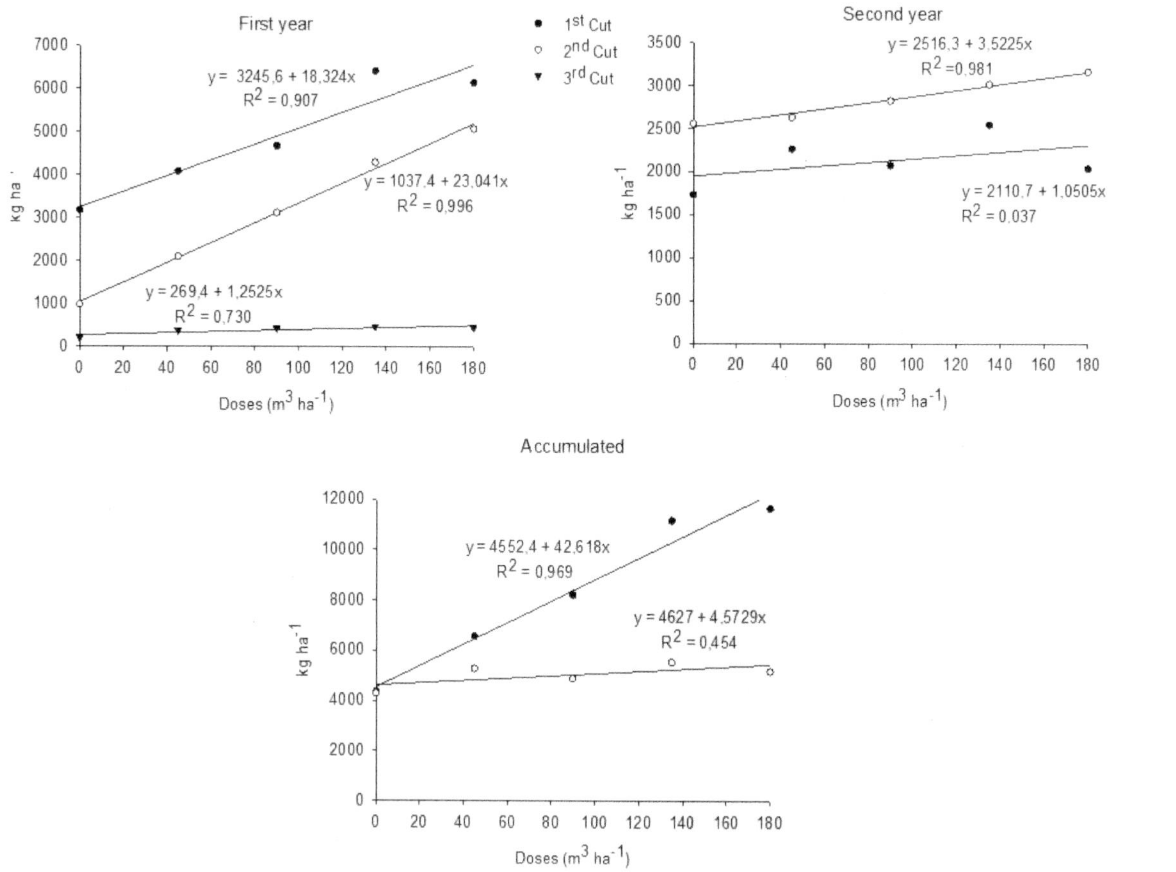

Figure 3. Dry matter yield in the first year (three cuttings) and second year (two cuttings) plotted against the accumulated yield of hermarthria grass treated with different rates of LWE in a pasture in the municipality of Fazenda Rio Grande, Paraná State, Brazil.

fertilized with 600 kg ha^{-1} $year^{-1}$ of N. Therefore, LWE may provide high yields in increments when applied to selected grass pastures.

The increase in dry matter productivity due to LWE application was linear for the first, second and third pasture cuttings (Figure 3). For each m^3 of LWE applied, dry matter was enhanced by 42.6 kg ha^{-1}, which corresponds to dry matter totals of 1.3, 18.3 and 23 kg ha^{-1} for the first, second and third cuttings, respectively. The equation in Figure 3 suggests that pasture dry matter productivity can increase nearly 2-fold when approximately 105 m^3 LWE is applied. In the second year, the slopes were 3.5 and 1.05 for the first and second cuttings, respectively.

Conclusion

The application of LWE led to increases in productivity (42.6 kg dry matter m^{-3} residue) and in pasture quality based on the N, P and K content in plants in the first year after application. Improvement of the soil occurred in terms of acidity and the availability of Ca^{2+}, Mg^{2+}, K^+ and P in the 0-10 cm layer, which indicates that LWE can improve acidity and nutrient content. The concentration of Ca in soil decreased after LWE application, and no change in C content was observed in plant tissue. These results indicate that the application of LWE is a feasible means of fertilization for pastures with acid, low fertility, and/or Cambisol soil.

Acknowledgements

The authors would like to thank the staff of Novozymes (Pedro Luiz Fernandes and Edna Genar Feliciano Maftum), Fabiane Machado Vezzani and the staff and students of the Department of Soils and Agricultural Engineering at UFPR. This project was conducted with financial support from Novozymes Latin America.

References

ALVIM, M. J.; XAVIER, D. F.; VERNEQUE, R. S.; BOTREL, M. A. Resposta do Tifton 85 a doses de nitrogênio e intervalos de cortes. **Pesquisa Agropecuária Brasileira**, v. 34, n. 3, p. 2345-2352, 1999.

APHA-American Public Health Association. **Standard methods for the examination of water and wastewater**. 20th ed. Washington, D.C.: AWWA, 1998.

ARAÚJO, R.; GOEDERT, W. J.; LACERDA, M. P. C. Qualidade de um solo sob diferentes usos e cerrado nativo. **Revista Brasileira de Ciência do Solo**, v. 31, n. 2, p. 1099-1108, 2007.

BRASIL. Ministério da Agricultura, Pecuária e Abastecimento. Secretária de Defesa Agropecuária. Instrução normativa n.º 15, de 2 de maio de 2001. **Diário Oficial da União**, Brasília, 17 de julho de 2001.

CANELLAS, L. P.; VELLOSO, A. C. X.; MARCIYEAR, C. R.; RAMALHO, J. F. G. P.; RUMJANEK, V. M.; REZENDE, C. E.; SANTOS, G. A. Propriedades químicas de um cambissolo cultivado com cana-de-açúcar, com preservação do palhiço e adição de vinhaça por longo tempo. **Revista Brasileira de Ciência do Solo**, v. 27, n. 2, p. 935-944, 2003.

CAVALLET, L. E.; LUCCHESI, L. A. C.; MORAES, A.; SCHIMIDT, E.; PERONDI, M. A.; FONSECA, R. A. Melhoria da fertilidade do solo decorrentes da adição de enzimas de água residuária da indústria de enzimas. **Revista Brasileira de Engenharia Agrícola e Ambiental**, v. 10, n. 3, p. 724-729, 2006.

CORREA, D. A.; SCHEFFER-BASSO, S. M.; FONTANELI, R. S. Efeito da fertilização nitrogenada na produção e composição química de uma pastagem natural. **Revista Brasileira de Agrociência**, v. 10, n. 1, p. 17-23, 2006.

COSTA, K. A. P.; FAQUIN, V.; OLIVEIRA, I. P.; SEVERIANO, E. C.; SIMON, G. A.; CARRIJO, M. S. Doses e fontes de nitrogênio na nutrição mineral do capim-marandu. **Ciência Animal Brasileira**, v. 10, n. 2, p. 115-123, 2009.

DURIGON, R.; CERETTA, C. A.; BASSO, C. J.; BARCELLOS, L. A. R.; PAVINATO, P. S. Produção de forragem em pastagem em pastagem natural com o uso de esterco líquido de suínos. **Revista Brasileira de Ciência do Solo**, v. 26, n. 2, p. 983-992, 2002.

EMBRAPA-Empresa Brasileira de Pesquisa Agropecuária. **Sistema brasileiro de classificação de solos**. Brasília: Embrapa Produção de Informação, 2006. p. 215.

GATIBONI, L. C.; KAMINSKI, J.; PELLEGRINI, J. B. R.; BRUNETTO, G.; SAGGIN, A.; FLORES, J. P. C. Influência da adubação fosfatada e da introdução de espécies forrageiras de inverno na oferta de forragem de pastagem natural. **Pesquisa Agropecuária Brasileira**, v. 35, n. 8, p. 1663-1668, 2000.

GEOHRING, L. D.; MCHUGH, O. V.; WALTER, M. T.; STEENHUIS, T. S.; AKHTAR, M. S.; WALTER, M. F. Phosphorus transport into subsurface drains by macropores after manure applications: implications for best management practices. **Soil Science**, v. 166, n. 12, p. 896-909, 2001.

IAPAR-Instituto Agronômico do Paraná. **Cartas climáticas do Estado do Paraná**. Londrina: Iapar, 1994.

IAPAR-Instituto Agronômico do Estado do Paraná. **Cartas climáticas do Estado do Paraná**. Ed. 2000. Available from: <http://www.iapar.br/Sma/Cartas_Climaticas/Cartas_Climaticas.htm>. Access on: June 23, 2008.

IBGE-Instituto Brasileiro de Geografia e Estatística. **Resultado do censo agropecuário 2010 para o Estado do Paraná** (resultado preliminar). Available from:<http://www.ibge.gov.br/estadosat/temas.php?sigla=pr&tema=censoagro>. Access on: Jan. 11, 2011.

KINGERY, W. L.; WOOD, C. W.; DELANEY, D. P.; WILIANS, J. C.; MULLINS, G. L. Impact of long-term land application of broiler litter on environmentally related soil properties. **Journal Environmental Quality**, v. 23, n. 1, p. 139-147, 1994.

MACEDO, W.; BRASIL, N. E.; PATELLA, J. F. Calcário na implantação em cobertura de leguminosas de inverno. **Pesquisa Agropecuária Brasileira**, v. 14, n. 2, p. 125-134, 1979.

MARQUES, R.; MOTTA, A. C. V. Análise química do solo para fins de fertilidade. In: LIMA, M. R. (Org.) **Manual de diagnóstico da fertilidade e manejo dos solos agrícolas**. Curitiba: Universidade Federal do Paraná, 2003. p. 82-101.

MAZZA, L. M.; FERRARO, F. P.; GARCIA, R. F.; CHEROBIM, V. F.; MOTTA, A. C. V.; MORAES, A. Efeito da adubação nitrogenada na composição química de uma pastagem de capim Mombaça (*Panicum maximum*) na região metropolitana de Curitiba. **Scientia Agraria**, v. 10, n. 4, p. 257-265, 2009.

PAVAN, M. A. Mobilização orgânica do calcário no solo através de adubo verde. In: PAULETTI, V.; SEGANFREDO, R. (Ed.). **Plantio direto**: atualização tecnológica. São Paulo: Fundação Cargill; Fundação ABC, 1999. p. 45-52.

PETRERE C.; ANGHINONI, I. Alteração de atributos químicos no perfil do solo pela calagem superficial em campo nativo. **Revista Brasileira de Ciência do Solo**, v. 25, n. 4, p. 885-895. 2001.

POWER, J.; SUTTON, A. Integrated animal waste management. **Task Force Report**, n. 128, p. 56-60, 1996. (Council for Agricultural Science and Technology).

RAIJ, B.; CANTARELLA, H.; QUAGGIO, J. A.; FURLANI, A. M. C. **Recomendações de adubação e calagem para o estado de São Paulo**. 2. ed. Campinas: Instituto Agronômico; Fundação IAC, 1996. p. 285.

RAIJ, B.; QUAGGIO, J. A.; CANTARELLA, H.; FERREIRA, M. E.; LOPES, A. S.; BATAGLIA, O. C. **Análise química do solo para fins de fertilidade**. Campinas: Fundação Cargill, 1987. p. 170.

REISSMAN, B. C.; MARTINS, A. P. L. Material vegetal e as rotinas laboratoriais nos procedimentos químico analíticos. **Scientia Agraria**, v. 8, n. 1, p. 1-17, 2007.

SHARPLEY, A. N.; SMITH, S. J.; BAIN, W. R. Nitrogen and phosphorus fate from long-term poultry litter applications to Oklahoma soils. **American Soil Science Society Journal**, v. 57, n. 1, p. 1131-1137, 1993.

SENGER, C. C. D.; SANCHEZ, L. M. B.; PIRES, M. B. G.; KAMINSKI, J. Teores minerais em pastagens do Rio

Grande do Sul. **Pesquisa Agropecuária Brasileira**, v. 31, n. 12, p. 897-904, 1996.

SILVA, R. M.; SIEWERDT, L.; SILVEIRA JÚNIOR, P.; SIEWERDT, F. Níveis e fracionamento da aplicação do nitrogênio nos teores de proteína bruta e macro-elementos na forragem de campo natural de Planosolo. **Revista Brasileira de Agrociência**, v. 2, n. 1, p. 165-170, 1996.

Soil surface spectral data from Landsat imagery for soil class discrimination

Marcos Rafael Nanni[1*], José Alexandre Melo Demattê[2], Marcelo Luiz Chicati[1], Peterson Ricardo Fiorio[3], Everson Cézar[1] and Roney Berti de Oliveira[1]

*[1]Departamento de Agronomia, Universidade Estadual de Maringá, Av. Colombo, 5790, 87020-900, Maringá, Paraná, Brazil. [2]Departamento de Solos e Nutrição de Plantas, Escola Superior de Agricultura "Luiz de Queiroz", Piracicaba, São Paulo, Brazil. [3]Departamento de Engenharia Rural, Escola Superior de Agricultura "Luiz de Queiroz", Piracicaba, São Paulo, Brazil. *Author for correspondence. E-mail: mrnanni@uem.br*

ABSTRACT. The aim of this study was to develop and test a method to determine and discriminate soil classes in the state of São Paulo, Brazil, based on spectral data obtained via Landsat satellite imagery. Satellite reflectance images were extracted from 185 spectral reading points, and discriminant equations were obtained to establish each soil class within the studied area. Sixteen soil classes were analyzed, and discriminant equations that comprised TM5/Landsat sensor bands 1, 2, 3, 4, 5, and 7 were established. The results showed that this methodology could effectively identify individual soil classes using discriminant analyses of the spectral data obtained from the surface. Success rates of > 40% were achieved for 14 of the 16 evaluated soil classes when applying the satellite image data. When the 10 soil classes containing the largest number of minimum cartographic areas were used, the hit rate increased to > 50%, for seven soil classes with a global hit rate of 52%. When the soil classes were grouped based on their parent materials, the hit rate increased to 70%. Thus, we concluded that the spectral method for soil classification was efficient.

Keywords: discriminant analysis, TM-Landsat, Brazilian soil classes, spectral response.

Introduction

In recent decades, Brazil has become a prominent worldwide leader in agricultural exports due to the expansion into new agricultural frontiers, implementation of technological improvements, and exceptional efforts to address adversities among growers. To continue this growth, Brazil must adopt suitable practices for using and managing previously uncultivated land areas.

Soil surveys that include appropriate and detailed cartographic representations allow us to obtain a wealth of information that, when properly managed, can enable technicians and growers to employ methodologies and establish strategies that extend and even increase the productive capacity of their lands. However, there are an insufficient number of professionals available to map such a large country, and there are still millions of hectares of land left for the rational and sustainable production of food (DALMOLIN, 1999).

Demattê (2001) stated that the study and implementation all available technologies regarding soil spectral analyses is of critical significance. Spectral behavior analyses of soils using remote sensors might provide answers to the problem of class discrimination. Such studies are based on the fact that each soil presents a spectral signature according to the absorption of specific wavelengths across the entire electromagnetic spectrum (BEN-DOR et al., 1999).

The fundamental theory of soil spectral analyses indicates that soils can be characterized according to their B horizon. This horizon is not detected using satellite data unless erosion has occurred. Therefore, how could spectral responses help in discriminating soil units? To answer this question, we begin with the assumption that many soil classes exhibit surface characteristics that differ from other classes and therefore can be used as a taxonomic identifier. This assumption has been combined with the use of aerial photographs for soil discrimination. This method does not detect B horizon values; however, it does infer information regarding the 'probable' soil class. A small number of published studies of soils and tropical soils have suggested that the use of satellite images can assist in the discrimination of surface information.

Because a spectral response is also an individual trait, the use of this technique allows for the separation of soil classes and may therefore aid in pedological surveys. Because soil properties are related to soil classes, and these soil properties are related to spectra, we can correlate spectra with soils to facilitate the detection of soil types. Therefore, the aim of this study was to develop and test a method for determining and distinguishing different soil classes in the southwestern region of São Paulo State, Brazil, based on spectral data obtained from satellite images.

Material and methods

The studied area is located in the southwestern region of the state of São Paulo and is delimited by the geographical coordinates 23°0'31.37" - 22°58'53.97" south latitude and 53°39'47.81" - 53°37'25.65" west longitude in a region known as the Paleozoic depression (IPT, 1981). As part of the hydrographic basin of the Tietê River (OLIVEIRA et al., 1992), the area is bordered by the Capivari river and spans an area of approximately 198 ha with a perimeter of 11,045.80 m (Figure 1).

Geologically, the studied area is situated in the Itararé Formation and is part of the Tubarão Group. These units were formed between the Upper Carboniferous Period and Middle Permian Epoch and occur in the state a complex association of several lithofacies, almost all of which are detritic that formed in rapid vertical and horizontal successions (IPT, 1981). The predominant lithologies consist of mineralogically immature sandstones with heterogeneous granulation ranging from feldspathic sandstones to arkoses; these units range from thin layers to banks that are tens of meters thick (VIDAL-TORRADO; LEPSCH, 1999).

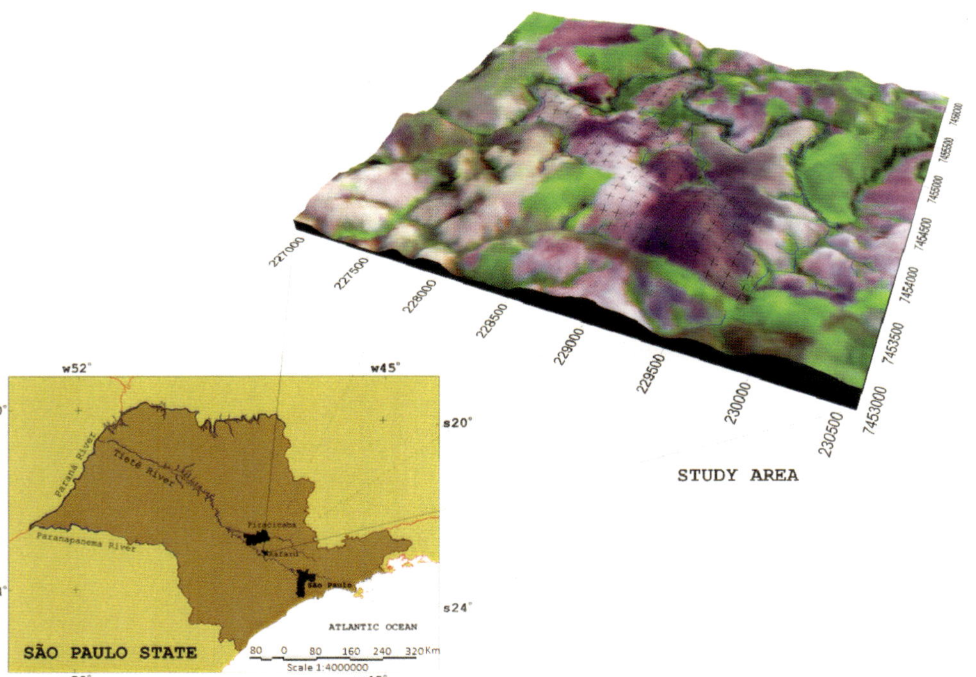

Figure 1. Study location and soil sampling grid representation (1 sample ha^{-1}).

The geographic information system "Sistema de Processamento de Informações Georreferenciadas (SPRING)" (INPE, 1999) was used to implement the cartographic study. This system has multiple functions and algorithms for processing georeferenced databases. A database was established using this system to incorporate the geospatialized information that was obtained from planimetric and altimetric charts and data that was collected in the field using GPS receivers and orbital images.

In the entire area, 185 augured points were selected (Figure 1). These points were marked in a grid fashion with one point per hectare (WOLKOWSKI; WOLLENHAUPT, 1994) following procedures used in soil surveys. Specifically, a synthetic approach was used (EMBRAPA, 1996) and the database was georeferenced with a differential global positioning system (DGPS) technique, which is a method used with GPS receivers to achieve higher precision with these data receptors in absolute positioning modes.

The collected samples were first dried in a forced ventilation oven at a constant temperature of 50°C for 48h and then sifted through a 2-mm sieve (air-dried soil). The textural soil groupings were conducted according to Embrapa (1999). The densitometric method was used to determine the total sand, silt and clay contents (CAMARGO et al., 1986). Organic matter (OM), active and residual acidity, pH and cation exchange capacity (CEC) determinations were performed according to Embrapa (1997). The sum of the cations (i.e., calcium, magnesium and potassium) (SC), base saturation (V%= [SC/CEC]*100), and aluminum saturation (m%= $[Al^{3+}/SC+Al^{3+}]$*100) were determined according to Van Raij and Quaggio (1989). Total iron (Fe), silicon (Si), and titanium (Ti) contents were determined by sulfuric acid digestion according to a methodology recommended by Embrapa (1997). The soil color was identified using the Munsell color chart (GRETAGMACBETH, 2000).

The physiographic unit boundaries were established after generating orbital image visual interpretations according to the recommendations of Donzeli et al. (1983) and Nanni and Rocha (1997). These boundaries were then used in the SPRING system after generating successive combinations of algorithms and filters to improve the visual quality as reported by Nanni and Rocha (1997). The soil classes were defined after analyzing the laboratory samples and open trenches at representative physiographic unit locations. The fieldwork procedures and the description and collection of the materials were performed according to the criteria established by Lemos and Santos (1996). The soil classes were defined according to the Brazilian classification system (EMBRAPA, 1999) and correlated to the soil taxonomy (COSTA; NANNI, 2004).

Spectral data acquisition and processing

The information extracted from the pixels was used to extract the gray value levels for each TM-Landsat-5 image point that corresponded to a sampling location in the field. The gray value levels for each band were then converted into reflectance values according to the procedures established by Markham and Barker (1986) and Thome et al. (1997). For Rayleigh scattering and ozone absorption corrections, the 5S irradiative transfer code simulation (TANRÉ et al., 1992; VERMOTE et al., 1997) was used and properly corrected for atmospheric effects (KAHLE et al., 1980). After the conversion and correction procedures, the zero gray level images corresponded to 0% reflectance, whereas a gray level of 255 corresponded to 100% soil reflectance (DEMATTÊ; NANNI, 2003).

To verify whether the points in the image that corresponded to the field-collected points characterized the exposed soils (i.e., without plant cover), the normalized difference vegetation index (NDVI) methodology was applied (DEMATTÊ et al., 2000; JACKSON, 1983; KAUTH; THOMAS, 1976).

Statistical analyses

The Statistical Analysis System (SAS, 1992) software package version 6 was used to manage the obtained data and perform the statistical analyses. To organize the database for statistical analyses and the evaluation of hypotheses, the reflectance values for each soil were fitted to a data matrix used for this analysis, which consisted of the six TM-Landsat-5 bands for each sampled terrain.

Completely randomized variance analyses were performed on this matrix, and the reflectance means of each soil per selected wavelength range were analyzed using the Tukey test at the $p \leq 0.01$ and $p \leq 0.05$ levels. The general linear models (GLM) method of the SAS program was used to test each selected wavelength range for the hypothesis where H_0: soil1=soil2=soil3=soil4=soil5. The rejection of this null hypothesis implies the acceptance of an alternative hypothesis where H_1: at least two soils are statistically different. The purpose of this study was to determine which variables would have a higher or lower potential to yield models that best explain the problem. Therefore, the STEPDISC method (SAS, 1992) was initially employed to select suitable bands. Discriminant analyses were then conducted to differentiate and characterize the soil with the goal of developing and testing the method for soil class determination based on either its spectral data or analytical characteristics. Thus, the DISCRIM method (SAS, 1992) was used. The parametric method was used to develop the discriminant linear functions.

Results and discussion

The area studied included a variety of parent materials consisting of reworked sandstone-diabase, sandstone-shale, diabase-shale materials, and a combination of all three. As a result of these geological conditions, 18 soil classes were established (Table 1).

The following soil subgroups and land areas dominated the study area: Paleudults and Paleudalfs, 58.12 ha (31.12%); Oxysols, 39.60 ha (21.21%); Dystrochrepts and Eutrochrepts, 36.92 ha (19.76%); Rhodic Paleudalfs, 19.08 ha (10.22%); Argiudolls, 12.76 ha (6.84%); Udorthent, 9.76 ha (10.22%); and Udifluvents, 10.52 ha (5.63%). The declivity conditions pre-established the possible presence of thin soils with depths ranging between 1.50 and 2.00 meters. Deeper soils were found in plains or gently rolling areas and reached up to tens of meters in depth. Because the cartographic land areas included in the strongly rolling and mountainous classes (with soils less than 0.80 cm in depth) were small, lower proportions of shallow soil classes were observed. Rocky outcrops were also included in the shallow soil class.

As expected, the soil classes that contained a small number of samples showed smaller amplitudes or variability of the attributes allowed by the class. The B1, B2, B3, B4, B5, and B7 values corresponded to the bands from the TM sensor (Table 1). This data bank was structured according to the following guidelines: (1) the soil spectral patterns from a given area are used to develop statistics and equations; (2) when a soil sample cannot be classified, its spectral information must be acquired; (3) the spectral information must be applied to all previous equations (Table 1); and (4) the highest obtained data on equations results will be the desired soil. However, such reasoning is valid only for the soils and conditions that were observed in this study.

Furthermore, this reasoning created the potential to characterize different regions in the country, thus allowing for faster preliminary identifications, as stated by Coleman and Montgomery (1990), Demattê and Garcia (1999), and Nanni et al. (2004).

According Gerbermann and Neher (1979), if a dataset is obtained via an automated technique, soil maps will be produced faster than with conventional methods. Table 2 summarizes the hit percentages for the classification of the soils that were examined using the discriminant equations compared to the conventional classification methodology.

Table 1. Discriminant equations obtained using the TM-Landsat evaluation bands for all soil classes.

Class	n[1]	Discriminant equations[2]
RUd[3]	7	$-34.9927 - 3.4605{*}B_1 - 0.14081{*}B_2 + 0.89447{*}B_3 + 2.48853{*}B_4 + 0.73691{*}B_5 - 0.43438{*}B_7$
MTfr	14	$-33.0937 - 6.4961{*}B_1 + 0.76258{*}B_2 + 0.67527{*}B_3 + 2.74335{*}B_4 + 1.43487{*}B_5 - 1.8539{*}B_7$
CE1	4	$-39.2013 - 6.70611{*}B_1 + 1.12848{*}B_2 + 0.9865{*}B_3 + 2.52815{*}B_4 + 1.3809{*}B_5 - 1.60193{*}B_7$
CD1	6	$-44.642 - 4.82595{*}B_1 + 0.63345{*}B_2 - 0.15483{*}B_3 + 3.4661{*}B_4 + 0.90695{*}B_5 - 0.53662{*}B_7$
CE3	3	$-31.4928 - 5.79768{*}B_1 + 0.2626{*}B_2 + 1.16703{*}B_3 + 2.52334{*}B_4 + 1.11786{*}B_5 - 1.36596{*}B_7$
CE2	6	$-31.2862 - 5.74454{*}B_1 + 1.0568{*}B_2 + 0.2449{*}B_3 + 2.95226{*}B_4 + 0.94872{*}B_5 - 1.15765{*}B_7$
CD2	12	$-47.5261 - 4.8144{*}B_1 + 0.61794{*}B_2 + 0.63277{*}B_3 + 2.83464{*}B_4 + 1.26189{*}B_5 - 1.06166{*}B_7$
LVe	20	$-30.4057 - 7.34009{*}B_1 + 0.58157{*}B_2 + 1.89452{*}B_3 + 2.13164{*}B_4 + 0.6073{*}B_5 - 0.69839{*}B_7$
LVAe	9	$-30.6845 - 6.71217{*}B_1 + 0.19287{*}B_2 + 1.98053{*}B_3 + 1.7553{*}B_4 + 0.73748{*}B_5 - 0.45603{*}B_7$
LVAd	5	$-38.5065 - 4.51743{*}B_1 + 0.48783{*}B_2 - 0.05451{*}B_3 + 3.19036{*}B_4 + 0.93884{*}B_5 - 0.6548{*}B_7$
PVAe1	27	$-50.0303 - 4.77088{*}B_1 + 0.45385{*}B_2 + 0.90233{*}B_3 + 2.78573{*}B_4 + 0.72144{*}B_5 - 0.08911{*}B_7$
PVAd	2	$-35.4309 - 6.44089{*}B_1 + 1.06144{*}B_2 + 0.76785{*}B_3 + 2.79228{*}B_4 + 0.83768{*}B_5 - 0.87981{*}B_7$
PVAe2	15	$-38.6123 - 4.66779{*}B_1 + 0.84416{*}B_2 + 0.49152{*}B_3 + 2.62668{*}B_4 + 0.79755{*}B_5 - 0.43845{*}B_7$
PVAe3	4	$-45.236 - 7.15908{*}B_1 + 0.79169{*}B_2 + 1.59665{*}B_3 + 2.48002{*}B_4 + 1.06265{*}B_5 - 0.92726{*}B_7$
Rle2	2	$-55.1678 - 8.65796{*}B_1 + 4.80298{*}B_2 - 1.77679{*}B_3 + 3.79686{*}B_4 + 1.42308{*}B_5 - 2.04336{*}B_7$
Rle1	9	$-30.0657 - 6.06055{*}B_1 + 0.15969{*}B_2 + 1.25555{*}B_3 + 2.17189{*}B_4 + 1.1738{*}B_5 - 1.15031{*}B_7$
NVef1	14	$-28.3508 - 7.2187{*}B_1 + 0.27142{*}B_2 + 1.96497{*}B_3 + 1.93981{*}B_4 + 1.06931{*}B_5 - 1.41957{*}B_7$
NVef2	4	$-22.2789 - 6.45176{*}B_1 + 0.35227{*}B_2 + 2.14137{*}B_3 + 1.49931{*}B_4 + 0.35021{*}B_5 - 0.36348{*}B_7$

[1]Number of individuals; [2]Discriminant equations obtained using the TM-Landsat analyses with the following bands: B_1 (450-520 nm); B_2 (520-600 nm); B_3 (630-690 nm); B_4 (760-900 nm); B_5 (1550-1750 nm); and B_7 (2080-2350 nm); Soil classes of studied area: Argissolo Vermelho-Amarelo distrófico, PVAd (Typic Paleudult); Argissolo Vermelho-Amarelo eutrófico textura arenosa/media, PVAe1 (Typic Paleudult); Argissolo Vermelho-Amarelo eutrófico abrúptico, PVAe3 (Arenic Abruptic Paleudalf); Argissolo Vermelho-Amarelo eutrófico textura média/argilosa, PVAe2 (Typic Paleudult); Cambissolos Háplicos Ta eutróficos lépticos substrato folhelhos da Formação Itararé, CE2 (Typic Eutrochrept); Cambissolos Háplicos Tb distróficos, CD2 (Typic Dystrochrept); Cambissolos Háplicos Tb distróficos e alumínicos substrato retrabalhamento de arenito e saprolito de folhelho da Formação Itararé, CD1 (Typic Dystrochrept); Cambissolos Háplicos Tb eutróficos típicos substrato diabásio, CE3 (Typic Eutrochrept); Cambissolos Háplicos Tb eutróficos típicos substrato regolito do retrabalhamento de arenito e diabásio, CE1 (Typic Eutrochrept); Chernossolo Argilúvico férrico saprolítico, MTfr (Typic Argiudoll); Latossolo Vermelho eutrófico típico – LVe (Typic Haplorthox); Latossolo Vermelho-Amarelo epieutrófico típico, LVAd (Typic Haplorthox); Latossolo Vermelho-Amarelo eutrófico típico, LVAe (Typic Haplorthox); Neossolo Litólicos eutróficos, Rle1 (Typic Udorthent); Neossolo Litólicos eutróficos chernossólicos e típicos, Rle2 (Typic Udorthent); Neossolos Flúvicos Tb distróficos típicos, RUd (Typic Udifluvent); Nitossolo Vermelho eutroférrico, NVef1 (Rhodic Paleudalf); and Nitossolo Vermelho eutroférrico latossólico, NVef2 (Rhodic Paleudalf).

Table 2. Discriminant analyses, the number of samples, and the percent of soil classifications for all soil classes in the studied area.

Class	Total samples	Correct		Incorrect [1]		
		Samples	%	Samples	%	Misidentified Soil Classes
RUd	7	4	57.14	3	42.86	CD1(2)[2],LVAd(1)
MTfr	14	9	64.29	5	35.71	CE2(2),LVe(2),PVAd(1)
CE1	4	3	75.00	1	25.00	LVAe
CE2	6	3	50.00	3	50.00	CD1(1),CE2(1),Rle1(1)
CE3	3	1	33.33	2	66.66	LVe(1),Rle1(1)
CD1	6	2	33.33	4	66.66	CD1(1),CD2(1),PVAe1(1),PVAe2(1)
CD2	12	0	0.00	12	100.00	CE1(2),CD1(2),CE2(1),LVAd(1), PVAe1(2),PVAe2(2),PVAe3(2)
LVe	20	9	45.00	11	55.00	MTfr(1),LVAe(2),NVef1(5),NVef2(3)
LVAe	9	4	44.44	5	55.56	PVAe3(1),NVef1(2),NVef2(2)
LVAd	5	2	40.00	3	60.00	CD1(2),PVAd(1)
PVAe1	27	16	59.26	11	40.74	RUd(1),CD1(4),CD2(1),LVAe(2), PVAe2(2),RLe1(1)
PVAe2	15	0	0.00	15	100.00	RUd(2),CD1(1),CD2(3),CE2(1),CE3(1), LVAd(1),PVAe1(3),Rle1(2),Rle2(1)
PVAe3	4	1	25.00	3	75.00	CE1(1),LVe(1),LVAe(1)
PVAd	2	0	0.00	2	100.00	CD2(1),LVe(1)
Rle1	9	2	22.22	7	77.78	MTfr(2),CE2(1),CE3(1),LVAe(3)
Rle2	2	1	50.00	1	50.00	CE2(2)
NVef1	14	10	71.43	4	28.57	CE3(2),LVe(1),LVAe(1)
NVef2	4	3	75.00	1	25.00	NVef1(1)

[1]Global error median=61.36%; [2]Number of incorrect classifications.

Of the 18 analyzed classes (Table 3), only five showed results hit values of > 60% within the class. Among these classes, the best results were achieved for CE1 and NVef2 with a 75% hit rate within each class. However, most of the classes showed inadequate results. Total failure was observed (i.e., no observations were auto classified using the established discriminant equations) for the soil classes CD2, PVAd and PVAe2 with six, two, and five individual sites, respectively.

In contrast to the total failure observed, the following aspects should be considered: (1) a misidentification occurred for those classes with similar physical surface characteristics, such as organic matter content, iron, clay, sand, and color, which would be difficult to distinguish even in the field; (2) classes with higher iron contents due to their parent material (diabase) showed infrequent or no misidentifications for confusion with the classes derived from sandstone; (3) the Dystrochrepts soil, which is a transitional class, showed the highest number of classification errors; and (4) in some classes with low numbers of sites, an error in a single point could result in a high rate of error for the class, which would contribute to an increase in the global error rate. For example, the Rle2 class (i.e., litholic soil with a shale substrate) with only two sites was problematic. Because one of the sites was classified as a Eutrochrepts soil with a shale substrate (CE2), the error rate was 50%. Considering that these soils were similar at the surface and in satellite images, the reported error (50%) was high and affected the global error for the entire analysis. This class showed the highest global median error value among all analyses performed for this study (61.36%).

In an attempt to eliminate this problem, only the 10 soil classes with the highest number of sites were evaluated. The classes that showed more than seven individual sites were selected following the procedure reported by Oliveira et al. (1982). To ensure that the statistical analyses could not exhibit a marked influence on the standard error of the average, a minimum of 7 samples per soil class were used.

Of the 18 identified classes in the study area, only 10 contained more than 7 observations, which were classified as the following: RUd, MTfr, CD1, CE2, CD2, LVe, LVAe, PVAe1, PVAe2, and NVef1. The obtained equations for the most abundant classes are shown in Table 3.

Table 4 summarizes the percentages of hits for the classification of the soils that were examined using the discriminant equations.

By limiting the number of soil classes, the global error was reduced to 48.25%. This reduction occurred for two primary reasons. The first reason was the elimination of classes with a small number of sites, which contributed to the increased classification error and the low hit percentages within each class. The second reason for the error reduction was the inclusion of those classes that were previously classified into one of the excluded classes. For example, the CD2 class had a hit rate within its own class of 0% based on an analysis that involved all classes. The hit rate increased to 33.33% when the analysis involved only the 10 classes with the highest number of sites.

Furthermore, the classes with similar surface characteristics caused misidentifications during the classification. Therefore, because the orbital data only referred to the surface soil layers, further discriminant analyses were conducted to group the soils according to their parent materials. The classes that were related to the parent material were designated as sand, shale, shale/diabase, and diabase.

The following soil classes were combined to compose the sand group: RUd, CE1, CD1, CD2, LVAd, PVAe1, PVAd, PVAe2, and PVAe3. The CE2 and Rle2 classes formed the shale group. The shale/diabase group contained the MTfr class, and the diabase group consisted of the CE3, LVe, LVAe, Rle1, NVef1, and NVef2 soil classes. The discriminant equations for these groupings are shown in Table 5.

Among the six bands that were evaluated using STEPDISC from the SAS system (Table 5), only the TM_B2 band was not used to assemble the discriminant equations. All of the other bands showed statistical significance at $p \leq 0.01$. After the discriminant equations were established, the classification data were generated as shown in Table 6.

Table 3. Discriminant equations obtained using TM-Landsat-evaluated bands for the 10 soil classes with the highest number of observations in the studied area.

Class	n[1]	Discriminant equations[2]
RUd	7	$-34.2896 - 3.55919 \star B_1 + 0.13371 \star B_2 + 0.63909 \star B_3 + 2.42146 \star B_4 + 1.18612 \star B_5 - 1.12938 \star B_7$
MTfr	14	$-34.6111 - 6.78654 \star B_1 + 1.50043 \star B_2 + 0.12797 \star B_3 + 2.84146 \star B_4 + 1.9366 \star B_5 - 2.66193 \star B_7$
CD1	6	$-44.6683 - 5.21448 \star B_1 + 1.28492 \star B_2 - 0.58129 \star B_3 + 3.47132 \star B_4 + 1.44051 \star B_5 - 1.42153 \star B_7$
CE2	6	$-32.2268 - 6.15409 \star B_1 + 1.8236 \star B_2 - 0.28028 \star B_3 + 3.01722 \star B_4 + 1.42041 \star B_5 - 1.92273 \star B_7$
CD2	12	$-47.9958 - 5.18662 \star B_1 + 1.20571 \star B_2 + 0.18731 \star B_3 + 2.88435 \star B_4 + 1.82645 \star B_5 - 1.98184 \star B_7$
LVe	20	$-29.8542 - 7.38688 \star B_1 + 1.14133 \star B_2 + 1.43477 \star B_3 + 2.09416 \star B_4 + 0.97403 \star B_5 - 1.24361 \star B_7$
LVAe	9	$-30.0024 - 6.66141 \star B_1 + 0.59336 \star B_2 + 1.62149 \star B_3 + 1.70646 \star B_4 + 1.09234 \star B_5 - 1.00113 \star B_7$
PVAe1	27	$-48.885 - 5.04352 \star B_1 + 0.92914 \star B_2 + 0.54099 \star B_3 + 2.74134 \star B_4 + 1.22831 \star B_5 - 0.90683 \star B_7$
PVAe2	15	$-38.464 - 5.05779 \star B_1 + 1.43548 \star B_2 + 0.07636 \star B_3 + 2.6489 \star B_4 + 1.27334 \star B_5 - 1.22066 \star B_7$
NVef1	14	$-28.3791 - 7.18731 \star B_1 + 0.77905 \star B_2 + 1.51134 \star B_3 + 1.94458 \star B_4 + 1.45759 \star B_5 - 1.99804 \star B_7$

[1]Number of individuals; [2]Discriminant equations obtained using TM-Landsat analyses with the following bands: B_1 (450-520 nm); B_2 (520-600 nm); B_3 (630-690 nm); B_4 (760-900 nm); B_5 (1550-1750 nm); and B_7 (2080-2350 nm).

Table 4. Discriminant analyses, the number of samples, and soil classification percentages for the 10 soil classes with the highest number of observations.

Class	Total of samples	Correct		Error[1]		
		Samples	%	Samples	%	Misidentified Soil Classes
RUd	7	6	85.71	1	14.29	CD1
MTfr	14	10	71.43	4	28.57	CE2(2)[2], LVe(2)
CE2	6?	3	50.00	3	50.00	MTfr(1),CD1(1),LVe(1)
CD1	6?	3	50.00	3	50.00	CD2(1),PVAe1(1),PVAe2(1)
CD2	12	4	33.33	8	66.67	RUd(1),CD1(2),CE2(1),PVAe1(2),PVAe2(2)
LVe	20	9	45.00	11	55.00	MTfr(1),LVAe(3),NVef1(7)
LVAe	9	5	55.56	4	44.44	LVe(1),PVAe1(1),NVef1(2)
PVAe1	27	15	55.56	12	44.44	RUd(2),CD1(4),CD2(2),LVAe(2),PVAe2(2)
PVAe2	15	1	6.67	14	93.33	RUd(3),MTfr(1),CD1(1),CE2(2),CD2(3),LVAe(1),PVAe1(3)
NVef1	14	9	64.29	5	35.71	RUd(1),MTfr(1),LVe(1),LVAe(2)

[1]Global error median=48.25%; [2]Number of error classifications.

Table 5. Discriminant equations obtained using TM-Landsat-evaluated bands for the parental material groups in the studied area.

Groups	n[1]	Discriminant equations[2]
Sandstones	78	$-36.45681 - 4.61984 \star B_1 + 1.00592 \star B_3 + 2.55491 \star B_4 + 0.65769 \star B_5 - 0.50672 \star B_7$
Shales	8	$-28.66557 - 5.15408 \star B_1 + 0.64817 \star B_3 + 2.91957 \star B_4 + 0.86388 \star B_5 - 1.26487 \star B_7$
Shales/Diabase	14	$-28.74755 - 6.00838 \star B_1 + 1.09110 \star B_3 + 2.52247 \star B_4 + 1.18156 \star B_5 - 1.68701 \star B_7$
Diabase	63	$-26.23708 + 1.87910 \star B_1 + 1.92418 \star B_3 + 0.71534 \star B_5 - 0.93404 \star B_7$

[1]Number of individuals; [2]Discriminant equations obtained using TM-Landsat analyses with the following bands: B_1 (450-520 nm); B_2 (520-600 nm); B_3 (630-690 nm); B_4 (760-900 nm); B_5 (1550-1750 nm); and B_7 (2080-2350 nm).

Table 6. The number of observations and percent of soils classified using TM-Landsat bands for the parent material groups.

Groups[1]	Sandstone	Shales/Diabase	Shales	Diabase	Total
Sandstone	60.00[2]	4.00	10.00	4.00	78.00
	76.92%	5.13	12.82	5.13	100.00
Shales/Diabase	0.00	10.00	2.00	2.00	14.00
	0.00	71.43%	14.29	14.29	100.00
Shales	0.00	1.00	6.00	1.00	8.00
	0.00	12.50	75.00%	12.50	100.00
Diabase	1.00	10.00	5.00	47.00	63.00
	1.59	15.87	7.94	74.60%	100.00
Total	61.00	25.00	23.00	54.00	163.00
Percent	37.42	15.34	14.11	33.13	100.00

[1]Identification of the soil class groupings with equal parental material; [2]Total frequencies for all classes.

A significant improvement was observed when the soil classes were grouped according to their parental material (Table 7). The global error for the Chi-square test was 25.51%, which was well below the 48.25% error for the classification using the 10 most populous classes and far better than the 61% error when all classes were used separately. Differences between the parent materials were found using the spectral data that were obtained by the TM Landsat sensor in the soil line analyses (HUETE, 1989) as described by Nanni and Demattê (2006). Nanni and Demattê (2006) found higher variations in soil lines for the soils classes that were developed in areas with sandstone and low iron levels relative to those that contained high amounts of clay and iron.

This fact observed by Nanni and Demattê (2006) opens new perspectives and possibilities for geological surveys because this method can set realistic boundaries between the different parent materials. This method confirms that the different parent materials can be discriminated according to their reflectance due to physical interactions with electromagnetic energy. To reinforce the discriminant analyses, a simulation was performed in which 80% of the sampled sites were used to generate a discriminant model that could be tested using the remaining 20% of the data. The sampled sites were randomly selected; i.e., the SAS system randomly selected the components that would take part in the discriminant analyses (80%) and those that would be used to test the obtained models (20%). The procedure was tested using 50 simulations with the system set to randomly choose 80% of the sampled sites, which generated the discrimination model that was used to test the remaining 20% of the sites. Only the 10 soil classes with the highest number of observations were used. Furthermore, classes with a reduced number of sites might not show the variability that exists within the class or might have characteristics that resemble another class. These issues could interfere with the safe and reliable analyses. In addition, as previously described, classes with a small number of sites greatly increased the global error.

After the simulation, the system yielded results that showed the data classification percentage within the model (Table 7).

The hit rate was low when 80% of the observations were applied to the model. Of the 5825 matches that were generated by the 50 simulations for the model-examined data, the model produced 3102 misclassifications with a global error of 53.3%, whereas the classifications matched in 2723 instances with an accuracy of 46.7%. Furthermore, 11% of the total data was lost (632 times). Hence, a difference between the numbers presented in Table 7 (5193) and the total number (5825) was observed. Despite this difference, the presented model was highly significant according to the Chi-square test ($p \leq 0.01$). The correlation between the classes that were observed and estimated by the model showed an r-square value of 83.8% as defined by the contingency coefficient.

Table 7. The number of observations and soil percentages classified using the 10 soil classes with the highest number of observations in the studied area and 80 % of the observations to develop the model.

Class[1]	RUd	MTfr	CD1	CE2	CD2	LVe	LVAe	PVAe1	PVAe2	NVef1	Total[2]
RUd	213.00[3]	0.00	39.00	3.00	11.00	0.00	0.00	2.00	7.00	0.00	275
	77.45[4]	0.00	14.18	1.09	4.00	0.00	0.00	0.73	2.55	0.00	
MTfr	0.00	376	0.00	88.00	0.00	76.00	1.00	0.00	1.00	12.00	554
	0.00	67.87	0.00	15.88	0.00	13.72	0.18	0.00	0.18	2.17	
CD1	2.00	0.00	110.00	2.00	47.00	0.00	0.00	41.00	36.00	0.00	238
	0.84	0.00	46.22	0.84	19.75	0.00	0.00	17.23	15.13	0.00	
CE2	0.00	49.00	26.00	123.00	0.00	25.00	7.00	0.00	0.00	1.00	231
	0.00	21.21	11.26	53.25	0.00	10.82	3.03	0.00	0.00	0.43	
CD2	42.00	18.00	73.00	16.00	192.00	0.00	0.00	63.00	74.00	0.00	478
	8.79	3.77	15.27	3.35	40.17	0.00	0.00	13.18	15.48	0.00	
LVe	0.00	57.00	0.00	4.00	0.00	396.00	124.00	0.00	0.00	229.00	810
	0.00	7.04	0.00	0.49	0.00	48.89	15.31	0.00	0.00	28.27	
LVAe	1.00	0.00	6.00	0.00	0.00	32.00	193.00	32.00	3.00	82.00	349
	0.29	0.00	1.72	0.00	0.00	9.17	55.3	9.17	0.86	23.5	
PVAe1	91.00	0.00	125.00	6.00	95.00	1.00	71.00	612.00	89.00	0.00	1090
	8.35	0.00	11.47	0.55	8.72	0.09	6.51	56.15	8.17	0.00	
PVAe2	99.00	48.00	44.00	52.00	104.00	0.00	25.00	115.00	114.00	0.00	601
	16.47	7.99	7.32	8.65	17.3	0.00	4.16	19.13	18.97	0.00	
NVef1	42.00	17.00	0.00	8.00	0.00	53.00	53.00	0.00	0.00	394.00	567
	7.41	3.00	0.00	1.41	0.00	9.35	9.35	0.00	0.00	69.49	
										Total[5]	5193
											100

[1]Classes with the highest number of observations; [2]Total frequencies in the model; [3]Frequency in the class; [4]% correct within the class; [5]Total frequencies for all classes.

Most of the errors occurred between classes that had very similar parent materials or surface layer properties. The only non-justifiable error was observed with the PVAe2 class, which was corrected classified only once in 554 trials (0.02%) by the model. Because the characteristics of these soils are quite distinct, this result indicates that the model requires adjustment.

Among all classes, RUd was the best classified with the smallest error and a 77.45% correct classifications The worst performance was observed for the PVAe2 class, which produced a 18.97% frequency of correct identification. As previously discussed, this class originated from a sandstone deposit overlying shale or diabase. Therefore, the characteristics of PVAe2 soils result in an intermediate classification between several soil classes within the area. Even so, the highest rate of misidentification occurred among the classes with high sand fractions at the surface, such as RUd, CD1, CD2, PVAe1 and LVAe classes. Nevertheless, classification errors occurred with the MTfr (7.99%), CE2 (8.65%) and LVAe (4.16%) classes, indicating that the model was not reliable for these classes.

Table 8 summarizes the match percentages for the classification of soils that were examined using the discriminant equations when only 20% of the observations were tested in the model generated using the remaining 80% of the data.

When using 20% of the classes for the application of the model established by the remaining 80% of the data, the hit rate was low, producing a large number of misidentifications among the classes and a low frequency of observations within the classes. Based on 50 simulations, the model produced 962 misclassifications of 1475 points with a global error of 65.2%, whereas the model correctly classified 34.8% of the data (513 points). Furthermore, 11% of the total data was lost (168 points). Hence, a difference between the frequencies shown in Table 8 (1307 test results) and the total frequency (1475 test results) was observed. In spite of this difference, the presented model was highly significant, as determined by the Chi-square test ($p \leq 0.01$). The correlation among the true and model-estimated classes showed an r-square value of 78.7%, as defined by the contingency coefficient.

However, even when testing 20% of the data, the classes that showed the highest rates of misidentification originated from similar parent materials. Thus, we were able to establish that the use of soil spectral curves for the discrimination of parent material classes showed satisfactory to good results when applying orbital-level data. For the discrimination of soil classes, the models that were generated with the obtained reflectance data at the orbital level showed satisfactory results for some classes and were inconsistent for others, indicating the need for further studies to improve the quality of the results.

Misclassifications occurred between similar soil data. CE and CD are moderate shallow soils that occur in a rolling relief. These soils were misclassified with other soil types under the same conditions (Table 2). Other misclassifications can also be explored, such as NVe1 is being confounded with NVe2. The only difference between these soil classes is that NVe1 is a highly clayey soil and NVe2 is a clayey soil. PVA soils also occur in rolling relief and can be misclassified as CE or CD. Thus, there are many cases where misclassifications may occur due to overlapping soil characteristics.

Table 8. Number of observations and soil percentages classified using the 10 soil classes with the highest number of observations in the studied area and 20% of the observations as developed by the model.

Class[1]	RUd	MTfr	CD1	CE2	CD2	LVe	LVAe	PVAe1	PVAe2	NVef1	Total[2]
RUd	26.00[3]	0.00	22.00	1.00	4.00	0.00	5.00	4.00	13.00	0.00	75
	34.67[4]	0.00	29.33	1.33	5.33	0.00	6.67	5.33	17.33	0.00	
MTfr	0.00	98.00	0.00	22.00	0.00	19.00	0.00	0.00	0.00	7.00	146
	0.00	67.12	0.00	15.07	0.00	13.01	0.00	0.00	0.00	4.79	
CD1	1.00	0.00	14.00	8.00	19.00	0.00	1.00	9.00	10.00	0.00	62
	1.61	0.00	22.58	12.9	30.65	0.00	1.61	14.52	16.13	0.00	
CE2	0.00	27.00	12.00	13.00	0.00	8.00	3.00	0.00	5.00	1.00	69
	0.00	39.13	17.39	18.84	0.00	11.59	4.35	0.00	7.25	1.45	
CD2	9.00	6.00	22.00	9.00	24.00	0.00	0.00	31.00	21.00	0.00	122
	7.38	4.92	18.03	7.38	19.67	0.00	0.00	25.41	17.21	0.00	
LVe	0.00	11.00	0.00	1.00	0.00	84.00	42.00	0.00	1.00	51.00	190
	0.00	5.79	0.00	0.53	0.00	44.21	22.11	0.00	0.53	26.84	
LVAe	1.00	0.00	1.00	0.00	0.00	17.00	39.00	5.00	1.00	37.00	101
	0.99	0.00	0.99	0.00	0.00	16.83	38.61	4.95	0.99	36.63	
PVAe1	23.00	0.00	37.00	3.00	31.00	0.00	18.00	135.00	13.00	0.00	260
	8.85	0.00	14.23	1.15	11.92	0.00	6.92	51.92	5.00	0.00	
PVAe2	30.00	11.00	16.00	24.00	23.00	1.00	12.00	28.00	4.00	0.00	149
	20.13	7.38	10.74	16.11	15.44	0.67	8.05	18.79	2.68	0.00	
NVef1	7.00	19.00	0.00	0.00	0.00	10.00	21.00	0.00	0.00	76.00	133
	5.26	14.29	0.00	0.00	0.00	7.52	15.79	0.00	0.00	57.14	
										Total[5]	1307
											100

[1]Classes with the highest number of observations; [2]Total frequency in the model; [3]Frequency in the class; [4]% correct within the class; [5]Total frequency for all classes.

Most of the results indicated that as a first analysis, this classification system could be important for assisting in the next step of soil classification. Discrepancies between the classifications and field results were due to several parameters, such as the relief and soil surface property, which are not related to the undersurface. In fact, the subsurface is more significant when classifying soils. This horizon contains less organic matter and influences soil management. In addition, the mineralogy is generally more accurate in the subsurface due to less weathering. For many reasons, the surface information provides important indications to assist the convergence of evidence and ultimately aid in soil classification. In the field, the surface of each soil type usually exhibits different appearances due to the dynamics of water, organic matter, relief, and humidity. All of these factors interfere with the spectral information and are responsible for the observed differences.

Therefore, we believe that the emergence of new sensors installed on orbital or sub-orbital platforms with a greater number of bands or spectral ranges will be very important for discriminating soils with a noticeable reduction of classification errors.

Conclusion

The surface information collected assisted in the discrimination of soils in the studied area using statistical analyses and the data obtained at the orbital level. Based on the discriminant analyses, the hit rate increased as the sample number decreased. The simulated statistical test was efficient for establishing errors and matches during the discriminant analyses.

Satellite-derived surface data can provide valuable and relevant information for soil class discrimination and assist in soil surveys. In addition, soil classes that were grouped according to their parent material were identified via discriminant analyses.

Acknowledgements

The authors thank the São Paulo Research Support Foundation (Fapesp) for the acquisition of the IRIS spectroradiometer (Process no. 95/6259 6), the Coordination for the Improvement of Higher Level Personnel (Capes) for providing funding for the first author's research, and the Brazilian National Research Council (CNPq) for providing funding for the second author's research (Process no. 98/01059 7).

References

BEN-DOR, E.; IRONS, J. R.; EPEMA, G. F. Soil reflectance. In: RENCZ, N. (Ed.). **Remote sensing for the earth sciences**: manual of remote sensing. New York: Wiley, 1999. p. 111-118.

CAMARGO, O. R.; MONIZ, A. C.; JORGE, J. A.; VALADARES, J. M. **Methods of chemical, mineralogical, and physical analysis of soils**. Campinas: IAC, 1986.

COLEMAN, T. L.; MONTGOMERY, O. L. Assessment of spectral characteristics for differentiating among soil categories in the southeastern United States. **Photogrammetric Engineering and Remote Sensing**, v. 52, n. 1, p. 1659-1663, 1990.

COSTA, A. C. S.; NANNI, M. R. **Brazilian classification**: encyclopedia of soil science. New York: Taylor and Francis, 2004.

DALMOLIN, R. S. D. Missing soil scientists in Brazil. **Boletim da Sociedade Brasileira de Ciência do Solo**, v. 24, n. 4, p. 13-15, 1999.

DEMATTÊ, J. A. M. The pedologist and precision agriculture. **Boletim da Sociedade Brasileira de Ciência do Solo**, v. 26, n. 1, p. 17-18, 2001.

DEMATTÊ, J. A. M.; GARCIA, G. J. Alteration of soil properties through a weathering sequence as evaluated by spectral reflectance. **Soil Science Society of America Journal**, v. 63, n. 2, p. 327-342, 1999.

DEMATTÊ, J. A. M.; NANNI, M. R. Weathering sequence of soils developed from basalt as evaluated by laboratory (IRIS), airborne (AVIRIS) and orbital (TM) sensors. **International Journal of Remote Sensing**, v. 24, n. 3, p. 4715-4738, 2003.

DEMATTÊ, J. A. M.; HUETE, A.; FERREIRA JR., L.; ALVES, M. C.; NANNI, M. R.; CERRI, C. E. P. Evaluation of tropical soils through ground and orbital sensors. **International Conference of Geospatial Information in Agriculture and Forestry 2**, v. 2, n. 1, p. 34-41, 2000.

DONZELI, P. L.; VALÉRIO FILHO, M.; NOGUEIRA F. P.; PEREZ FILHO, A.; KOFFLER, N. F. Satellite images and radar on the physiographic setting of standards applied to soils. **Revista Brasileira de Ciência do Solo**, v. 7, n. 1, p. 89-94, 1983.

EMBRAPA-Empresa Brasileira de Pesquisa Agropecuária. **Criteria and standards for soil surveys**. Rio de Janeiro: Embrapa, 1996.

EMBRAPA-Empresa Brasileira de Pesquisa Agropecuária. **Manual of methods for soil analysis**. Rio de Janeiro: Embrapa, 1997.

EMBRAPA-Empresa Brasileira de Pesquisa Agropecuária. **Brazilian system of soil classification**. Brasília: Embrapa, 1999.

GERBERMANN, A. H.; NEHER, D. D. Reflectance of varying mixtures of a clay soil and sand. **Photogrammetric Engineering and Remote Sensing**, v. 45, n. 8, p. 1145-1151, 1979.

GRETAGMACBETH. **Munsell soil color charts – year 2000, revised**. New York: USA, 2000.

HUETE, A. R. Soil influences in remotely sensed vegetation-canopy spectra. In: ASRAR, G. (Ed.). **Theory and application of optical remote sensing**. New York: Wiley Interscience, 1989. p. 107-141.

INPE-Instituto Nacional de Pesquisas Espaciais. **Spring tutorial**: basic Spring. São José dos Campos: INPE, 1999.

IPT-Instituto de Pesquisas Tecnológicas. **Mapa Geológico do Estado de São Paulo**. São Paulo: Brasil, 1981.

JACKSON, R. D. Spectral indices in space. **Remote Sensing of Environmental**, v. 13, n. 1, p. 409-421, 1983.

KAHLE, A. B.; MADURA, D. P.; SOHA, J. M. Middle infrared multispectral aircraft scanner data analysis for geological applications. **Applied Optics**, v. 19, n. 1, p. 2279-2290, 1980.

KAUTH, R. J.; THOMAS, J. S. The tasseled cape graphic description of the spectral temporal development of agriculture crops as seen by Landsat. **Symposium on Machine Processing of Remotely Sensed Data**, v. 1, n. 1, p. 41-51, 1976.

LEMOS, R. C.; SANTOS, R. D. **Manual of description and soil samples from the field**. Campinas: SBCS, 1996.

MARKHAM, B. L.; BARKER, J. L. Landsat MSS and TM post-calibration dynamic ranges, exoatmospheric reflectances and at-satellite temperatures. **EOSAT Technical Notes**, n. 1, 1986.

NANNI, M. R.; DEMATTÊ, J. A. M. Spectral reflectance methodology in comparison to traditional soil analysis. **Soil Science Society of America Journal**, v. 70, n. 2, p. 393-407, 2006.

NANNI, M. R.; ROCHA, H. O. Integration of GIS technology, remote sensing and multivariate analysis in the delimitation of physiographic units for pedological mapping. **USP Scientific Series: IG report**, v. 28, n. 1, p. 129-143, 1997.

NANNI, M. R.; DEMATTÊ, J. A. M.; FIORIO, P. R. Soil discrimination analysis by spectral response in the ground level. **Pesquisa Agropecuária Brasileira**, v. 39, n. 10, p. 995-1006, 2004.

OLIVEIRA, J. B.; JACOMINE, P. K. T.; CAMARGO, M. N. **General soil classes from Brazil**: a guide to assist recognition. São Paulo: Funep/Unesp, 1992.

OLIVEIRA, J. B.; MENCK, J. R. F.; BARBIERI, J. L. **Semidetailed pedological survey of São Paulo State**: Araras quadrangle. Campinas: IAC, 1982.

SAS-Statistical Analysis System. **User's guide, version 6.0**. Cary: Statistical Analysis System Institute, 1992.

TANRÉ, D.; HOLBEN, B. N.; KAUFMAN Y. J. Atmospheric correction algorithm for NOAA-AVHRR products: theory and application. **IEEE Transaction on Geoscience and Remote Sensing**, v. 30, n. 1, p. 231-248, 1992.

THOME, K. B.; MARKHAM, B. L.; BARKER, J. L.; SLATER, P.; BIGGAR, S. Radiometric calibration of landsat. **Photogrammetric Engineering and Remote Sensing**, v. 63, n. 7, p. 853-858, 1997.

VAN RAIJ, B.; QUAGGIO, J. A. **Methods of soil analysis for fertility**. Campinas: IAC, 1989.

VERMOTE, E. F.; TANRÉ, D.; DEUZÉ, J. L.; HERMAN, L.; MORCRETTE, J. J. Second simulation of the satellite signal in the solar spectrum, 6S: an overview. **IEEE Transaction on Geoscience and Remote Sensing**, v. 35, n. 1, p. 675-686, 1997.

VIDAL-TORRADO, P.; LEPSCH, I. F. Soil/parent material relations and pedogenesis on a slope dominated by clayey oxidic soils over sandstone at the S. Paulo State peripheral depression, southeastern Brazil. **Revista Brasileira de Ciência do Solo**, v. 23, n. 2, p. 357-369, 1999.

WOLKOWSKI, R. P.; WOLLENHAUPT, N. C. Grid soil Sampling. **Better Crops**, v. 78, n. 4, p. 6-9, 1994.

Permissions

 Every chapter published in this book has been scrutinized by our experts. Their significance has been extensively debated. The topics covered herein carry significant findings which will fuel the growth of the discipline. They may even be implemented as practical applications or may be referred to as a beginning point for another development.

The contributors of this book come from diverse backgrounds, making this book a truly international effort. This book will bring forth new frontiers with its revolutionizing research information and detailed analysis of the nascent developments around the world.

We would like to thank all the contributing authors for lending their expertise to make the book truly unique. They have played a crucial role in the development of this book. Without their invaluable contributions this book wouldn't have been possible. They have made vital efforts to compile up to date information on the varied aspects of this subject to make this book a valuable addition to the collection of many professionals and students.

This book was conceptualized with the vision of imparting up-to-date information and advanced data in this field. To ensure the same, a matchless editorial board was set up. Every individual on the board went through rigorous rounds of assessment to prove their worth. After which they invested a large part of their time researching and compiling the most relevant data for our readers.

The editorial board has been involved in producing this book since its inception. They have spent rigorous hours researching and exploring the diverse topics which have resulted in the successful publishing of this book. They have passed on their knowledge of decades through this book. To expedite this challenging task, the publisher supported the team at every step. A small team of assistant editors was also appointed to further simplify the editing procedure and attain best results for the readers.

Apart from the editorial board, the designing team has also invested a significant amount of their time in understanding the subject and creating the most relevant covers. They scrutinized every image to scout for the most suitable representation of the subject and create an appropriate cover for the book.

The publishing team has been an ardent support to the editorial, designing and production team. Their endless efforts to recruit the best for this project, has resulted in the accomplishment of this book. They are a veteran in the field of academics and their pool of knowledge is as vast as their experience in printing. Their expertise and guidance has proved useful at every step. Their uncompromising quality standards have made this book an exceptional effort. Their encouragement from time to time has been an inspiration for everyone.

The publisher and the editorial board hope that this book will prove to be a valuable piece of knowledge for researchers, students, practitioners and scholars across the globe.

List of Contributors

André Fernão Martins de Andrade
Instituto Federal de Educação, Ciência e Tecnologia do Rio de Janeiro, Pinheiral, Rio de Janeiro, Brazil

Nelson Moura Brasil do Amaral Sobrinho
Departamento de Solos, Instituto de Agronomia, Universidade Federal Rural do Rio de Janeiro, Seropédica, Rio de Janeiro, Brazil

Fabiana Soares dos Santos
Departamento de Engenharia de Agronegócios, Universidade Federal Fluminense, Volta Redonda, Rio de Janeiro, Brazil

Marcio Osvaldo Lima Magalhães
Departamento de Agronomia, Universidade do Estado de Mato Grosso, Rodovia MT-358, km 7, 78300-000, Tangará da Serra, Mato Grosso, Brazil

Alfredo Tolón-Becerra
Departamento de Engenharia Rural, Universidade de Almería, Almería, Andaluzia, Spain

Leilane da Silva Lima
Programa de Pós-graduação em Agronomia-Ciência do Solo, Universidade Federal Rural do Rio de Janeiro, Rio de Janeiro, Rio de Janeiro, Brazil

Ivanildo Amorim de Oliveira, José Marques Júnior, Renato Eleotério de Aquino and Ludmila de Freitas
Departamento de Solos e Adubos, Faculdade de Ciências Agrárias e Veterinárias, Universidade Estadual Paulista"Julio de Mesquita Filho", Via de Acesso Prof. Paulo Donato Castellane, s/n^{o}, 14884-900, Jaboticabal, São Paulo, Brazil

Milton César Costa Campos
Departamento de Agronomia, Universidade Federal do Amazonas, Humaitá, Amazonas, Brazil

Antônio Sérgio Ferraudo
Departamento de Ciências Exatas, Faculdade de Ciências Agrárias e Veterinárias, Universidade Estadual Paulista "Júlio de Mesquita Filho", Jaboticabal, São Paulo, Brazil

Elcio Liborio Balota, Oswaldo Machineski and Priscila Viviane Truber
Laboratório de Microbiologia do Solo, Área de Solos, Instituto Agronômico do Paraná, Londrina, Paraná, Brazil

Selene Cristina de Pierri Castilho and Laura Fernandes Simões da Silva
Programa de Pós-graduação em Solos e Nutrição de Plantas, Escola Superior de Agricultura "Luiz de Queiroz", Universidade de São Paulo, Piracicaba, São Paulo, Brazil

Miguel Cooper
Escola Superior de Agricultura "Luiz de Queiroz", Universidade de São Paulo, Avenida Pádua Dias, 11, 13418-900, Piracicaba, São Paulo, Brazil

Jorge Luiz Moretti de Souza, Kharyn Freitas Fezer, Bruno César Gurski and Daniela Jerszurki
Departamento de Solos e Engenharia Agrícola, Universidade Federal do Paraná, Rua dos Funcionários, 1540, 80035-050, Curitiba, Paraná, Brazil

Paulo Eugênio Pachechenik
Escola de Ciências Agrárias e Medicina Veterinária, Pontifícia Universidade Católica, Curitiba, Paraná, Brazil

Jessé Rodrigo Fink, Alberto Vasconcellos Inda and Cimélio Bayer
Departamento de Solos, Universidade Federal do Rio Grande do Sul, Av. Bento Gonçalves, 7712, Porto Alegre, Rio Grande do Sul, Brazil

José Torrent and Vidal Barrón
Departamento de Agronomía, Universidad de Córdoba, Córdoba, Spain

Letícia d'Agosto Miguel Fonseca
Instituto Nacional de Pesquisas Espaciais, Av. dos Astronautas, 12227-010, São José dos Campos, São Paulo, Brazil

João Luiz Lani and Elpídio Inácio Fernandes Filho
Departamento de Solos, Centro de Ciências Agrárias, Universidade Federal de Viçosa, Viçosa, Minas Gerais, Brazil

Gerson Rodrigues dos Santos and Adriana Maria Rocha Trancoso Santos
Departamento de Estatística, Universidade Federal de Viçosa, Viçosa, Minas Gerais, Brazil

Williams Pinto Marques Ferreira
Empresa Brasileira de Pesquisa Agropecuária, Embrapa Café, Viçosa, Minas Gerais, Brazil

Pedro Alexandre Varella Escosteguy, Rafael Mariath Cherubini, Cristiano Pauletto and Simone Meredith Scheffer-Basso
Faculdade de Agronomia e Medicina Veterinária, Universidade de Passo Fundo, BR-285, km 171, 99001-970, Passo Fundo, Rio Grande do Sul, Brazil

Bernardo da Silva Tissot
Programa de Pós-graduação em Agronomia, Faculdade de Agronomia e Medicina Veterinária, Universidade de Passo Fundo, Passo Fundo, Rio Grande do Sul, Brazil

Luiz Fernando Pires and Nivea Maria Piccolomini Dias
Laboratório de Física Aplicada a Solos e Ciências Ambientais, Universidade Estadual de Ponta Grossa, Ponta Grossa, Paraná, Brazil

Cezar Francisco Araujo-Junior
Instituto Agronômico do Paraná, Pólo de Londrina, Londrina, Paraná, Brazil

Moacir de Souza Dias Junior
Departamento de Ciência do Solo, Universidade Federal de Lavras, Lavras, Minas Gerais, Brazil

Elifas Nunes de Alcântara
Empresa de Pesquisa Agropecuária de Minas Gerais, Lavras, Minas Gerais, Brazil

Araína Hulmann Batista, Antônio Carlos Vargas Motta, Carlos Bruno Reissmann, Iara Lang Martins and Maurício Hashimoto
Programa de Pós-graduação em Ciência do Solo, Universidade Federal do Paraná, Rua dos Funcionários, 1540, 80035-050, Curitiba, Paraná, Brazil

Thays Schneider
Universidade Estadual de Ponta Grossa, Ponta Grossa, Paraná, Brazil

Fábio Lúcio Santos and Domingos Sárvio Magalhães Valente
Departamento de Engenharia Agrícola, Universidade Federal de Viçosa, Av. P. H. Rolfs, s/n, 36570-000, Viçosa, Minas Gerais, Brazil

Valquíria Aparecida Mendes de Jesus
Programa de Pós-graduação em Fitotecnia, Universidade Federal de Viçosa, Viçosa, Minas Gerais, Brazil

Murilo Rezende Machado
Instituto Chico Mendes de Conservação da Biodiversidade, Floresta Nacional do Rio Preto, Zona Rural, 29960-000, Conceição da Barra, Espírito Santo, Brazil

Rodrigo Camara and Marcos Gervasio Pereira
Universidade Federal Rural do Rio de Janeiro, Seropédica, Rio de Janeiro, Brazil

Paulo de Tarso Barbosa Sampaio and João Baptista Silva Ferraz
Instituto Nacional de Pesquisas da Amazônia, Manaus, Amazonas, Brazil

José Flávio Neto, Eduardo da Costa Severiano, Kátia Aparecida de Pinho Costa, Wellingthon Silva Guimarães Junnyor, Wainer Gomes Gonçalves and Renata Andrade
Instituto Federal de Educação, Ciência e Tecnologia Goiano, Rod. Sul Goiana, Km 01, 75901-970, Campûs Rio Verde, Rio Verde, Goiás, Brazil

Muhammad Mahmood-ul-Hassan and Eajaz Rafique and Abdul Rashid
Land Resources Research Institute, National Agricultural Research Centre, Islamabad 45500, Pakistan

Ademir Sergio Ferreira de Araújo, Luciano Moura Lima and Vilma Maria dos Santos
Laboratorio de Qualidade do Solo, Centro de Ciências Agrárias, Universidade Federal do Piauí, Campus da Socopo, 64000-000, Teresina, Piauí, Brazil

Wanderley José de Melo
Faculdade de Ciências Agrárias e Veterinárias, Universidade Estadual de São Paulo, Jaboticabal, São Paulo, Brazil

Fabio Fernando de Araujo
Universidade do Oeste Paulista, Presidente Prudente, São Paulo, Brazil

Aline Marques Genú
Departamento de Agronomia, Universidade Estadual do Centro-Oeste, Rua Simeão Camargo Varela de Sá, 3, 85040-080, Guarapuava, Paraná, Brazil

José Alexandre Melo Demattê
Departamento de Ciência do Solo, Escola Superior de Agricultura "Luiz de Queiroz", Universidade de São Paulo, Piracicaba, São Paulo, Brazil

José de Oliveira Melo Neto and Antônio Marciano da Silva
Instituto de Ciência e Tecnologia, Universidade Federal de Alfenas, Cidade Universitária, Rodovia José Aurélio Vilela, BR-267, km 533, 11999, 37715-400, Poços de Caldas, Minas Gerais, Brazil

Carlos Rogério de Mello, Marcelo Ribeiro Viola and Sílvia de Nazaré Monteiro Yanagi
Departamento de Engenharia, Universidade Federal de Lavras, Lavras, Minas Gerais, Brazil

José Márcio de Mello
Departamento de Ciência Florestal, Universidade Federal de Lavras, Lavras, Minas Gerais, Brazil

Aline Marques Genú
Departamento de Agronomia, Universidade Estadual do Centro-Oeste, Rua Simeão Camargo Varela de Sá, 3, 85040-080, Guarapuava, Paraná, Brazil

Dar Roberts
Department of Geography, University of California, Santa Barbara, California, United State of America

José Alexandre Melo Demattê
Departamento de Ciência do Solo, Escola Superior de Agricultura "Luiz de Queiroz", Universidade de São Paulo, Piracicaba, São Paulo, Brazil

Rafael Thomaz de Aquino Moura
Secretaria de Educação da Cidade do Recife, Av. Afonso Olindense, 1513, 50.810-000, Várzea, Recife, Pernambuco, Brazil

Marlon da Silva Garrido
Universidade Federal do Vale do São Francisco, Joazeiro, Bahia, Brazil

Carla da Silva Sousa
Instituto Federal de Educação, Ciência e Tecnologia Baiano, Campus Uruçuca, Uruçuca, Bahia, Brazil

Romulo Simões Cezar Menezes and Everardo Valadares de Sá Barretto Sampaio
Departamento de Energia Nuclear, Universidade Federal de Pernambuco, Recife, Pernambuco, Brazil

Maria Inês Lopes de Oliveir
Programa de Pós-graduação em Agronomia, Universidade Estadual de Londrina, Londrina, Paraná, Brazil

Didier Brunet, Danielle Mitja and Michel Brossard
Institut de Recherche pour le Développement, Unité Eco&Sols, BP 64501, 34394, Montpellier cedex 5, France

Wellington dos Santos Cardoso
Faculdade da Terra de Brasília, Brasília, Distrito Federal, Brazil

Norton Pólo Benito
Núcleo Temático de Segurança Biológica, Embrapa Recursos Genéticos e Biotecnologia, Brasília, Distrito Federal, Brazil

Maria de Fátima Guimarães
Departamento de Agronomia, Centro de Ciências Agrárias, Universidade Estadual de Londrina, Londrina, Paraná, Brazil

Cácio Luiz Boechat
Departamento de Ciência do Solo, Universidade Federal do Rio Grande do Sul, Av. Bento Gonçalves, 7712, 91540-000, Porto Alegre, Rio Grande do Sul, Brazil

Jorge Antonio Gonzaga Santos
Centro de Ciências Agrárias Biológicas e Ambientais, Universidade Federal do Recôncavo da Bahia, Cruz das Almas, Bahia, Brazil

Adriana Maria de Aguiar Accioly
Empresa Brasileira de Pesquisa Agropecuária, Centro Nacional de Pesquisa de Mandioca e Fruticultura, Cruz das Almas, Bahia, Brazil

Osmar Henrique de Castro Pias and Cimélio Bayer
Departamento de Ciência do Solo, Universidade Federal do Rio Grande do Sul, Av. Bento Gonçalves, 7712, 91540-000, Porto Alegre, Rio Grande do Sul, Brazil

Maurício Roberto Cherubin and José Paulo Molin
Escola Superior de Agricultura "Luiz de Queiroz", Universidade de São Paulo, Piracicaba, São Paulo, Brazil

Claudir José Basso and Antônio Luis Santi
Universidade Federal de Santa Maria, Frederico Westphalen, Rio grande do Sul, Brazil

Marcio Osvaldo Lima Magalhães, Nelson Moura Brasil do Amaral Sobrinho, Everaldo Zonta and Izabella Bezerra Coutinho
Departamento de Solos, Universidade Federal Rural do Rio de Janeiro, BR-465, km 7, 23890-000, Seropédica, Rio de Janeiro, Brazil

Alfredo Tolón Becerra and Xavier Bolívar Lastra-Bravo
Rural Engineering Department, University of Almería, Almería, Spain

Cid Tacaoca Muraishi, Marlene Cristina Alves and Aparecido da Silva Júnior
Departamento de Fitossanidade, Faculdade de Engenharia, Universidade Estadual Paulista "Júlio de Mesquita Filho", Av. Brasil, 56, Cx.Postal 31, 15385-000, Ilha Solteira, São Paulo, Brazil

Zigomar Menezes de Souza
Departamento de Água e Solo, Faculdade de Engenharia Agrícola, Universidade Estadual de Campinas, Campinas, São Paulo, Brazil

André Faé Giostri, Luis Fernando Montrucchio Bond, Antônio Carlos Vargas Motta and Jair Alves Dionísio
Programa de Pós-graduação em Ciência do Solo, Universidade Federal do Paraná, Rua dos Funcionários, 1540, 80035-050, Curitiba, Paraná, Brazil

Ida Chapaval Pimentel
Setor de Ciências Biológicas, Departamento de Patologia Básica, Centro Politécnico, Universidade Federal do Paraná, Curitiba, Paraná, Brazil

Paulo César de Faccio Carvalho
Programa de Pós-graduação em Produção Vegetal, Universidade Federal do Paraná, Curitiba, Paraná, Brazil

Marcos Rafael Nanni, Marcelo Luiz Chicati, Everson Cézar and Roney Berti de Oliveira
Departamento de Agronomia, Universidade Estadual de Maringá, Av. Colombo, 5790, 87020-900, Maringá, Paraná, Brazil

José Alexandre Melo Demattê
Departamento de Solos e Nutrição de Plantas, Escola Superior de Agricultura "Luiz de Queiroz", Piracicaba, São Paulo, Brazil

Peterson Ricardo Fiorio
Departamento de Engenharia Rural, Escola Superior de Agricultura "Luiz de Queiroz", Piracicaba, São Paulo, Brazil

Index